Life at Interfaces and Under Extreme Conditions

Developments in Hydrobiology 151

Series editor
H. J. Dumont

Life at Interfaces and Under Extreme Conditions

Proceedings of the 33rd European Marine Biology Symposium, held at Wilhelmshaven, Germany, 7–11 September 1998

Edited by

Gerd Liebezeit, Sabine Dittmann & Ingrid Kröncke

Reprinted from Hydrobiologia, volume 426 (2000)

Springer Science+Business Media, B.V.

Library of Congress Cataloging-in-Publication Data

A C.I.P. Catalogue record for this book is available from the Library of Congress.

ISBN 978-94-010-5808-7 ISBN 978-94-011-4148-2 (eBook)
DOI 10.1007/978-94-011-4148-2

Printed on acid-free paper

Contents

Hydrobiologia **426:** vii, 2000.

Preface

The 33rd European Marine Biology Symposium held in Wilhelmshaven, Germany, from September 7 to 11, 1998, covered two main topics: 'Life at Interfaces' and 'Life under Extreme Conditions'. It assembled about 180 marine biologists from 16 countries to present and discuss recent developments in the field.

Interfaces between media, be it the air–water or the sediment–water interfaces or organisms themselves, pose considerable problems to marine organisms attempting to make a living at these boundaries. In view of the present discussion on the deleterious effects of antifouling agents on marine ecosystems, these aspects especially will regain increased attention by the scientific community.

Although marine organisms are well adapted to – by human standards – extreme conditions of temperature, salinity or pressure, they nevertheless have to develop techniques to survive under these conditions.

In the present volume of *Hydrobiologia* a number of authors address various aspects of these two topics. Locations range from intertidal areas to the deep sea while both macro- and meiofaunal organisms were investigated with respect to distribution patterns and effects of variable temperatures, pressures and salinities. Aspects of fouling induction and prevention are also addressed.

Although the papers assembled in this volume do not represent the full array of material presented at the conference they provide a representative cross section of both oral and poster contributions.

The editors are indebted to J. Beukema, D. Blome, G. Cadée, N. Dankers, D. Desbruyeres, E. Flach, R. Hartnoll, M. Huettel, I. Karakassis, A. Mackensen, T. Noji, H. L. Rees, C. Reid, J. Ryland, K. Schaumann, S. Schumacher, M. Scholler, M. Thiel, M. Türkay, A. Underwood and 16 colleagues who wished to remain anonymous for reviewing the manuscripts. Ummo Fooken expertly undertook the final checking of the manuscripts. We are also indebted to the Federal Ministry for Education and Research, Merck, Germany, the Oldenburgische Landesbank and the Nordwest-Oelleitungs-GmbH for financial and other support of the conference.

GERD LIEBEZEIT
SABINE DITTMANN
INGRID KRÖNCKE

Hydrobiologia **426:** 1–24, 2000.
G. Liebezeit, S. Dittmann & I. Kröncke (eds), Life at Interfaces and Under Extreme Conditions.

Organic matter diagenesis at the oxic/anoxic interface in coastal marine sediments, with emphasis on the role of burrowing animals

Erik Kristensen
Institute of Biology, Odense University, SDU, DK-5230 Odense M, Denmark
E-mail: ebk@biology.ou.dk

Key words: oxic/anoxic interfaces, diagenesis, carbon, oxygen, bioturbation, irrigation

Abstract

The present paper reviews the current knowledge on diagenetic carbon transformations at the oxic/anoxic interface in coastal marine sediments. Oxygen microelectrodes have revealed that most coastal sediments are covered only by a thin oxic surface layer. The penetration depth of oxygen into sediments is controlled by the balance between downward transport and consumption processes. Consumption of oxygen is directly or indirectly caused by respiration of benthic organisms. Aerobic organisms have the enzymatic capacity for complete oxidation of organic carbon. Anaerobic decay occurs stepwise, involving several types of bacteria. Large organic molecules are first fermented into small moieties. These are then oxidized completely by anaerobic respirers using a sequence of electron acceptors: Mn^{4+}, NO_3^-, Fe^{3+}, SO_4^{2-} and CO_2. The quantitative role of each electron acceptor depends on the sediment type and water depth. Since most of the sediment oxygen uptake is due to reoxidation of reduced metabolites, aerobic respiration is of limited importance. It has been suggested that sediments contain three major organic fractions: (1) fresh material that is oxidized regardless of oxygen conditions; (2) oxygen sensitive material that is only degraded in the presence of oxygen; and (3) totally refractory organic matter. Processes occurring at the oxic/anoxic boundaries are controlled by a number of factors. The most important are: (1) temperature, (2) organic supply, (3) light, (4) water currents, and (5) bioturbation. The role of bioturbation is important because the infauna creates a three-dimensional mosaic of oxic/anoxic interfaces in sediments. The volume of oxic burrow walls may be several times the volume of oxic surface sediment. The infauna increases the capacity, but not the overall organic matter decay in sediments, thus decreasing the pool of reactive organic matter. The increase in decay capacity is partly caused by injection of oxygen into the sediment, and thereby enhancing the decay of old, oxygen sensitive organic matter several fold. Finally, some future research directions to improve our understanding of diagenetic processes at the oxic/anoxic interface are suggested.

Introduction

The boundary between oxic and anoxic zones is a well defined and sharp interface in most aquatic environments; particularly in the sediment. Oxygen is the energetically most favorable electron acceptor for microbial respiration (Fenchel et al., 1998), but the high consumption rate combined with low solubility in water usually prevents deep penetration of oxygen into coastal sediments. The lack of available oxygen may have serious implications for the biotic community and, thus, rates of organic matter diagenesis in sediments (Kristensen et al., 1995; Fenchel, 1996a, b).

Macrofaunal structures, such as burrows formed by bottom-dwelling animals, represent an important mosaic of physico-chemical and biological microenvironments in most coastal sediments. The surface area available for diffusive solute exchange, as well as the areas of oxic/anoxic boundaries, are considerably increased by the presence of irrigated burrows (Kristensen, 1984; Fenchel, 1996a). Accordingly, the activities of burrowing and irrigating infauna alter the one-dimensional diagenetic stratification into a three-dimensional, complex and time-dependent stratification with effects on microbial communities deep in the sediment (Aller, 1982; Kristensen, 1988).

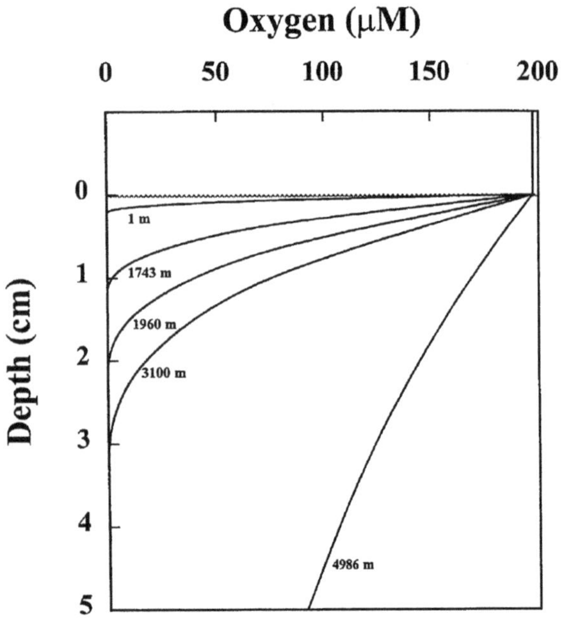

Figure 1. Oxygen penetration into marine sediments at water column depths from 1 to about 5000 m. The horizontal line indicates the sediment–water interface. The oxygen saturation level is, for simplicity, fixed to 197 μM in all environments. Modified from Jørgensen & Revsbech (1985) and Glud et al. (1994).

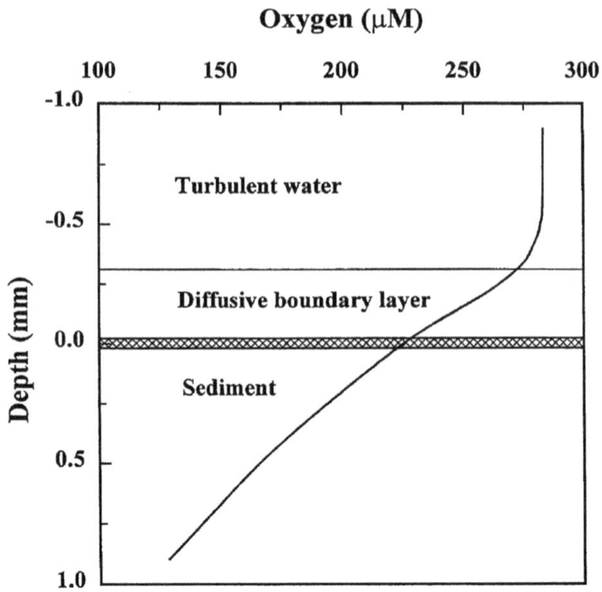

Figure 2. Hypothetical oxygen profile in a coastal marine sediment. A diffusive boundary layer of 0.3 mm thickness separates the sediment from the turbulent overlying water. The horizontal line indicates the interface between the boundary layer and the turbulent water phase. The cross-hatched horizontal bar represents the sediment–water interface.

The present paper reviews the current knowledge on dynamics of oxic/anoxic interfaces with respect to carbon transformations in coastal marine sediments. The distribution patterns of oxic/anoxic interfaces are discussed and related to the factors affecting the dominating oxic and anoxic diagenetic processes. The quantitative role of deep oxygen penetration caused by animal burrows on the overall rate of organic matter decomposition is evaluated. The general discussion is supplemented with relevant case studies and conceptual models.

Oxygen distribution and oxic/anoxic interfaces

Marine sediments are reducing environments covered only by a thin oxic surface layer. Sediments in productive shallow coastal waters are generally characterized by oxygen penetration depths of millimeters compared with cm or dm scales in oceanic sediments underlying a deep oligotrophic water column (Figure 1) (Reimers et al., 1986; Glud et al., 1994). The penetration depth of oxygen is controlled by the balance between downward transport of oxygen from above and by consumption processes of all benthic organisms and their metabolic products within the sediment. The transport of oxygen in sediments is driven by molecular diffusion and water currents or bioturbation induced advective forces (Huettel & Gust, 1992a), whereas the consumption processes are driven by microbially mediated oxidation of organic matter and reduced inorganic metabolites (Jørgensen, 1983).

The rate of benthic oxygen uptake may be hampered by a mm thick diffusive boundary layer above the sediment–water interface (Jørgensen & Revsbech, 1985; Archer et al., 1989). The diffusive boundary layer is a viscous film of water at the sediment–water interface created by the internal friction of water close to a solid surface (Figure 2). Molecular diffusion is the principal mechanism for mass transport within the diffusive boundary layer (Santschi et al., 1983). As eddy diffusion is reduced under low water current regimes, the diffusive boundary layer can create a barrier between the sediment and the overlying water, thus reducing the oxygen concentration at the sediment surface compared with the stirred overlying water. Consequently, the thickness of the diffusive boundary layer may control the influx and penetration depth of oxygen into the sediment, particularly when oxygen uptake is high (Jørgensen & Revsbech, 1985).

The oxidized zone that extends just below the upper oxic zone in sediments is frequently denoted the

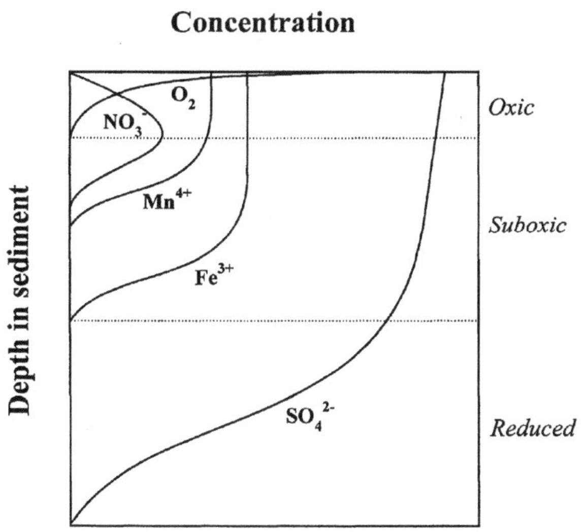

Figure 3. Idealized presentation of vertical profiles of oxidized compounds (electron acceptors) in a marine sediment. The depth of the 'oxic' zone is determined by the penetration of oxygen. The 'suboxic' zone contains a number of electron acceptors, i.e. nitrate, oxidized manganese and iron, and the position of the lower bounday is usually defined by the penetration depth of oxidized iron. Sulfate is the dominating electron acceptor in the 'reduced' zone. When sulfate is depleted, methane and carbon dioxide (not shown) are the dominating compounds for diagenetic processes.

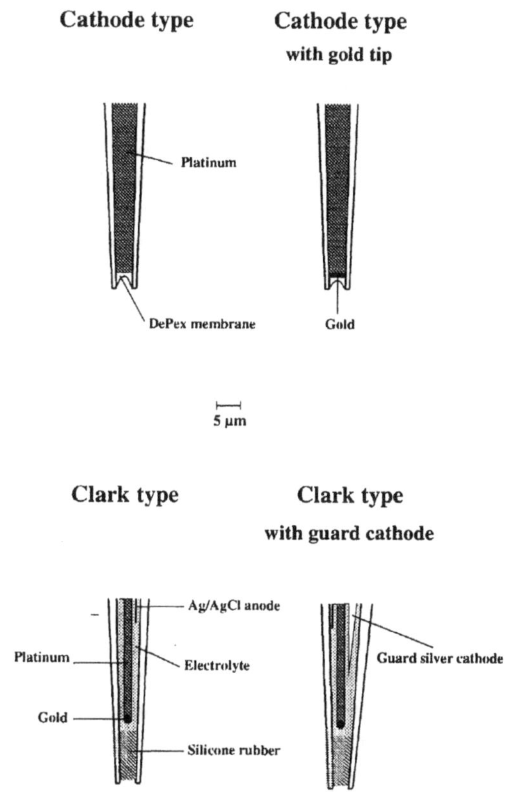

Figure 4. Schematic drawings of oxygen microelectrode tips. The four electrode types indicate 10 years (1980–1990) of evolution in the Revsbech-electrode design, i.e. from the simple cathode type to the Clark type with a guard cathode. Modified from Revsbech et al. (1983) and Revsbech (1989).

'suboxic zone' (Froelich et al., 1979). The suboxic zone is characterized by high concentrations of oxidized inorganic compounds such as nitrate, manganese oxides and iron oxyhydroxides (Jørgensen, 1983), and appears visually as a light brown upper layer of most sediments (Figure 3). The reduced zone extends below the suboxic zone and is often characterized by the presence of sulfides produced by bacterial sulfate reduction, either in precipitated form as iron sulfides or in dissolved form as free sulfide (Chanton et al., 1987). Under conditions of high sediment oxygen uptake combined with stagnant conditions in the overlying water, the suboxic zone may disappear and the oxic/anoxic interface with free sulfide present just below the oxic zone moves upwards to the surface or even into the overlying water (Stigebrandt & Wulff, 1987; Kemp et al., 1992; Møller, 1996).

Determination of oxygen penetration depth

It is important to know the exact penetration depth and concentration of oxygen for the understanding of microbial processes at the oxic/anoxic interface in sediments. Before Revsbech et al. (1980) introduced oxygen microelectrodes in ecological research, the oxic surface layer in sediments was assumed to be identical with the brown oxidized surface layer; i.e. the layer having positive redox potentials. However, the use of oxygen microelectrodes has shown that oxygen penetration depth generally is less than 10% of the oxidized layer thickness (Revsbech & Jørgensen, 1986). The redox potential of the remainder (suboxic) layer is kept positive by occasional oxygen input and the presence of considerable amounts of oxidized iron and manganese compounds.

The first cathode type of oxygen microelectrode with external Ag/AgCl reference electrode was developed around 1980 (Revsbech et al., 1980), since then a number of improvements and new developments of electrode design have been made (Figure 4). The cathode electrodes of Revsbech et al. (1980) were made of 0.1 mm platinum wire enclosed by a thin glass casing. The platinum tip was electrolytically etched in saturated KCN to a diameter of 1–4 μm and covered with a polystyrene membrane. The final diameter of the electrode tip was less than 10 μm. The cathode microelectrode was later improved by coating the platinum tip with gold before application of the membrane in order to increase electrode stability and signal quality (Revsbech et al., 1983). A considerable improvement was the development of a combined microsensor that is a small version of the conventional Clark electrode (Revsbech & Ward, 1983). In this microsensor, the gold coated cathode is situated behind an electrically insulating membrane of silicone rubber which is extremely permeable to oxygen. The cathode is bathed in an electrolyte solution of 1 M KCl into which an Ag/AgCl reference electrode is immersed. Finally, the stability of the 'Revsbech' microelectrode was improved be inserting an internal guard cathode that removes all oxygen diffusing towards the sensor from the internal electrolyte (Revsbech, 1989).

More recently, a new fiber-optic oxygen microsensor (microoptrode) was developed (Klimant et al., 1995). The microoptrode is made by immobilizing an oxygen-quenchable fluorophore at the tapered tip of an optical fiber with a 15–40 μm core diameter. An optoelectronic system is used to illuminate the fluorophore (blue) and to detect the fluorescent light (red) from the fiber tip. The intensity of fluorescent red light proportionally increases with decreasing oxygen concentration. In contrast to oxygen microelectrodes, the microoptrodes are relatively easy to make, do not consume oxygen and show no stirring dependence. The optrode principle has recently been used to develop planar optrodes (fluorophore coated PVC sheets) for measuring fine scale two-dimensional oxygen distributions in sediments (Glud et al., 1996).

During measurements, the microelectrode (-optrode) tip is introduced into the substratum by a micromanipulator at steps of 25–100 μm with a precision better than 10 μm (Revsbech & Jørgensen, 1986). For *in situ* measurements in oceanic environments, microelectrodes have been successfully mounted on benthic landers of various designs (Reimers, 1987; Gundersen & Jørgensen, 1990).

Oxic and anoxic diagenesis

Organic matter diagenesis

Organic matter is degraded (mineralized) in sediments by an array of aerobic and anaerobic microbial processes with a concurrent release of inorganic nutrients (Figure 5). The actual rates of decay depend primarily on organic matter quality (i.e. the content of protein, cellulose, lignin etc.), age (decomposition stage) and temperature (season) (Fenchel et al., 1998). The chemical composition of organic matter in marine environments can be generalized by the following formula:

$$(CH_2O)_x(NH_3)_y(H_3PO_4)_z,$$

where x, y and z may vary strongly depending on the origin and age of the material. For marine organic matter (e.g. phytoplankton) having the Redfield composition: $x = 106$, $y = 16$, and $z = 1$.

A number of organisms including bacteria, fungi and micro- and macrofauna are responsible for the aerobic degradation of organic carbon (Fenchel et al., 1998). Almost all of these have the enzymatic capacity to perform a total mineralization of organic substrates. Organic matter is, therefore, completely metabolized by a single organism to H_2O, CO_2 and inorganic nutrients using oxygen as electron acceptor according to the following stoichiometry:

$$(CH_2O)_x(NH_3)_y(H_3PO_4)_z + xO_2 \rightarrow xCO_2 + yNH_3 + zH_3PO_4 + xH_2O. \quad (1)$$

However, due to an efficient energy metabolism, a large fraction of the metabolized organic matter ends up as cell material. A unique feature of aerobic decomposition is the formation and consumption of reactive oxygen-containing radicals such as superoxide anion ($\cdot O_2^-$), hydrogen peroxide (H_2O_2) and hydroxyl radicals ($\cdot OH$). These are capable of of breaking bonds and depolymerize relatively refractory organic compounds like lignin (Canfield, 1994). As the oxic (oxygen containing) zone in coastal sediments usually is limited to a thin uppermost layer, a large fraction of the organic matter is buried in a more or less decomposed form into anoxic layers. Here, anaerobic decomposition is accomplished by mutualistic consortia of bacteria because no single type of anaerobic bacterium seems capable of complete mineralization (Fenchel et al., 1998).

Anaerobic decomposition occurs stepwise, involving several different functional types of bacteria (Figure 5). First, the large and normally complex polymeric organic molecules stepwise are split into water soluble monomers (amino acids, monosaccharides and fatty acids) by hydrolysis and fermentation under the production of energy and release of inorganic nutrients (Kristensen & Hansen, 1995), e.g. mixed propionate and acetate formation:

$$8(CH_2O)_x(NH_3)_y(H_3PO_4)_z \rightarrow xCH_3CH_2COOH + xCH_3COOH + 3xCO_2 + 3xH_2 + yNH_3 + zH_3PO_4. \quad (2)$$

The small organic acids are then oxidized completely to H_2O and CO_2 by a number of respiring mi-

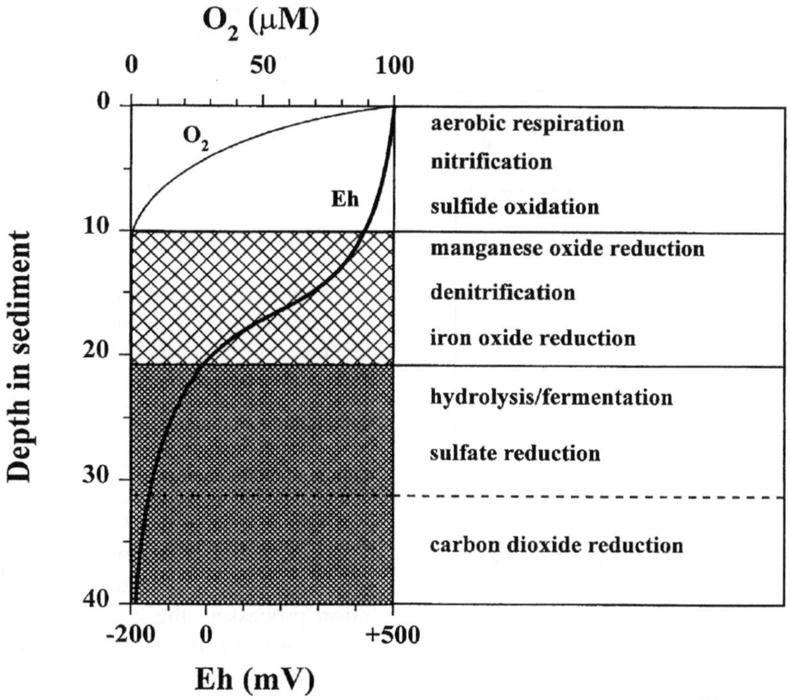

Figure 5. The idealized vertical distribution of diagenetic processes in marine sediments. The oxic zone is illustrated by an oxygen profile (white zone), the suboxic zone is shown as the layer where the redox discontinuity is evident (light cross-hatched), the reduced zone is shown as the layer where E_h is below zero (dark cross-hatched). The depth scale is arbitrary.

croorganisms using a variety of inorganic compounds as electron acceptors. The individual anaerobic respiration processes generally occur in a sequence with depth in the sediment according to the availability of electron acceptors: $Mn^{4+} \rightleftarrows NO_3{}^-$, Fe^{3+}, $SO_4{}^{2-}$ and CO_2 respiration (Figure 5). The actual sequence is determined by the ability of each electron acceptor to receive electrons, and thus the energy output per degraded organic carbon atom (Fenchel et al., 1998), e.g. nitrate respiration (denitrification) is favored energetically compared to sulfate reduction. The suboxic zone contains the most potent anaerobic electron acceptors, Mn^{4+}, $NO_3{}^-$ and Fe^{3+}. The transition from one electron acceptor to the other downwards in the sediment occurs when the most favorable is exhausted. “When the best is gone, one has to accept something less good”. However, some vertical overlap may occur between the various zones. Only two examples of anaerobic degradation stoichiometries, denitrification and sulfate reduction, will be presented here:

$$\text{Denitrification}: \quad CH_3COOH + 1.6NO_3^- + 1.6H^+ \rightarrow 2CO_2 + 0.7N_2 + 2.8H_2O. \quad (3)$$

$$\text{Sulfate reduction}: \quad CH_3COOH + SO_4^{2-} \rightarrow 2CO_2 + S^{2-} + 2H_2O. \quad (4)$$

The strict vertical distribution of electron acceptors as depicted in Figure 5 is an over-simplification of the true spatial distribution. The influence of sediment inhomogeneities, such as worm burrows, on porewater profiles and vertical distribution of microbial processes has been clearly documented (Aller, 1982). Furthermore, patches associated with e.g. fecal pellets are known to create anaerobic microniches, where anaerobic processes such as denitrification and sulfate reduction occur in otherwise oxic surface sediments (Jørgensen, 1977; Jahnke, 1985; Brandes & Devol, 1995).

Nevertheless, the usually observed decreasing degradation rate with depth in sediments is not primarily caused by the less efficient electron acceptors in the deeper layers, but rather by the decreasing quality of organic matter (lability or degradability) with depth (Canfield, 1994). Even within a few mm thick oxic zone, the decreasing degradability may be evident as a considerable reduction in volume specific oxygen uptake with depth (Jensen et al., 1993; Figure 6). However, anaerobic bacteria appear more limited than aerobic organisms in their ability to depolymerize certain large complex molecules. These include among others saturated hydrocarbons (Schink, 1988), certain

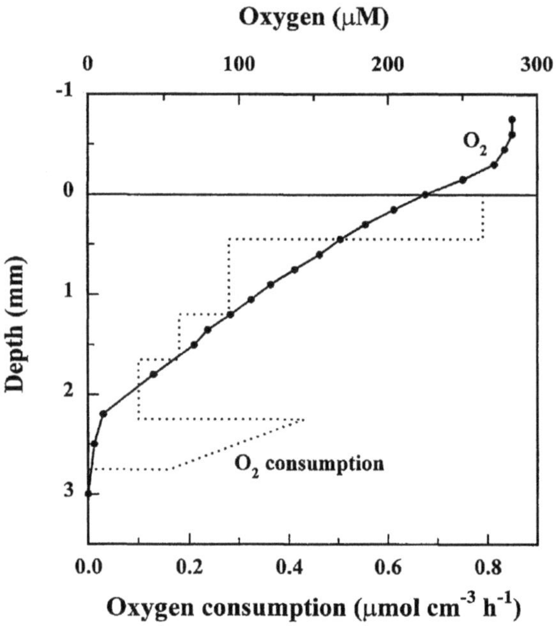

Figure 6. A sediment oxygen profile (full line) with estimated volume specific rates of oxygen consumption (dotted line). Oxygen consumption was estimated from changes in curvature of the oxygen profile. The oxygen consumption down to 2 mm is due to aerobic respiration, whereas the peak between 2 and 3 mm depth is caused be reoxidation of reduced inorganic metabolites. The horizontal line indicates the sediment–water interface. Modified from Jensen et al. (1993).

Table 1. Importance of different carbon oxidation pathways in sediments from different depths in the Skagerrak. Rates are given in mmol m^{-2} d^{-1}. Numbers in brackets indicate the fraction (%) of total carbon oxidation by each pathway. From Canfield et al. (1993a)

	Carbon oxidation		
	190 m	380 m	695 m
Respiration type			
Oxygen	2.1 (13.6)	1.7 (17.4)	0.4 (3.6)
Nitrate	0.5 (3.2)	0.4 (3.8)	0.6 (5.7)
Manganese	0.0 (0.0)	0.0 (0.0)	9.9 (90.7)
Iron	5.1 (32.1)	5.2 (50.9)	0.0 (0.0)
Sulfate	8.1 (51.1)	2.9 (27.9)	<0.1 (<1.0)

pigments (Sun et al., 1993a) and lignin (Benner et al., 1984).

Separation of total benthic metabolism into respiration types

Quantification of sediment respiration processes and separation of total benthic metabolism into individual processes is problematic. Only total metabolism (measured as total O_2 or CO_2 flux; Kristensen & Hansen, 1999), sulfate reduction (measured by ^{35}S assay; Fossing & Jørgensen, 1989) and nitrate respiration (measured by ^{15}N assay; Nielsen, 1992) can be determined directly on undisturbed sediment. The remainder, O_2, Mn^{4+} and Fe^{3+} respiration, has to be deduced indirectly from the shape of porewater profiles, from reaction rates in sediment incubation or by simple subtraction. By subtracting measured rates of sulfate reduction from measured rates of total sediment O_2 uptake, oxygen respiration and sulfate reduction have generally been estimated to contribute by about 50% each to total benthic metabolism in coastal sediments (Jørgensen, 1983). In many organic-rich sediments, however, sulfate reduction have been estimated to account for up to 70–100% (Mackin & Swider, 1989; Kristensen et al., 1998). Although sulfate reduction is one of the least favorable respiration processes, the high concentration of SO_4^{2-} in seawater (300–1000 times higher than O_2) is responsible for its deep vertical distribution and thus high quantitative importance (Jørgensen, 1983). Nitrate, manganese and iron respiration, on the other hand, are generally of limited quantitative significance for carbon oxidation (few percent) in most shallow coastal sediments, but may in certain environments account for a significant fraction (Jørgensen & Sørensen, 1985; Aller, 1990; Canfield et al., 1993a).

By the use of a combination of porewater and solid phase analysis, as well as a series of sediment incubations, Canfield et al. (1993a, b) provided one of the first experimentally derived attempts to quantify rates of organic carbon mineralization by various electron acceptors in sediments. They concluded that oxygen respiration accounted for only between 4 and 17% of the total organic carbon oxidation in sediments from the Skagerrak (between Denmark and Norway) (Table 1). Nitrate respiration (denitrification) was limited to a narrow zone just below the oxygen penetration depth and was responsible for only 3–6% of the total carbon oxidation. The processes of Fe respiration, Mn respiration and sulfate respiration dominated organic carbon mineralization, but their relative significance varied depending on the sediment and water column depth. In deep areas with high concentrations of Mn(IV), more than 90% of the total respiration was due to Mn(IV) reduction. With lower Mn(IV) concentrations more typical of coastal sediments, Fe(III) reduction in shallower areas accounted for 32–51% and sulfate reduction for 28–51% of the total carbon oxidation with no contribution from Mn(IV) reduction.

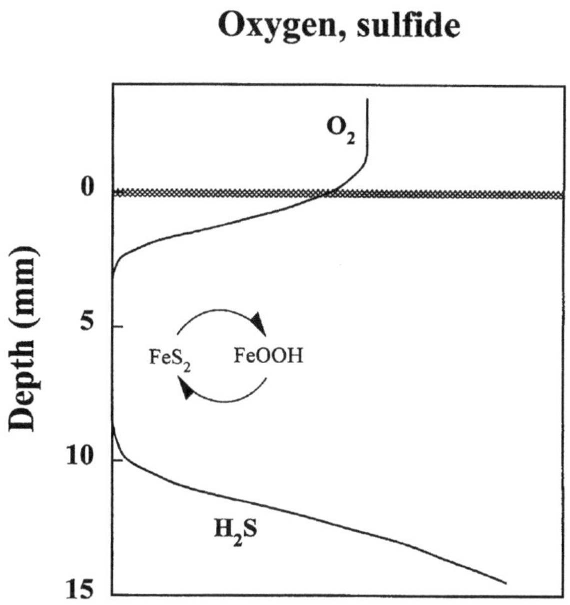

Figure 7. Schematic presentation of oxygen and sulfide profiles in a marine sediment. Sulfide precipitation and oxidation is mediated by a Fe(II)/Fe(III) shunt. Modified from Jørgensen (1989).

Canfield et al. (1993a) suggested that the importance of oxygen respiration in many coastal sediments has previously been overestimated, whereas metal oxide reduction has probably been well underestimated.

Reoxidation of metabolites

Most of sediment oxygen uptake is not caused by aerobic respiration, but is rather due to reoxidation of reduced inorganic metabolites (e.g. H_2S) close to the oxic/anoxic interface (Figure 6). A large fraction (up to 85%) of the sulfide produced by sulfate reduction in marine sediments is not trapped permanently due to reactions with iron and other metals, but is continuously diffusing upwards to be reoxidized in the surface sediment (Thamdrup et al., 1994). About 50% or more of the sediment oxygen uptake is usually consumed directly or indirectly by oxidation of sulfide (Jørgensen, 1989). Reoxidation can be a pure chemical process, but it is usually mediated by microorganisms. The oxidation is usually not a direct reaction between sulfide and oxygen, because the two compounds rarely meet (Figure 7). Instead, a number reaction appears to take place mostly in the oxidized, but anoxic (suboxic) sediment layers through oxidation by nitrate, manganese(IV) and iron(III). An example of sulfide oxidation with iron where pyrite acts as an intermediate is as follows (Howarth, 1984):

$$\text{Sulfide oxidation}: \quad 2FeOOH + 2H_2S + 2H^+ \rightarrow FeS_2 + Fe^{2+} + 4H_2O. \qquad (5)$$

$$\text{Iron oxidation}: \quad 4FeS_2 + 15O_2 + 10H_2O \rightarrow 4FeOOH + 8SO_4^{2-} + 16H^+. \qquad (6)$$

Only at very high organic loading of the sediment or during bottom water hypoxia, when the pools of oxidized metals are depleted, sulfide may reach the oxic zone where chemoautotrophs, particularly the filamentous *Beggiatoa* spp., proliferate and grow (Jørgensen & Revsbech, 1983):

$$\text{The `}\textit{Beggiatoa}\text{' oxidation}: \quad HS^- + 2O_2 \rightarrow SO_4^{2-} + H^+. \qquad (7)$$

Although sulfide is a potential poison for many organisms, it is an energy-rich substrate for the sulfide oxidizers near the oxic/anoxic interface. About 75% of the energy released during organic matter oxidation by sulfate reduction is stored in the produced sulfide. Accordingly, Howarth (1984) estimated that from the the total degradation of organic matter in the sediment, 15–35% of the energy is released upon reoxidation of sulfide at the sediment surface.

Rates of organic matter decay under oxic and anoxic conditions

A controversy

There has been considerable debate within the last decade about the rate and efficiency of aerobic *versus* anaerobic decomposition and the role of these processes in carbon preservation (e.g. Canfield, 1994). Organic matter that escapes decomposition is obviously buried and preserved in marine sediments, but the role of bottom water oxygen concentration on this process is less obvious. It has been suggested that oxygen has relatively little influence on preservation of organic carbon in sediments and that high-productivity events and sedimentation rates are more important (Henrichs & Reeburgh, 1987; Calvert & Pedersen, 1992). Others have argued that organic carbon burial efficiencies are higher in euxinic sediments (e.g. Black Sea) than in oxic sediments of the same accumulation rate (Canfield, 1989; Hartnett et al., 1998). As convincing evidence is available in support of both views, they both can be accepted as correct or, more likely, the truth may be a compromise. A number of

Table 2. Examples of recently published aerobic:anaerobic decay ratios (A/AN) in marine environments. The original 'material' used and its 'decomposition stage' is presented together with information on the incubation 'medium' and the specific 'compound' examined

Material	Decomp. stage	Medium	Compound	A/AN	Reference
Macroalgae	Fresh	Sediment	POC	~1	Hansen & Blackburn (1991)
Reag. grade	Fresh	Water	Amino acids	1.3–4.5	Lee (1992)
Reag. grade	Fresh	Water	Acetate	0.8–4.0	Lee (1992)
Reag. grade	Fresh	Water	Glucose	~1	Lee (1992)
Diatom	Fresh	Sediment	POC	2.7	Sun et al. (1993b)
Diatom	Fresh	Sediment	Chl. a	2.5–7.0	Sun et al. (1993a)
Diatom	Fresh	Water	POC	3.7	Harvey et al. (1995)
Diatom	Fresh	Water	PON	4.4	Harvey et al. (1995)
Diatom	Fresh	Water	Lipids	2.5	Harvey et al. (1995)
Diatom	Fresh	Water	Proteins	1.4	Harvey et al. (1995)
Diatom	Fresh	Water	Carbohydrates	4.7	Harvey et al. (1995)
Barley hay	Fresh	Sediment	POC	~1	Kristensen et al. (1995)
Diatom	Aged	Sediment	POC	~10	Kristensen et al. (1995)
Diatom	Fresh	Sediment	POC	~1	Andersen, 1996
Diatom	Aged	Sediment	POC	5.2	Andersen (1996)
Reag. grade	Fresh	Sediment	Fatty acids	2–4	Sun et al. (1997)
0–1 cm sed.	Fresh	Sediment	POC	~1	Hulthe et al. (1998)
17–20 cm sed.	Aged	Sediment	POC	3.6	Hulthe et al. (1988)

experimental studies using various techniques have attempted to determine organic matter decomposition under oxic and anoxic conditions (Table 2). Most studies using fresh organic matter (e.g. diatoms) found that rates of decomposition under aerobic conditions do not greatly exceed those under anaerobic conditions (Hansen & Blackburn, 1991; Kristensen et al., 1995; Andersen, 1996). Major exceptions are hydrolysis resistant materials such as lignin (Benner et al., 1984), complex lipids (Harvey et al., 1995) and photosynthetic pigments (Sun et al., 1993a). Nevertheless, a number of recent studies have revealed that aerobic decomposition of old and partly degraded organic matter may exceed anaerobic rates by up to a factor 10 (Kristensen et al., 1995; Andersen, 1996; Hulthe et al., 1998).

In accordance with these results, Hedges & Keil (1995) suggested an alternative hypothesis of carbon preservation in marine sediments. Since only a fraction (<10%) of the total organic matter in marine sediments is present as discrete particles, a major portion of the organic matter must be sorbed to mineral particles (Keil et al., 1994). The observation that higher organic loadings in surficial marine sediments are reduced to a monolayer equivalent at depth suggests that intrinsically reactive organic matter is stabilized with respect to decomposition when associated with mineral grain surfaces (Mayer, 1994). The sediment particles are in many cases highly irregular and appear to have a major fraction of their surface area in mesopores less than 10 nm in width (Mayer, 1994). Although a direct cause for the stability to decomposition of monolayer coatings has not been established, a number of mechanisms have been proposed. Mayer (1994) argued that hydrolytic enzymes are excluded from or inactivated within confined mesopores. Another explanation is that the activity of enzymes is sterically limited by close association of substrate molecules to mineral surfaces via chemisorptive bonds (Hedges & Keil, 1995).

However, in the deep-sea and in turbidites where sediment layers may be exposed to oxygen for a long time (>1000 years), the organic content is considerably less than expected from the monolayer-equivalent theory. This led Hedges & Keil (1995) to propose a model for oxic degradation of sedimentary organic matter. They assumed that all sediments receive only three organic components: 1. hydrolyzable (fresh) organic matter that is completely mineralized regardless of redox conditions; 2. oxygen-sensitive organic matter (monolayer coatings) that degrades slowly in the presence of oxygen, but not at all under anoxic con-

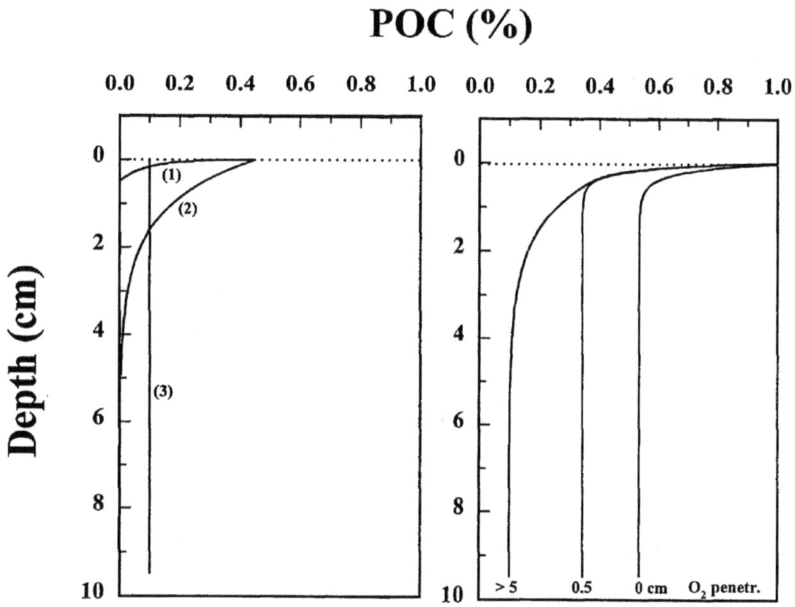

Figure 8. Hypothetical depth dependent degradation of organic matter in marine sediments. *Left panel*: Degradation pattern of three sedimentary organic pools, (1) hydrolyzable (fresh) organic matter that is completely mineralized regardless of redox conditions (exhausted at 0.5 cm depth), (2) oxygen-sensitive organic matter (monolayer coatings) that degrades slowly in the presence of oxygen (then exhausted at 5 cm depth), but not at all under anoxic conditions (not shown), and (3) totally refractory organic matter. *Right panel*: Depth distribution of combined organic matter composed of the three pools at different oxygen penetration depths in the sediment. Modified from Hedges & Keil 1995.

ditions, and 3. totally refractory organic matter (Figure 8). The sorptive protection of fraction (2) may be slowly disrupted under oxic conditions because H_2O_2 produced by aerobic organisms is sufficiently small and agressive to alleviate the mesopore or steric protection of monolayer coatings. In deeply oxygenated sediments that deposit slowly, fraction (2) has ample time to degrade completely (deep-sea and turbidites). As more rapid sediment accumulation and shallower oxygen penetration depth contribute to shorter oxygen exposure times at more landward sites, the fraction (2) is preserved at progressively shallower sediment depths.

Based on the facts and thoughts given above, Hulthe et al. (1998) supplemented the speculative scenario of Hedges & Keil (1995) for the effect of oxygen on degradation/preservation of organic matter in continental shelf sediments. Oxic and anoxic degradation rates are similar for fresh organic matter (newly deposited phytodetritus and zooplankton fecal pellets) in surficial sediments before adsorption to mineral grains (fraction (1) of Hedges & Keil, 1995). With time and deeper burial into the sediment, the fraction of degrading material that is adsorbed to mineral surfaces increases (fraction (2) of Hedges & Keil, 1995). Degradation by anoxic bacteria is retarded because the organic material is protected from exoenzymatic attack. Oxic rates of decomposition are now faster than anoxic rates.

Experimental evidence

In a recent series of experiments, Kristensen et al. (1995) and Kristensen & Holmer (submitted) examined the decomposition rates of fresh and aged organic matter under oxic, suboxic and anoxic conditions. Two different ^{14}C-labeled plant materials, diatoms (*Skeletonema costatum*) and barley straw (*Hordeum vulgare*) were used. Aged material was obtained by allowing both substrates to pre-decompose aerobically in seawater for 40–50 days to 33–50% of the original radioactivity. Each of the materials was then mixed into intertidal sediment and spread in a 1.5 mm layer on the bottom of oxic and anoxic recirculating seawater chambers. One anoxic series was incubated both with and without addition of 2 mM nitrate. All dissolved ^{14}C-pools being produced were sampled at 3-day intervals. After 3–4 weeks incubation, some of the chambers were switched from aerobic to anaerobic and vise versa for another 1–2 weeks.

Decomposition of fresh diatoms decreased rapidly from high initial rates to more steady levels after about

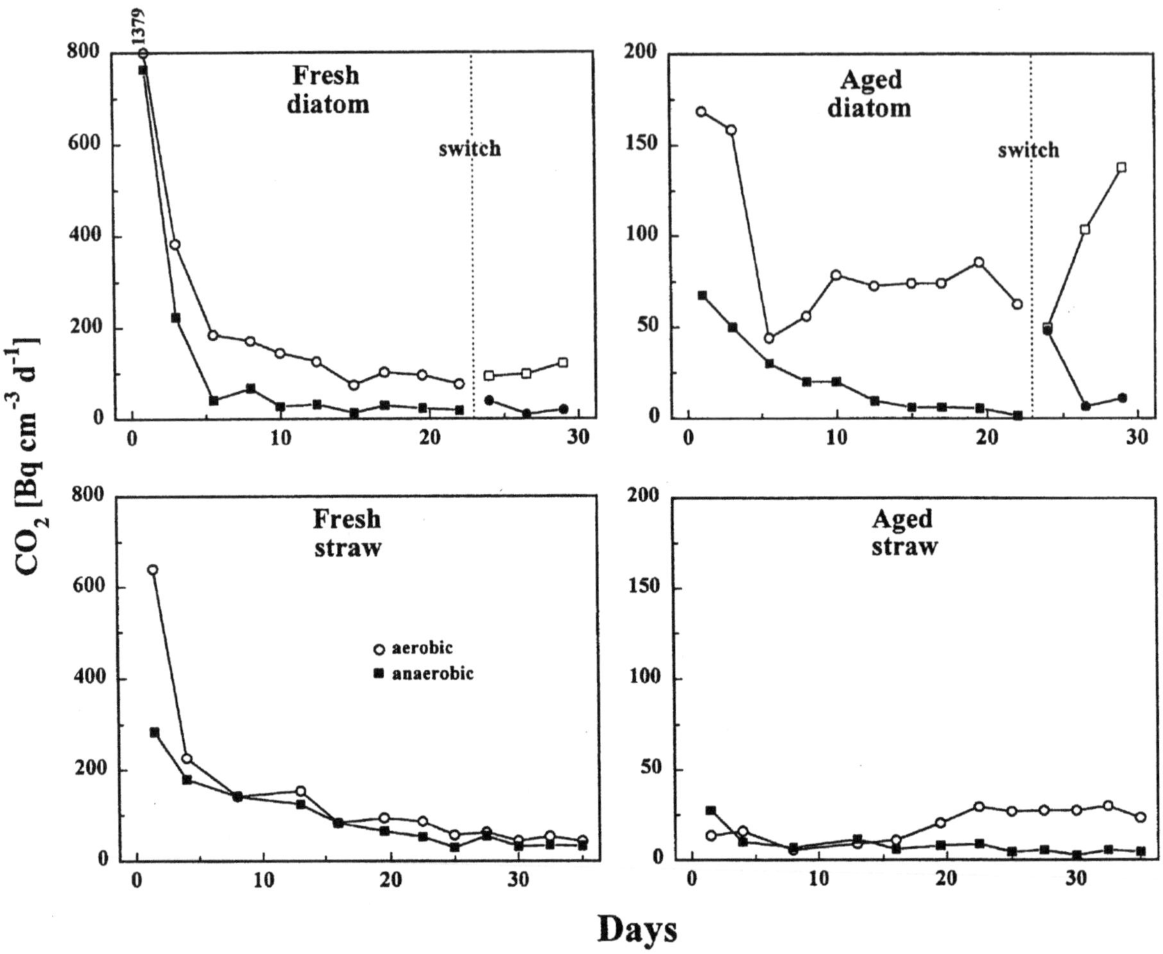

Figure 9. Aerobic and anaerobic decay ($^{14}CO_2$ release) of ^{14}C-labelled diatoms (upper panels) and barley straw (lower panels) incubated in thin-layer, flow-through (seawater) systems. Both fresh (left panels) and pre-decomposed (right panels) materials were used. The dotted vertical lines in the upper panels indicate the time when aerobic systems were switched to anaerobic conditions and vice versa.

1 week (Figure 9). The decay was always faster in the presence than in the absence of oxygen, but only by less than 40% during the first few days. However, these experiments may not provide the true decay pattern of fresh diatoms because the diatom material may have lost labile cell contents due to cell rupture during freezing before being used. The decay of aged diatoms decreased gradually throughout the experiment at about ten times higher rates under aerobic than anaerobic conditions. When redox conditions were switched, the aerobic (former anaerobic) treatments exhibited the highest decay rates. The barley straw treatments behaved differently. Here, the initial decrease in decay rate was less dramatic than for diatoms. Fresh straw was degraded at similar rates irrespective of the presence of oxygen. Aged straw also exhibited similar initial rates irrespective of redox conditions, but with a gradual divergence after 2 weeks, eventually reaching five times higher aerobic than anaerobic rates. The divergence in rates may have been delayed several days in the present experiment because all aged treatments accidentally turned anaerobic for 1 week (day 4–9) due to a pump failure. The results obtained by these experiments clearly show that decay of aged organic matter is at least five times faster when mediated by aerobic microbial communities than by anaerobic assemblages with sulfate reduction as terminal oxidation process (Figure 9). Enzymatic hydrolysis related to oxygen is also more efficient than suboxic decomposition with nitrate as the terminal respiration process. Thus, respiration under suboxic (with nitrate) and reduced (with sulfate) conditions appears equally efficient, or even faster with the latter electron acceptor (Figure 10).

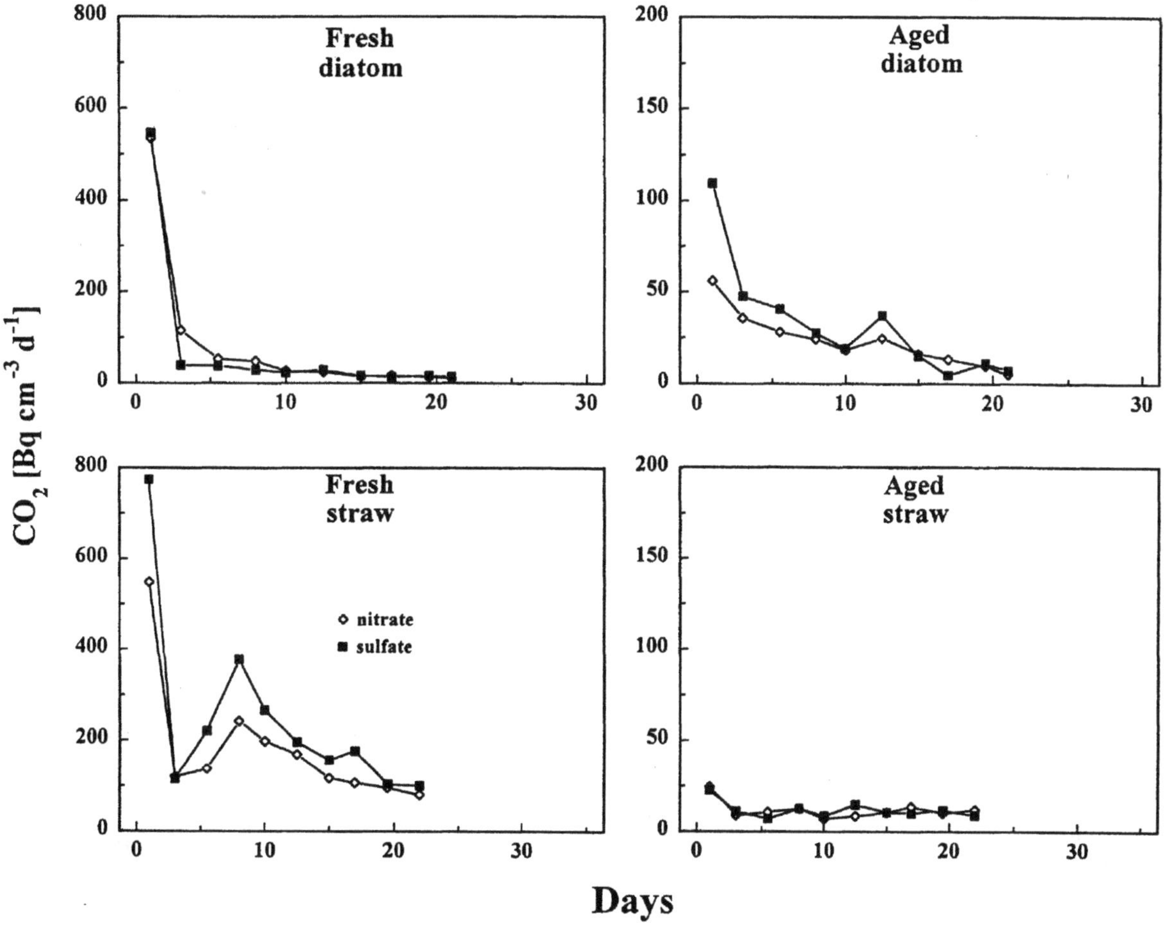

Figure 10. Decay ($^{14}CO_2$ release) by nitrate and sulfate respiration of ^{14}C-labelled diatoms (upper panels) and barley straw (lower panels) incubated in thin-layer, flow-through (seawater) systems. Both fresh (left panels) and pre-decomposed (right panels) materials were used.

As virtually no DO^{14}C was produced in any of the treatments, the limiting step of organic matter decay appears to be the initial hydrolysis of particulate matter into dissolved forms rather than the terminal respiration process. Fresh organic matter is degraded equally fast under aerobic and anaerobic conditions as long as the easily leachable and hydrolyzable fractions of the organic particles are not exhausted. The removal and decay of these fractions occur much faster for the diatom than the straw material.

Factors controlling processes at oxic/anoxic boundaries

The balance between downward transport and consumption and thus the penetration depth of oxygen in coastal sediments is controlled by a number of factors, of which many are of oscillating nature. The most important are:

1. temperature;
2. organic supply;
3. light;
4. water currents;
5. bioturbation.

Only the first four factors are dealt with in this chapter, whereas bioturbation will be treated in detail later.

Temperature

Temperature is a master factor controlling rates of chemical and biological processes in the marine environment. The temperature dependence of chemical and biological processes can be quantified by the apparent activation energy, E_a, according to the Arrhenius

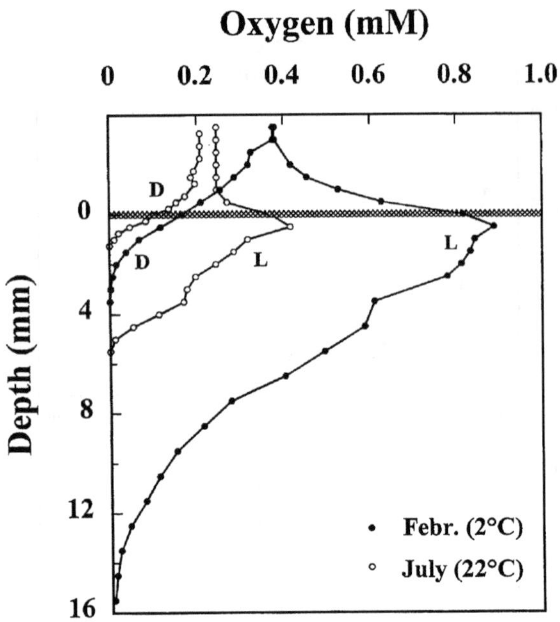

Figure 11. Oxygen profiles in a sandy coastal sediment (1 m water depth) measured in daylight (L) and in darkness (D) during winter (February) and summer (July).

equation or by the quotient of rate increase following a 10° increase in temperature, Q_{10}. E_a values can be converted to Q_{10} values in the temperature range from T to T+10K according to:

$$Q_{10} = \exp(E_a \times 10\text{K} \times [R \times T(T + 10\text{K})]^{-1}),$$

where R is the gas constant (Thamdrup et al., 1998). A number of studies have shown that seasonal variations of oxygen consumption in marine sediments is significantly related to temperature with a Q_{10} of 2–3 (Kristensen, 1993; Banta et al., 1995; Kristensen et al., 1998; Thamdrup et al., 1998). Since molecular diffusion increases only by 30–40% for a temperature increase of 10° within the range of 0–30°C, the observed high temperature dependence of oxygen uptake must be mediated by microbial (and chemical) processes. The high temperature dependence of sediment processes is further substantiated by the fact that a 10° increase in temperature (within the range of 0–30°C) decreases the solubility of oxygen in water by 17–23%, i.e. the highest oxygen uptake is reached when the availability of oxygen is lowest. As a consequence, the penetration depth and concentration of oxygen in sediments is lower during warm than cold periods (Figure 11), whereas the gradient driving the diffusive uptake at the sediment–water interface is steeper (Rasmussen & Jørgensen, 1992).

Table 3. The influence of phytoplankton deposition on bacterial and oxygen variables in a sandy sediment. Bacterial variable were integrated over the 0–63 mm sediment layer. Three treatments are presented: (1) starved sediment (no algal addition) averaged over 130 days; (2) single pulse of 24 g C m^{-2} and averaged over 39 days; (3) continuous weekly additions of 8 g C m^{-2} and averaged over 130 days (after Van Duyl et al., 1992)

	Starved	Pulse 24 g C m^{-2}	Weekly 8 g C m^{-2}
Bacterial biomass [mg C m^{-2}]	816	1036	1374
Bacterial production [mg C m^{-2} d^{-1}]	205	320	358
Oxygen consumption [mmol O_2 m^{-2} d^{-1}]	3.66	11.8	29.0
Oxygen penetration depth [mm]	13.4	11.4	10.2

Organic supply

Organic matter in sediments are considered a complex mixture of more or less labile compounds, and the overall microbial decay can be described as first-order decay using the multi-G model of Berner (1980):

$$G_T(t) = \sum_{i=l}^{n} G_i[\exp(-k_i t)] + G_{NR},$$

where G_T is the total decomposable (reactive) organic pool, k_i is the first-order decay constant of the type i pool, G_i is the amount of the type i pool, G_{NR} is a non-metabolizable pool. The pool, G_1, will be microbially degraded with a high k_1 before the pool, G_2 with a lower k_2, etc. The value of G_i, since it refers to a group of organic compounds, is independent of the microbial decomposition process, whereas the value of k_i will vary with the process. In most cases sedimentary organic matter can be divided into two decomposable fractions (G_1 and G_2) of considerably different reactivity (k_1 typically 5–10 yr^{-1} and k_2 less than 1 yr^{-1}), and a non-metabolizable fraction (G_{NR}) (Westrich & Berner, 1984). The non-metabolizable organic pool in marine sediments is either intrinsically stable or stabilized through monolayer-equivalent sorption of organic matter to mineral surfaces (Mayer, 1994; Hedges & Keil, 1995).

It is generally accepted that benthic oxygen uptake is directly related to the input of labile organic matter to the sediment (primarily G_1 material), and that no simple relationship can be obtained between oxygen uptake and total organic content (G_T) of the sediment due to unpredictable amounts of the G_{NR} fraction (Suess, 1980; Jørgensen, 1983; Cai & Reimers, 1995; Kristensen et al., 1998). As a consequence, oxygen penetration into sediment is highly dependent on the input of labile organic matter (Table 3) (Van Duyl et al., 1992; Kristensen & Hansen, 1995), which may vary with season (Kristensen, 1993). The limited input of labile organic matter to deep pelagic sediments is substantiated by the general increase in oxygen penetration depth observed with water depth (Figure 1)(Schlüter, 1991; Glud et al., 1994), and the negative empirical relationship between organic carbon deposition (normalized to annual primary production in the photic zone) and water depth as reported by Suess (1980).

Variations in oxygen penetration caused by seasonal fluctuations in fresh detritus influx are most conspicuous in coastal regions experiencing spring phytoplankton blooms (Jørgensen, 1996). In deep pelagic sediments, on the other hand, the seasonal signal may be dampened by pelagic degradation processes (e.g. zooplankton grazing) (Smith & Baldwin, 1984).

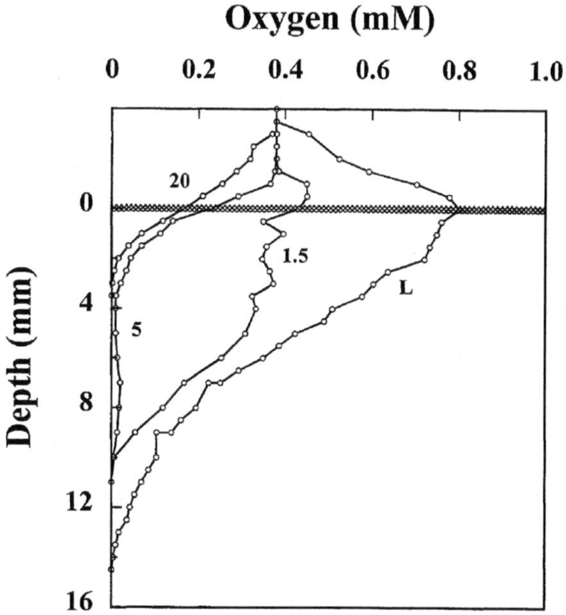

Figure 12. Oxygen profiles in a sandy coastal sediment (1 m water depth) measured in light, 1.5, 5 and 20 h after the light was turned off. Measurements were done in February (2°C).

Light

The direct influence of light on oxygen conditions in sediments is limited to shallow localities where light is sufficient to maintain a positive net photosynthesis by benthic primary producers. Such sediments are often inhabited by a dense population of benthic microalgae (e.g. pennate diatoms) or cyanobacterial mats utilizing the abundance of both light and nutrients occurring at the sediment–water interface (Colijn & Jonge, 1984). The chemical microenvironment created by the benthic primary producers may have more pronounced impacts on coastal ecosystems than their photosynthetic production of organic matter.

Benthic photosynthesis can cause large diurnal oscillations in chemical parameters like oxygen and pH at the sediment–water interface (Revsbech et al., 1983). Although light only penetrates a few mm into shallow coastal sediments (Kühl & Jørgensen, 1994), the benthic primary producers in the narrow photosynthetic zone may increase the oxygen concentration at the sediment–water interface several times the atmospheric saturation level (Figure 11). Such oversaturation may in extreme cases result in bubble formation (Revsbech et al., 1981). Temporal changes in oxygen concentration near the sediment surface as a response to light/dark shifts occur rapidly (Figure 12). The actual response time varies with sediment type, microphytobenthic community and temperature. Thus, in contrast to the concentration changes in the time scale of hours shown in Figure 12 for a sandy sediment during winter (2°C), Revsbech et al. (1986) found that a peak concentration at 0.2 mm depth of about 1000 μM during light exposure in a cyanobacterial mat overlying an organic-rich sediment (21°C) was reduced to zero after only 80 s.

As a result of the wide variation in oxygen concentrations found in the upper few mm photic zone of the sediment, the diffusion based penetration depth is also affected. The penetration depth of oxygen may increase up to 10-fold (Figure 12) by a shift from dark to light exposure (Revsbech et al., 1986). The diurnal up and down movement of the oxic/anoxic interface in photosynthetically active sediments may have serious implications for the zonation of aerobic and anaerobic microbial processes in the upper sediment layers. It is, therefore, advantageous for microorganisms associated with interfaces to be motile in order to follow the interface when it moves. Many sulfide-oxidizing and ammonium-oxidizing bacteria are in fact motile (Austin, 1988). Most of the non-motile heterotrophic bacteria living at or near the oxic/anoxic interface are instead facultative anaerobes with the capacity for both oxic and suboxic respiration (Fenchel et al., 1998).

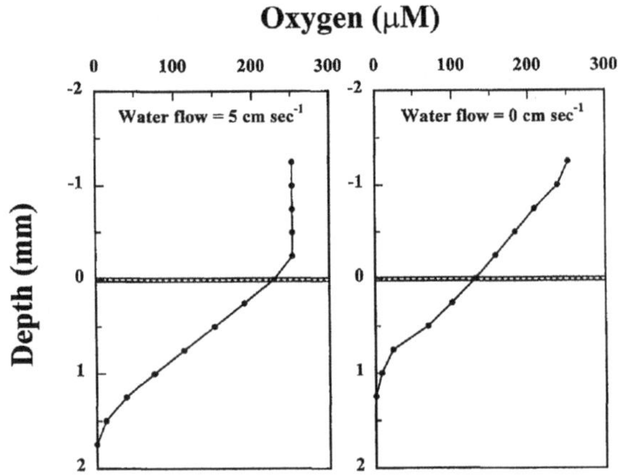

Figure 13. Oxygen profiles in a marine sediment under different flow regimes of the overlying water. *Left panel*: high water flow. *Right panel*: no water flow. The horizontal cross-hatched bar represents the sediment–water interface. Modified from Revsbech & Jørgensen (1986).

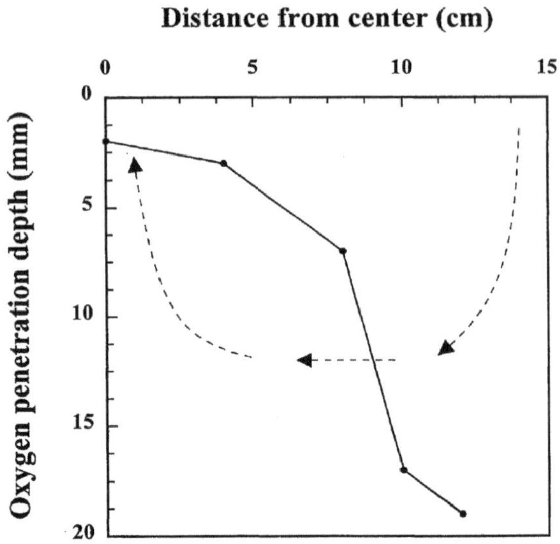

Figure 14. Oxygen penetration depth in a permeable sediment as a function of distance from the center of a cylindrical chamber with circular water movement. Dashed arrows indicate the direction of advective porewater movements. Modified from Huettel & Gust (1992b).

Water currents

The actual thickness of the diffusive boundary layer and thus the penetration depth of oxygen into sediments is determined by the water flow velocity and the roughness of the sediment surface. When the water flow above the sediment is increased, the boundary layer is reduced in thickness (Figure 13), thus facilitating diffusion dependent solute flux across the sediment–water interface (Jørgensen & Revsbech, 1985). At a constant water flow regime, however, the boundary layer generally increases with sediment particle size (surface roughness). Thus, the impact of water flow on solute flux is positively correlated with the degree of surface roughness.

Huettel & Gust (1992a) showed that biogenic microtopography, termed bioroughness (e.g. burrow excavations, feeding traces and fecal pellets) affect interfacial solute fluxes by up to one order of magnitude at high water flow velocities. Small-scale horizontal pressure gradients generated by flowing water around biogenic structures create advective porewater flows exceeding diffusive transport considerably. The water current induced advective porewater flow is highly dependent on the permeability of the sediment, and thus directly related to median grain size (Huettel & Gust, 1992b). Advective porewater flow has implications for flux measurements in permeable sediments. The use of cylindrical core tubes with circular water flow during sediment incubations may increase solute fluxes by more than a factor of 5 than expected under laminar, unidirectional flow (Huettel & Gust, 1992b). Causes are flow generated pressure gradients which create advective porewater flushing by forcing overlying water into the sediment close to the walls, pushing porewater up through the interface in the core center (Figure 14).

Benthic animals and oxic/anoxic interfaces

The burrow environment and the role of irrigation

Marine sediments underlying oxygenated waters are perforated with tubes and burrows formed by bottom-dwelling animals such as polychaetes, crustaceans and bivalves (Figure 15). These structures influence the geometry of reaction rates and solute (e.g. oxygen) distribution in the sediment creating a mosaic of microenvironments. The hypothetical vertical distribution of microbial processes (Figure 5) based on the availability of electron acceptors (Figure 3) is influenced by macrobenthos in a number of ways, e.g.:

1. Material is translocated continuously between reaction zones by feeding, burrowing and tube construction.
2. New reactive substrates in the form of mucus secretions are introduced into the sediment independent of sedimentation processes.
3. Tubes and burrows are irrigated with oxic surface water by ventilation activities of their inhabitants.

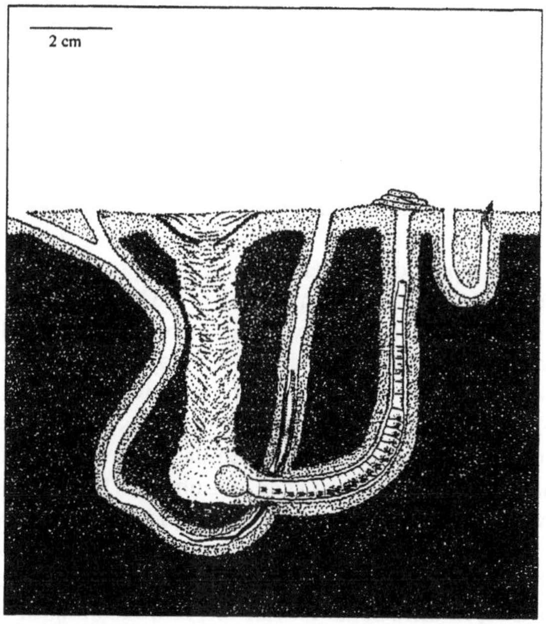

Figure 15. Drawing of the infaunal community in a shallow coastal sediment. From left, the polychaetes *Nereis diversicolor* and *Arenicola marina*, and the crustacean *Corophium colutator*. The heavily dotted part of the sediment is reduced and the lightly dotted sediment is oxidized.

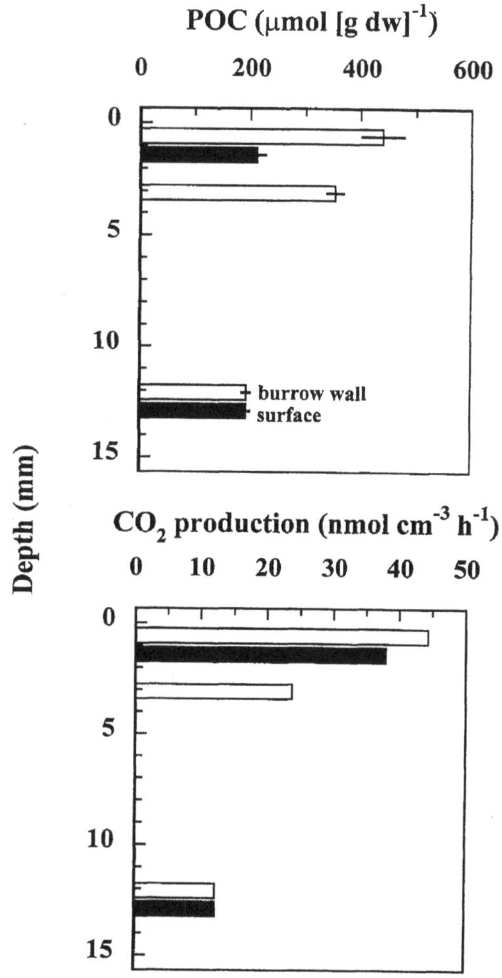

Figure 16. Particulate organic carbon content (POC, *upper panel)* and carbon mineralization rates (CO_2 production, *lower panel*) in a 15 mm deep radial profile around burrows of the polychaete, *Nereis diversicolor* (open bars) and a 15 mm deep vertical profile at the sediment surface (solid bars).

The extent to which these effects are realized depends on the functional groups of animals present, their abundance, taxonomic peculiarities and the size of individuals. Functional groups are defined by feeding type, life habit and mobility.

Tube or burrow structures differ in size, appearance and composition according to the functional group and size of the various infaunal species. They vary from mm-sized (small oligochaetes and polychaetes) to dm- or m-sized (large polychaetes and crustaceans) vertical or horizontal structures with variable degree of branching (Hertweck, 1986; Davey, 1994; Fenchel, 1996a; Ziebis et al., 1996). The wall lining of tubes and burrows usually consists of mucoid, membranous, parchment-like secretions encrusted with sand or shell debris (Defretin, 1971; Kristensen et al., 1991b). The lining is in most cases highly enriched in organic matter compared with the surrounding sediment (Figure 16), but its biodegradability is highly dependent on the chemical composition and structure of the secreted material. For example, the protein rich mucopolysaccharide secretions produced by burrow-dwelling infaunal animals, like polychaetes of the genus *Nereis*, are a readily degradable substrate for microbial growth (Figure 16; Aller & Aller, 1986; Reichardt, 1988). Structures, like the fibrous, leathery cerianthin tubes of the infaunal sea anemone, *Ceriantheopsis americanus*, on the other hand, are degraded at rates of less than 1% of those usually found for fresh planktonic debris (Kristensen et al., 1991b).

The permeability of tube and burrow linings to solute diffusion can be an important determinant of the chemical and biological composition of the surrounding sediment and the tube or burrow habitat. Aller (1983) found that the diffusive permeability of linings from eight infaunal species of marine invertebrates are 10–40% of that in free solution. The permeability of linings can, therefore, affect sedimentary solute distribution differently depending on the types of controlling reactions. The concentration of solutes subject

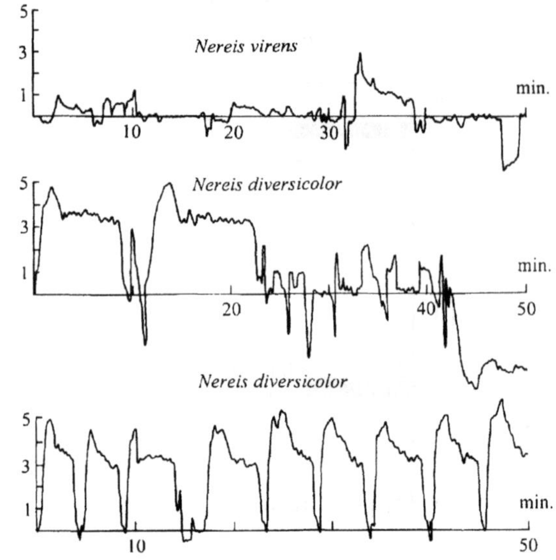

Figure 17. Examples of ventilation patterns of nereid polychaetes. *Upper panel*: *Nereis virens*. *Middle panel*: *Nereis diversicolor* during a non filter-feeding period. *Lower panel*: *Nereis diversicolor* during a filter-feeding period. The worms used in all three traces were of the same size (about 0.5 g wet weight).

to zero-order reactions are greatly influenced by lining permeability, but net fluxes across the lining are not. The opposite is true for solutes subject to first or higher order reactions.

Most infaunal animals actively ventilate or irrigate their burrows with oxygen-rich overlying water. The water current is driven by peristaltic or undulatory body movements in most polychaetes, by pleopods in crustaceans and by cilia in most bivalves. The renewal of burrow water serves important transport functions, such as supply of oxygen and other oxidized compounds (electron acceptors) at depth in the sediment and removal of metabolites (e.g. sulfide and ammonium). Burrow irrigation may therefore be an important factor controlling microbial processes in the sediment (Kristensen, 1988; Aller & Aller, 1998). Many studies have reported on infaunal ventilation patterns (Gust & Harrison, 1981; Kristensen, 1989; Riisgård, 1991; Forster & Graf, 1995). These have revealed that most infaunal animals show intermittent ventilation, interrupted by periods of rest, in a more or less rhythmic fashion. Kristensen (1989) found that the average duration of ventilation periods in the deposit-feeding polychaete, *Nereis virens*, is 5–8 min followed by a rest period of about 30 min; ventilation ocurring about 20% of the time (Figure 17). The total water ventilated by a population of *N. virens* (700 ind per m^2) at 16°C is then about 100 l m^{-2} d^{-1}. The closely related suspension-feeding species, *N. diversicolor*, behaves differently with almost continuous ventilation at high rates only interrupted by rest periods of a few minutes (Riisgård, 1991). The total amount of water pumped by populations of this active species (2400 ind per m^2) at 17°C is up to 9800 l m^{-2} d^{-1}.

The supply of oxygen in burrows is primarily dependent on the ventilation activity of the burrow inhabitants. The intermittent ventilation pattern observed for the majority of infaunal species may promote very variable oxygen conditions in the burrows. For nereid polychaetes, the oxygen level approaches that of the surface water during active ventilation periods, but during resting periods oxygen consumption by the burrow inhabitant and wall microbes rapidly exhaust the oxygen (Figure 18). The radial geometry of burrows together with highly reactive linings are responsible for a rapid diffusional loss of oxygen, resulting in a variable and generally low oxygen penetration into the wall sediment. Fenchel (1996a) found that the oxic zone around burrows of *N. diversicolor* typically extends 1–2 mm from the wall, which corresponds to between 40 and 70% of the oxic layer thickness of the surface sediment. Based on these observations, Fenchel (1996a) developed a simple model to describe the ratio between the thickness of the oxic zone around burrows (L_b) and of the oxic zone at the surface (L_s):

$$L_b/L_s = (-r + (r^2 + 2rL_s)^{\frac{1}{2}})/L_s,$$

where r is the radius of the burrow. The model assumes that the oxygen uptake of the sediment at the surface is identical to that surrounding the burrow and that the oxygen concentration at the surface and in the burrow are identical (realistically not true in most cases). The model output (Figure 19) shows that the thickness of the oxic zone around burrows always is thinner than at the sediment surface, but that the difference decreases with increasing r and increases with increasing L_s.

The quantitative role of burrows for aerobic microbial sediment processes cannot be evaluated from the temporal variability and penetration depth of oxygen in the wall alone. These data should be combined with a quantification of burrow wall areas, i.e. the surface-area-specific increase in sediment–water interfaces caused by burrows. A number of studies have attempted, by the use of a variety of techniques, to determine the surface area of burrow structures. For nereid polychaetes, the surface area representing burrow walls have been reported to exceed that of the

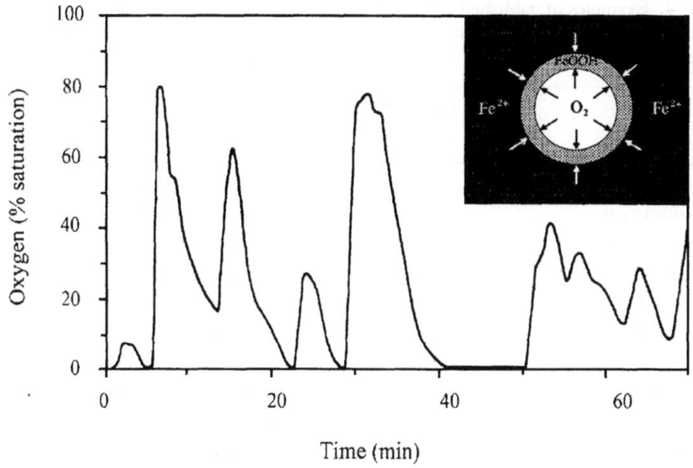

Figure 18. Temporal pattern of oxygen concentration in the middle of a *Nereis virens* burrow measured by a needle oxygen electrode. The inserted figure shows a cross section of a nereid burrow indicating the radial diffusion geometry involving oxygen and iron. Dissolved Fe^{2+} diffuses from the surrounding reduced sediment against the burrow and is concentrated in the wall. Oxygen diffuses rapidly from the burrow into the wall sediment where it drives the oxidation of Fe^{2+} which is precipitated as FeOOH.

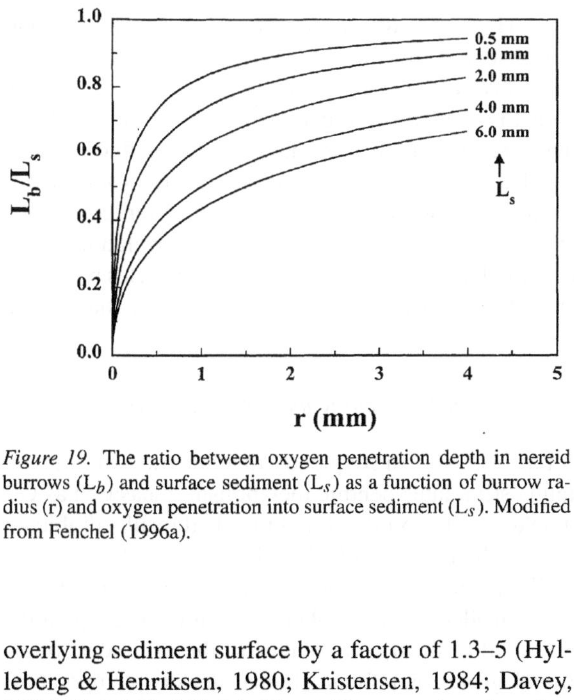

Figure 19. The ratio between oxygen penetration depth in nereid burrows (L_b) and surface sediment (L_s) as a function of burrow radius (r) and oxygen penetration into surface sediment (L_s). Modified from Fenchel (1996a).

overlying sediment surface by a factor of 1.3–5 (Hylleberg & Henriksen, 1980; Kristensen, 1984; Davey, 1994; Fenchel, 1996a). The wide range is a function of worm density and size distribution. By combining these data with the oxygen penetration model of Fenchel (1996a), assuming an average burrow radius $r = 2$ mm and an oxygen penetration into surface sediment $L_S = 2$ mm, the ratio of oxic sediment volume associated with burrows relative to the volume of oxic surface layer is between 0.9 and 3.3. Accordingly, the volume of oxic burrow-wall sediment may be several times the volume of oxic surface sediment.

Oxygen availability in the burrow environment is important for the macrofaunal inhabitant, but it also affects the associated meio- and microorganisms. The abundance of these organisms alongside infaunal burrows is normally quantitatively and qualitatively different from both the ambient anoxic and oxic surface sediment (Aller & Yingst, 1978; Wetzel et al., 1995; Fenchel, 1996b). For example, the density of meiofauna in burrow walls is usually high; in some instances higher than in the surface and in others lower, but generally higher than in the ambient sediment (Aller & Yingst, 1978; Reise, 1981). Thus, biogenic structures are expected to harbor 10–50% of the total abundance in the sediment. The diversity of meiofauna is highest at the sediment surface and generally much lower in both burrow walls and ambient anoxic sediment, where nematodes dominate (Kristensen, 1988). These observations indicate that the burrow environment is the habitat for specific biological assemblages, possibly induced by the unpredictable chemical conditions.

The chemical environment of burrow walls, e.g. narrow redox zonations, steep chemical gradients and presence of labile organic matter, is the basis for a very dynamic bacterial community. Reichardt (1988) found that the wall lining of burrows of the polychaete, *Arenicola marina*, has higher bacterial abundance and production than surface sediment, ambient anoxic sediment and fecal casts. Also, microheterotrophic activity and concentrations of hydrolytic enzymes were highest in the wall lining. Despite the generally low and variable oxygen concentrations, the activity of

Table 4. Examples of published values for enhancement of benthic metabolism in sediment inhabited by burrow-dwelling macrobenhos. The enhancement is given as percent difference in oxygen uptake between faunal-inhabited and defaunated sediment

Species	O_2 flux enhancement [%]	Reference
Nereis virens	74	Kristensen (1985)
Nereis virens	152	Kristensen & Blackburn (1987)
Nereis virens	74–84	Andersen & Kristensen (1988)
Nereis virens	72–89	Banta et al. (1999)
Nereis diversicolor	25–35	Kristensen et al. (1992)
Nereis diversicolor	100–140	Hansen & Kristensen (1997)
Nereis diversicolor	38–122	Kristensen & Hansen (1999)
Nephtys spp.	225	Hansen & Blackburn (1992)
Arenicola marina	141–271	Banta et al. (1999)
Penaeus setiferus	60– 90	Vetter & Hopkinson (1985)
Echinocardium chordatum	10–108	Van Duyl et al. (1992)

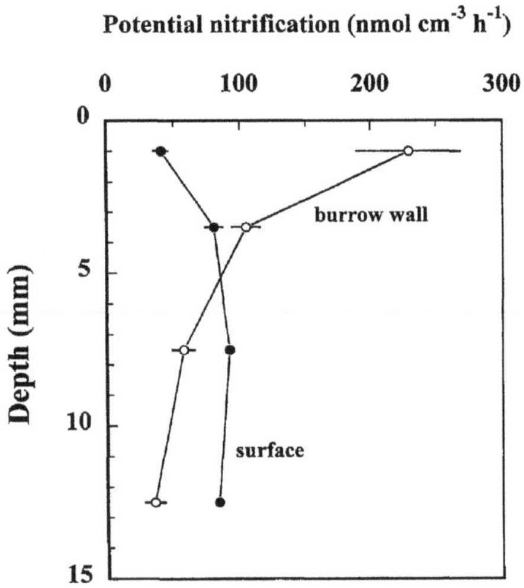

Figure 20. Rates of potential nitrification in a 15 mm deep radial profile around burrows of *Nereis virens* (burrow wall) and a 15 mm deep vertical profile at the sediment surface (surface). Modified from Kristensen et al. (1985).

certain aerobic processes is enhanced considerably in wall linings. Mayer et al. (1995) compiled published information on nitrification potentials in burrow linings of a variety of infaunal animals. They found that wall linings of most species have nitrification potentials significantly greater (2–20 times) than those of ambient anoxic sediment from the same depth interval. The nitrification potential in wall linings is also significantly greater (1.5–61 times) than that of surface sediment. They concluded that macrofaunal tubes and burrows tend to be sites of enhanced nitrification potential, and that this enhancement varies among species due to variations in irrigation behavior. However, Kristensen et al. (1985) found that the observed high nitrification potential in wall linings of *Nereis virens* burrows (Figure 20) is correlated with the content of mucus and fine particles. Nitrifiers are commonly associated with the fine-particle and high-organic fraction of sediments despite their chemoautotrophic metabolism (Henriksen & Kemp, 1988). It has been suggested that this correlation is due to the pH-buffering capacity of such particles retaining an optimum pH for nitrifiers (Fenchel & Blackburn, 1979).

Implications for organic matter diagenesis

It has been shown frequently that burrow-dwelling infauna stimulates benthic metabolism, measured as O_2 uptake or CO_2 production. Published results (Table 4) show enhancements of 25–271% compared with artificially defaunated sediments. The degree of stimulation is controlled by a number of factors, e.g. quantity and quality of organic matter in the sediment, functional group of infauna, faunal density, temperature and season (Aller, 1982; Kikuchi, 1987; Kristensen & Blackburn, 1987; Kristensen et al., 1991a; Forster & Graf, 1995). Besides faunal respiration and stimulation of microbial activity, a significant part of the flux enhancement is caused by porewater flushing, i.e. removal of porewater CO_2 or sulfide (with subsequent reoxidation), because most experiments have been conducted with artificial batch systems. Faunal induced flux enhancements measured in non-steady state batch systems are most likely overestimates of

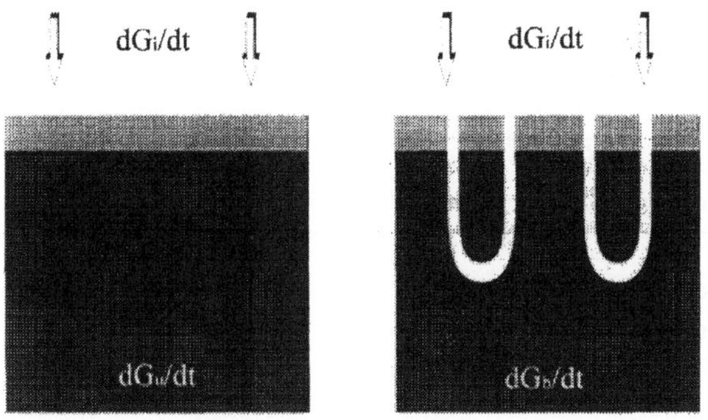

Figure 21. A hypothetical marine sediment with two adjoining sites. *Left*: without benthic macrofauna. *Right*: with a normal density of deposit-feeding infauna. Arrows indicate sedimentation of reactive organic matter at a rate of dG_i/dt. Bulk sediment organic matter decomposition is dG_u/dt at the fauna-free site and dG_b/dt at the faunated site.

conditions in the field and rather represent the enhanced metabolic capacity. This proposition is difficult to test under natural conditions because defaunated sediments underlying oxic water columns are rare. Instead, we must apply a speculative scenario including the current knowledge on animal effects.

The frame of our scenario is a hypothetical marine sediment environment, with no temporal (seasonal and diurnal) variability, which is supplied with organic carbon from above by deposition on the surface. The environment consists of two adjoining sites: into one a deposit-feeding infauna is introduced whereas the other remains defaunated. They both receive the same quantity and quality of organic input (dG_i/dt) (Figure 21). Thus, the deposit-feeding infauna is assumed not to increase organic carbon deposition. In the presence of filter-feeders, on the other hand, the organic input may be increased considerably (Christensen et al., 1999). The bulk decomposition rate within the sediment is assumed to be first-order with respect to reactive organic carbon content: $dG/dt = k\ G$, where k is the first-order decomposition coefficient (decomposition capacity) and G is the bulk pool of reactive organic carbon. The infauna is assumed to enhance the capacity for bulk benthic metabolism, and thus to increase the decomposition coefficient compared with the defaunated sediment, i.e. $k_b > k_u$. Initially (just after introduction of fauna), the total metabolism of the faunated ($dG_b/dt = k_b G_u$) system is k_b/k_u times higher than that of the defaunated system ($dG_u/dt = k_u G_u$) because the reactive organic pool G_u is similar in both systems. However, at steady state the amount of organic carbon mineralization must be equivalent to that being supplied (ignoring permanent burial), irrespective of the presence of animals: $dG_b/dt = dG_u/dt = dG_i/dt$. The steady state pool of reactive organic carbon in the faunated sediment must then be lower than in the defaunated sediment in proportion to the ratio between decomposition coefficients: $G_b = k_u/k_b\ G_u$. Basically, the difference between the two systems is an increased decomposition capacity (i.e. the decomposition coefficient, k) in the faunated system and thus a decreased total pool of reactive organic carbon at steady state compared with a defaunated system, while the bulk decomposition rate and flux are similar at both sites. In cases where filter-feeders increase organic input to the sediment, the steady state flux may be increased, but the organic pool can be higher than, similar to or lower than in the defaunated sediment depending on the functional type, density and size of animals (Christensen et al., 1999).

A number of mechanisms has been suggested to be responsible for the faunal induced enhancement of microbial metabolism and capacity for organic matter degradation in sediments, e.g. redistribution of particles, enhanced porewater transport and secretions of labile mucus alongside burrow walls (Aller, 1982; Kristensen, 1988; Aller & Aller, 1998). The animals themselves may also contribute significantly to the total benthic metabolism by feeding, assimilation and respiration (Andersen & Kristensen, 1988; Kristensen et al., 1991a). However, none of the microbial stimulations have yet been quantified experimentally. Here, a conceptual model is derived to explain the potential role of oxygen on enhancement of microbial decomposition capacity within burrows. The model describes the initial situation presented for the hypothetical sediment environment discussed above (Figure 21). Assume that all diagenetic processes occur in the upper 20 cm of the sediment column and that the oxic surface zone is 3 mm thick. Accordingly, the oxic sediment accounts for 1.5% of the active sediment volume. At the oxic surface zone, all organic matter is fresh and labile, whereas it is old and partly degraded in the anoxic zone. There is no depth dependent change in degradability with sediment depth in either zone. The total benthic CO_2 production is 100 mmol m^{-2} d^{-1}, of which 25 mmol m^{-2} d^{-1} is due to oxic respiration in the upper 3 mm. So, the volume specific carbon mineralization is 22 times higher in oxic (8.33 μmol cm^{-3} d^{-1}) than anoxic (0.38 μmol cm^{-3} d^{-1}) sediment (Table 5).

The deposit-feeding polychaete, *Nereis diversicolor*, is introduced at a density of 2000 ind m^{-2} (average length of 5 cm). These produce U-shaped burrows with a total length of 20 cm and a diameter of

Table 5. Conceptual model for determining the role of enhanced degradation of old organic matter exposed to oxygen along burrow walls of the polychaete, *Nereis diversicolor*, at a population density of 2000 ind m^{-2}. For further information consult the text

	Sediment volume [cm^3 m^{-2}]	Specific activity [μmol cm^{-3} d^{-1}]	Total activity [mmol m^{-2} d^{-1}]
Defaunated			
Oxic surface	3000	8.33	25.0
Anoxic subsurface	197 000	0.38	75.0
Total	200 000	0.50 ($\sim k_u$)	100.0 ($\sim k_u G_u$)
Faunated			
Oxic surface	2850	8.33	23.8
Burrow cavity	5027	0.00	0.0
Oxic burrow wall	14 627	3.81	55.7
Anoxic subsurface	177 496	0.38	67.4
Total	200 000	0.73 ($\sim k_b$)	146.9 ($\sim k_b G_u$)

0.4 cm. Continuous ventilation is assumed for simplicity and the oxic zone around burrows is 2 mm thick. Accordingly, the burrow cavity accounts for 2.5% and the oxic wall sediment for 7.3% of the total sediment volume. Based on the information given earlier, mineralization of the old and partly degraded organic matter around burrows is assumed to be enhanced by up to a factor of 10 when exposed to oxygen (3.81 μmol cm^{-3} d^{-1}). Accordingly, the total CO_2 production by the bioturbated sediment system is 147 mmol m^{-2} d^{-1}, equivalent to an increase of 47% (or k_b/k_u = 1.47) caused by the presence of *Nereis diversicolor* (Table 5). This increase is in the low range of published enhancement of benthic metabolic capacity due to these animals in batch systems, but accounts for a considerable fraction. The remainder must be caused by other, yet not quantified, mechanisms (Aller & Aller, 1998). However, this simple model example illustrates that enhancement of decomposition of old organic matter along the oxic walls of infaunal burrows should be considered an important contributor to the increased capacity for decomposition in sediments caused by benthic animals.

Future research directions

Much effort has been devoted within the last couple of decades to elucidate rates and controlling factors for biogeochemical processes occurring at and around the oxic/anoxic inteface in marine sediments. We are now able to determine the physical location of the interface with a precision of few micrometers due to the development of oxygen microelectrodes. Extensive work has also provided considerable knowledge on the distribution of various oxic, suboxic and anoxic respiration types within sediments. We have, however, a crucial lack of experimental capacity regarding various respiration types in marine sediments. Intensive methodological work is needed to develop methods to quantify oxygen, manganese and iron respiration directly in undisturbed sediment. So far, this has proved to be a difficult task, but nothing is impossible – new thinking is urgently needed here. We might use or develop alternative techniques which have not yet been considered in this context.

Our knowledge of factors determining preservation of organic carbon in sediments is still limited, and a controversy has evolved regarding the role of oxygen on rates of organic matter degradation. More work is certainly needed to fully elucidate the role of oxygen and other electron acceptors for decomposition rate of organic matter in order to understand the underlying mechanisms for carbon preservation. The work of Don Canfield, John Hedges, Larry Mayer and others has provided essential insight into the complexity of the problems, but we still need to know the processes involved in full detail before the, in many respects, important process of carbon preservation is understood.

Activities of burrow-dwelling fauna appears to be one of the most important factors controlling organic matter diagenesis in sediments. Bob Aller and coworkers improved our understanding of the basic role of burrow-dwelling infauna on sediment processes two decades ago. These early studies did not fully solve the problems related to benthic infauna, although they made a giant jump forward – now is the time to continue. Thus, the underlying mechanisms for the commonly observed enhancement of organic matter decomposition in the presence of animals are still not fully understood. We still rely on largely untested hypothesis raised some 20 years ago. Is it the redistribution of partcles, removal of inhibitory metabolites, secretions of labile mucus along burrow walls, direct effects by animal ingestion and metabolism, or the introduction of oxygen into otherwise anaerobic sediment that enhances diagenetic processes in sediments? In the present review, I provide some ideas which need to be confirmed (or rejected) experimentally.

In our effort to study bioturbation effects, we should expand the work to include more animals species or functional groups of infauna – not only in coastal waters, but also in deeper parts of the ocean. Virtually nothing is known on biotubation effects in continental slope and deep sea areas (e.g. in the vicinity of hydrothermal vent zones).

References

Aller, J. Y. & R. C. Aller, 1986. Evidence for localized enhancement of biological activity associated with tube and burrow structures in deep-sea at the HEBBLE site western North Atlantic. Deep Sea Res. 33: 755–790.

Aller, R. C., 1982. The effects of macrobenthos on chemical properties of marine sediment and overlying water. In McCall, P. L. & P. J. S. Tevesz (eds), Animal-Sediment Relations. Plenum, New York: 53–102.

Aller, R. C., 1983. The importance of the diffusive permeability of animal burrow linings in determining marine sediment chemistry. J. mar. Res. 41: 299–322.

Aller, R. C., 1990. Bioturbation and manganese cycling in hemipelagic sediments. Phil. Trans. r. Soc., Lond. A 331: 51–68.

Aller, R. C. & J. Y. Aller, 1998. The effect of biogenic irrigation intensity and solute exchange on diagenetic reaction rates in marine sediments. J. mar. Res. 56: 905–936.

Aller, R. C. & J. Y. Yingst, 1978. Biogeochemistry of tube-dwellings: a study of the sedentary polychaete *Amphitrite ornata* (Leidy). J. mar. Res. 36: 201–254.

Andersen, F. Ø., 1996. Fate of organic carbon added as diatom cells to oxic and anoxic marine sediment microcosms. Mar. Ecol. Prog. Ser. 134: 225–233.

Andersen, F. Ø. & E. Kristensen, 1988. The influence of macrofauna on estuarine benthic community metabolism – a microcosm study. Mar. Biol. 99: 591–603.

Archer, D., S. Emerson & C. R. Smith, 1989. Direct measurement of the diffusive sublayer at the deep sea floor using oxygen microelectrodes. Nature 340: 623–626.

Austin, B., 1988. Marine Microbiology. Cambridge Univ. Press, Cambridge. 222 pp.

Banta, G. T., A. E. Giblin, J. E. Hobbie & J. Tucker, 1995. Benthic respiration and nitrogen release in Buzzards Bay, Massachusetts. J. mar. Res. 53: 107–135.

Banta, G. T., M. Holmer, M. H. Jensen & E. Kristensen, 1999. The effects of two polychaete worms, *Nereis diversicolor* and *Arenicola marina*, on aerobic and anerobic decomposition in a sandy marine sediment. Aquat. Microb. Ecol. 19: 189–204.

Benner, R., A. E. Maccubbin & R. E. Hodson, 1984. Preparation, characterization and microbial degradation of specifically radiolabelled [^{14}C] lignocelluloses from marine and freshwater macrophytes. Apl. envir. Microbiol. 47: 381–389.

Berner R. A., 1980. Early Diagenesis, a Theoretical Approach. Princeton University Press, New Jersey: 241 pp.

Brandes, J. A. & A. H. Devol, 1995. Simultaneous nitrate and oxygen respiration in coastal sediments: evidence for discrete diagenesis. J. mar. Res. 53: 771–797.

Cai, W.-J. & C. E. Reimers, 1995. Benthic oxygen flux, bottom water oxygen concentration and core top organic carbon content in the deep northeast Pacific Ocean. Deep Sea Res. I 42: 1681–1699.

Calvert, S. E. & T. F. Pedersen, 1992. Organic carbon accumulation and preservation in marine sediments: how important is anoxia? In Whelan, J. K. & J. W. Farrington (eds), Organic Matter: Productivity, Accumulation and Preservation in Recent and Ancient Sediments. Columbia Univ. Press, New York: 231–263.

Canfield, D. E., 1989. Sulfate reduction and oxic respiration in marine sediments: implications for organic carbon preservation in euxinic environments. Deep-Sea Res. 36: 121–138.

Canfield, D. E., 1994. Factors influencing organic carbon preservation in marine sediments. Chem. Geol. 114: 315–329.

Canfield, D. E., B. B. Jørgensen, H. Fossing, R. Glud, J. Gundersen, N. B. Ramsing, B. Thamdrup, J. W. Hansen, L. P. Nielsen & P. O. J. Hall, 1993a. Pathways of organic carbon oxidation in three continental margin sediments. Mar. Geol. 113: 27–40.

Canfield, D. E., B. Thamdrup & J. W. Hansen, 1993b. The anaerobic degradation of organic matter in Danish coastal sediments: iron reduction, manganese reduction and sulfate reduction. Geochim. Cosmochim. Acta 57: 3867–3884.

Chanton, J. P., C. S. Martens & M. B. Goldhaber, 1987. Biogeochemical cycling in an organic-rich coastal marine basin. 7. Sulfur mass balance, oxygen uptake and sulfide retention. Geochim. Cosmochim. Acta 51: 1187–1199.

Christensen, B., A. Vedel & E. Kristensen, 1999. Carbon and nitrogen fluxes in sediment inhabited by suspension-feeding (*Nereis diversicolor)* and non suspension-feeding (*Nereis virens*) polychaetes. Mar. Ecol. Prog. Ser. (in press).

Colijn, F. & V. N. de Jonge, 1984. Primary production of microphytobenthos in the Ems-Dollard estuary. Mar. Ecol. Prog. Ser. 14: 185–196.

Davey, J. T., 1994. The architecture of the burrow of *Nereis diversicolor* and its quantification in relation to sediment–water exchange. J. exp. mar. Biol. Ecol. 179: 115–129.

Defretin, R., 1971. The tubes of polychaete annelids. In Florkin, M. & E. H. Stotz (eds), Comprehensive Biochemistry, Vol. 26.2. Elsevier, Amsterdam: 713–747.

Van Duyl, F. C., A. J. Kop, A. Kok & A. J. J. Sandee, 1992. The impact of organic matter and macrozoobenthos on bacterial and oxygen variables in marine sediment boxcosms. Neth. J. Sea Res. 29: 343–355.

Fenchel, T., 1996a. Worm burrows and oxic microniches in marine sediments. 1. Spatial and temporal scales. Mar. Biol. 127: 289–295.

Fenchel, T., 1996b. Worm burrows and oxic microniches in marine sediments. 2. Distribution patterns of ciliated protozoa. Mar. Biol. 127: 297–301.

Fenchel, T & T. H. Blackburn, 1979. Bacteria and Mineral Cycling, Academic Press, London: 225 pp.

Fenchel, T., G. M. King & T. H. Blackburn, 1998. Bacterial Biogeochemistry: the Ecophysiology of Mineral Cycling. Academic Press, San Diego: 307 pp.

Forster, S & G. Graf, 1995. Impact of irrigation on oxygen flux into the sediment: intermittent pumping by *Callianassa subterranea* and 'piston pumping' by *Lanice conchilega*. Mar. Biol. 123: 335–346.

Fossing, H. & B. B. Jørgensen, 1989. Measurement of bacterial sulfate reduction in sediments: evaluation of a single-step chromium reduction. Biogeochemistry 8: 205–222.

Froelich P. N., G. Klinkhammer, M. L. Bender, N. A. Luedtke, G. R. Heath, D. Cullen, P. Dauphin, D. Hammond & B. Hartman, 1979. Early oxidation of organic matter in pelagic sediments of the eastern equatorial Atlantic: suboxic diagenesis. Geochim. Cosmochim. Acta 43: 1075–1095.

Glud, R. N., J. K. Gundersen, B. B. Jørgensen, N. P. Revsbech & H. D. Schulz, 1994. Diffusive and total oxygen uptake of deep-sea sediments in the eastern South Atlantic Ocean: *in situ* and laboratory measurements. DeepSea Res. I 41: 1767–1788.

Glud, R. N., N. B. Ramsing, J. K. Gundersen & I. Klimant, 1996. Planar optrodes: a new tool for fine scale measurements of two-dimensional O_2 distribution in benthic communities. Mar. Ecol. Prog. Ser. 140: 217–226.

Gundersen, J. K. & B. B. Jørgensen, 1990. Microstructure of diffusive boundary layers and the oxygen uptake of the sea floor. Nature 345: 604–607.

Gust, G. & J. T. Harrison, 1981. Biological pumps at the sediment–water interface: mechanistic evaluation of the alpheid shrimp *Alpheus mackayi* and its irrigation pattern. Mar. Biol. 64: 71–78.

Hansen, L. S. & T. H. Blackburn, 1991. Aerobic and anaerobic mineralization of organic material in marine sediment microcosms. Mar. Ecol. Prog. Ser. 75: 283–291.

Hansen, L. S. & T. H. Blackburn, 1992. Mineralization budgets in sediment microcosms: effect of the infauna and anoxic conditions. FEMS Microbiol. Ecol. 102: 33–43.

Hansen, K. & E. Kristensen, 1997. Impact of macrofaunal recolonization on benthic metabolism and nutrient fluxes in a shallow marine sediment previously overgrown with macroalgal mats. Estuar. coast. shelf Sci. 45: 613–628.

Hartnett, H. E., R. G. Keil, J. I. Hedges & A. H. Devol, 1998. Influence of oxygen exposure time on organic carbon preservation in continental margin sediments. Nature 391: 572–574.

Harvey, H. R., J. H. Tuttle & J. T. Bell, 1995. Kinetics of phytoplankton decay during simulated sedimentation: changes in biochemical composition and microbial activity under oxic and anoxic conditions. Geochim. Cosmochim. Acta 59: 3367–3378.

Hedges, J. I. & R. G. Keil, 1995. Sedimentary organic matter preservation: an assessment and speculative synthesis. Mar. Chem. 49: 81–115.

Henrichs, S. M. & W. S. Reeburgh, 1987. Anaerobic mineralization of marine sediment organic matter: rates and the role of anaerobic processes in the oceanic carbon economy. Geomicrobiol. J. 5: 191–237.

Henriksen, K. & W. M. Kemp, 1988. Nitrification in estuarine and coastal marine sediments. In Blackburn, T. H. & J. Sørensen (eds), Nitrogen Cycling in Coastal Marine Environments. John Wiley & Sons Ltd., Chichester: 207–249.

Hertweck, G., 1986. Burrows of the polychaete *Nereis virens* Sars. Senckenberg. marit. 17: 319–331.

Howarth, R. W., 1984. The ecological significance of sulfur in the energy dynamics of salt marsh and coastal marine sediments. Biogeochemistry 1: 5–27.

Huettel, M. & G. Gust, 1992a. Impact of bioroughness on interfacial solute exchange in permeable sediments. Mar. Ecol. Prog. Ser. 89: 253–267.

Huettel, M. & G. Gust, 1992b. Solute release mechanisms from confined sediment cores in stirred benthic chambers and flume flows. Mar. Ecol. Prog. Ser. 82: 187–197.

Hulthe, G., S. Hulth & P. O. J. Hall, 1998. Effect of oxygen on degradation rate of refractory and labile organic matter in continental margin sediments. Geochim. Cosmochim. Acta 62: 1319–1328.

Hylleberg, J. & K. Henriksen, 1980. The central role of bioturbation in sediment mineralization and element re-cycling. Ophelia, Suppl. 1: 1–16.

Jahnke, R., 1985. A model of microenvironments in deep-sea sediments: formation and effects on porewater profiles. Limnol. Oceanogr. 30: 956–965.

Jensen, K., N. P. Revsbech & L. P. Nielsen, 1993. Microscale distribution of nitrification activity in sediment determined with a shielded microsensor for nitrate. Apl. envir. Microbiol. 59: 3287–3296.

Jørgensen, B. B., 1977. Bacterial sulfate reduction within reduced microniches of oxidized marine sediments. Mar. Biol. 41: 7–17.

Jørgensen, B. B., 1983. Processes at the sediment–water interface. In Bolin, B. & R. B. Cook (eds), The Major Biogeochemical Cycles and Their Interactions. SCOPE 21, Stockholm: 477–509.

Jørgensen, B. B., 1989. Biogeochemistry of chemoautotrophic bacteria. In Schlegel, H. G. & B. Bowien (eds), Autotrophic Bacteria. Science Tech Publ. & Springer-Verlag, Madison: 117–146.

Jørgensen, B. B., 1996. Material flux in the sediment. In Jørgensen, B. B. & K. Richardson (eds), Eutrophication in Coastal Marine Ecosystems (Coastal and Estuarine Studies 52). American Geophysical Union, Washington DC: 115–135.

Jørgensen, B. B. & N. P. Revsbech, 1983. Colorless sulfur bacteria *Beggiatoa* spp. and *Thiovulum* spp. in O_2 and H_2S microgradients. Apl. envir. Microbiol. 45: 1261–1270.

Jørgensen, B. B. & N. P. Revsbech, 1985. Diffusive boundary layers and the oxygen uptake of sediments and detritus. Limnol. Oceanogr. 30: 111–122.

Jørgensen, B. B. & J. Sørensen, 1985. Seasonal cycles of O_2, NO_3^- and SO_4^{2-} reduction in estuarine sediments: the significance of a NO_3^- reduction maximum in spring. Mar. Ecol. Prog. Ser. 24: 65–74.

Keil, R. G., D. B. Montlucon, F. G. Prahl & J. I. Hedges, 1994. Sorptive preservation of labile organic matter in marine sediments. Nature 370: 549–552.

Kemp, W. M., P. A. Sampou, J. Garber, J. Tuttle & W. R. Bounton, 1992. Seasonal depletion of oxygen from bottom waters of Chesapeake Bay: roles of benthic and planktonic respiration and physical exchange processes. Mar. Ecol. Prog. Ser. 85: 137–152.

Kikuchi, E., 1987. Effect of the brackish deposit-feeding polychaetes *Notomastus* sp. (Capitellidae) and *Neanthes japonica* (Izuka) (Nereidae) on sedimentary O_2 consumption and CO_2 production rates. J. exp. mar. Biol. Ecol. 114: 15–25.

Klimant, I., V. Meyer & M. Kühl, 1995. Fiber-optic oxygen microsensors, a new tool in aquatic biology. Limnol. Oceanogr. 40: 1159–1165.

Kristensen, E., 1984. Effect of natural concentrations on exchange of nutrients between a polychaete burrow in estuarine sediment and overlying water. J. exp. mar. Biol. Ecol. 75: 171– 190.

Kristensen, E., 1985. Oxygen and inorganic nitrogen exchange in a *Nereis virens* (Polychaeta) bioturbated sediment–water system. J. Coast. Res. 1: 109–116.

Kristensen, E., 1988. Benthic fauna and biogeochemical processes in marine sediments: microbial activities and fluxes. In Blackburn, T. H. & J. Sørensen (eds), Nitrogen Cycling in Coastal Marine Environments. John Wiley & Sons Ltd., Chichester: 275–299.

Kristensen, E., 1989. Oxygen and carbon dioxide exchange in the polychaete *Nereis virens*: influence of ventilation activity and starvation. Mar. Biol. 101: 381–388.

Kristensen, E., 1993. Seasonal variations in benthic community metabolism and nitrogen dynamics in a shallow, organic poor Danish lagoon. Estuar. coast. shelf. Sci. 36: 565–586.

Kristensen, E., S. I. Ahmed & A. H. Devol, 1995. Aerobic and anaerobic decomposition of organic matter in marine sediment: which is fastest? Limnol. Oceanogr. 40: 1430–1437.

Kristensen, E., R. C. Aller & J. Y. Aller, 1991b. Oxic and anoxic decomposition of tubes from the burrowing sea-anemone *Ceriantheopsis americanus*: implications for sediment carbon and nitrogen dynamics. J. mar. Res. 49: 589–617.

Kristensen, E., F. Ø. Andersen & T. H. Blackburn, 1992. Effects of benthic macrofauna and temperature on degradation of macroalgal detritus: the fate of organic carbon. Limnol. Oceanogr. 37: 1404–1419.

Kristensen, E. & T. H. Blackburn, 1987. The fate of organic carbon and nitrogen in experimental marine sediment systems: influence of bioturbation and anoxia. J. mar. Res. 45: 231–257.

Kristensen, E. & K. Hansen, 1995. Decay of plant detritus in organic-poor marine sediment: production rates and stoichiometry of dissolved C and N compounds. J. mar. Res. 53: 675–702.

Kristensen, E. & K. Hansen, 1999. Transport of carbon dioxide and ammonium in bioturbated (*Nereis diversicolor*) coastal, marine sediments. Biogeochemistry 45: 147–168.

Kristensen, E., M. H. Jensen & T. K. Andersen, 1985. The impact of polychaete (*Nereis virens* Sars) burrows on nitrification and nitrate reduction in estuarine sediments. J. exp. mar. Biol. Ecol. 85: 75–91.

Kristensen, E., M. H. Jensen & R. C. Aller, 1991a. Direct measurement of dissolved inorganic nitrogen exchange and denitrification in individual polychaete (*Nereis virens*) burrows. J. mar. Res. 49: 355–377.

Kristensen, E., M. H. Jensen & K. M. Jensen, 1998. Sulfur dynamics in sediments of Königshafen. In Gätje, C. & K. Reise (eds), Ökosystem Wattenmeer – Austausch-, Transport- und Stoffumwandlungsprozesse. Springer-Verlag, Berlin: 233–256.

Kühl, M. & B. B. Jørgensen, 1994. The light field of microbenthic communities: radiance distribution and microscale optics of sandy coastal sediments. Limnol. Oceanogr. 39: 1368–1398.

Lee, C., 1992. Controls on organic carbon preservation: the use of stratified water bodies to compare intrinsic rates of decomposition in oxic and anoxic systems. Geochim. Cosmochim. Acta 56: 3323–3335.

Mackin, J. E. & K. T. Swider, 1989. Organic matter decomposition pathways and oxygen consumption in coastal marine sediments. J. mar. Res. 47: 681–716.

Mayer, L. M., 1994. Surface area control of organic carbon accumulation in continental shelf sediments. Geochim. Cosmochim. Acta 58: 1271–1284.

Mayer, M. S., L. Schaffner & W. M. Kemp, 1995. Nitrification potentials of benthic macrofaunal tubes and burrow walls: effects of sediment NH_4^+ and animal irrigation behavior. Mar. Ecol. Prog. Ser. 121: 157–169.

Møller, J. S., 1996. Water masses, stratification and circulation. In Jørgensen, B. B. & K. Richardson (eds), Eutrophication in Coastal Marine Ecosystems (Coastal and Estuarine Studies 52). American Geophysical Union, Washington DC: 51–66.

Nielsen, L. P., 1992. Denitrification in sediment determined from nitrogen isotope pairing. FEMS Microbiol. Ecol. 86: 357–362.

Rasmussen, H. & B. B. Jørgensen, 1992. Microelectrode studies of seasonal oxygen uptake in a coastal sediment: role of molecular diffusion. Mar. Ecol. Prog. Ser. 81: 289–303.

Reichardt, W., 1988. Impact of bioturbation by *Arenicola marina* on microbiological parameters in intertidal sediments. Mar. Ecol. Prog. Ser. 44: 149–158.

Reimers, C. E., 1987. An *in situ* microprofiling instrument for measuring interfacial pore water gradients: methods and oxygen profiles from the North Pacific Ocean. Deep Sea Res. 34: 2019–2035.

Reimers, C. E., K. M. Fischer, R. Merewether, K. L. Smith Jr. & R. A. Jahnke, 1986. Oxygen microprofiles measured *in situ* in deep ocean sediments. Nature 320: 741–744.

Reise, K., 1981. High abundance of small zoobenthos around biogenic structures in tidal sediments of the Wadden Sea. Helgolënder wiss. Meeresunters. 34: 413–425.

Revsbech, N. P., 1989. An oxygen microsensor with a guard cathode. Limnol. Oceanogr. 34: 474–478.

Revsbech, N. P. & B. B. Jørgensen, 1986. Microelectrodes: Their use in microbial ecology. Adv. Microb. Ecol. 9: 293–352.

Revsbech, N. P., B. B. Jørgensen, T. H. Blackburn & Y. Cohen, 1983. Microelectrode studies of the photosynthesis and O_2, H_2S and pH profiles of a microbial mat. Limnol. Oceanogr. 28: 1062–1074.

Revsbech, N. P., B. B. Jørgensen & O. Brix, 1981. Primary production in sediments measured by oxygen microprofiles $H^{14}CO_3^-$ fixation, and oxygen exchange methods. Limnol. Oceanogr. 26: 717–730.

Revsbech, N. P., B. Madsen & B. B. Jørgensen, 1986. Oxygen production and consumption in sediments determined at high spatial resolution by computer simulation of oxygen microelectrode data. Limnol. Oceanogr. 31: 293–304.

Revsbech, N. P., J. Sørensen, T. H. Blackburn & J. P. Lomholt, 1980. Distribution of oxygen in marine sediments measured with microelectrodes. Limnol. Oceanogr. 25: 403–411.

Revsbech, N. P. & D. M. Ward, 1983. Oxygen microelectrode that is insensitive to medium chemical composition: use in an acid microbial mat dominated by *Cyanidium caldarium*. Apl. envir. Microbiol. 45: 755–759.

Riisgård, H. U., 1991. Suspension feeding in the polychaete *Nereis diversicolor*. Mar. Ecol. Prog. Ser. 70: 29–37.

Santschi, P. H., P. Bower, U. P. Nyffeler, A. Azvedo & W. S. Broecker, 1983. Estimates of the resistance of chemical transport posed by the deep-sea boundary layer. Limnol. Oceanogr. 28: 899–912.

Schink, B., 1988. Principles and limits of anaerobic degradation: Environmental and technological aspects. In Zehnder, A. J. B. (ed.), Biology of Anaerobic Microorganisms. John Wiley & Sons Ltd., New York: 771–846.

Schlüter, M., 1991. Organic carbon flux and oxygen penetration into sediments of the Weddell Sea: indicators for regional differences in export production. Mar. Chem. 35: 569–579.

Smith, K. L., Jr. & R. J. Baldwin, 1984. Seasonal fluctuations in deep-sea sediment community oxygen consumption: central and eastern North Pacific. Nature 307: 624–625.

Stigebrandt, A. & F. Wulff, 1987. A model for the dynamics of nutrients and oxygen in the Baltic proper. J. mar. Res. 45: 729–759.

Suess, E., 1980. Particulate organic carbon flux in the oceans – surface productivity and oxygen utilization. Nature 288: 260–263.

Sun, M.-Y., C. Lee & R. C. Aller, 1993b. Anoxic and oxic degradation of ^{14}C-labeled chloropigments and a ^{14}C-labeled diatom in Long Island Sound sediments. Limnol. Oceanogr. 38: 1438–1451.

Sun, M.-Y., C. Lee & R. C. Aller, 1993a. Laboratory studies of oxic and anoxic degradation of chlorophyll-*a* in Long Island Sound sediments. Geochim. Cosmochim. Acta 57: 147–158.

Sun, M.-Y., S. G. Wakeham & C. Lee, 1997. Rates and mechanisms of fatty acid degradation in oxic and anoxic coastal marine sediments of Long Island Sound, New York, U.S.A. Geochim. Cosmochim. Acta 61: 341–356.

Thamdrup, B, H. Fossing & B. B. Jørgensen, 1994. Manganese, iron and sulfur cycling in a coastal marine sediment, Aarhus Bay, Denmark. Geochim. Cosmochim. Acta 58: 5115–5130.

Thamdrup, B., J. W. Hansen & B. B. Jørgensen, 1998. Temperature dependence of aerobic respiration in a coastal sediment. FEMS Microbiol. Ecol. 25: 189–200.

Vetter, E. F. & C. S. Hopkinson, Jr., 1985. Influence of white shrimp (*Penaeus setiferus*) on benthic metabolism and nutrient flux in a coastal marine ecosystem: Measurements *in situ*. Contr. Mar. Sci. 28: 95–107.

Westrich, J. T. & R. A. Berner, 1984. The role of sedimentary organic matter in bacterial sulfate reduction: the G model tested. Limnol. Oceanogr. 29: 236–249.

Wetzel, M. A., P. Jensen & O. Giere, 1995. Oxygen/sulfide regime and nematode fauna associated with *Arenicola marina* burrows: new insights in the thiobios case. Mar. Biol. 124: 301–312.

Ziebis, W., S. Forster, M. Huettel & B. B. Jørgensen, 1996. Complex burrows of the mud shrimp *Callianassa truncata* and their geochemical impact in the sea bed. Nature 382: 619–622.

Hydrobiologia **426:** 25–36, 2000.
G. Liebezeit, S. Dittmann & I. Kröncke (eds), Life at Interfaces and Under Extreme Conditions.

Variation in abundances of intertidal populations: consequences of extremities of environment

A.J. Underwood & M.G. Chapman
Centre for Research on Ecological Impacts of Coastal Cities, Marine Ecology Laboratories, A11, University of Sydney, NSW 2006 Australia. E-mail: aju@bio.usyd.edu.au

Key words: rocky shore, intertidal, spatial variability, temporal variability, barnacles, limpets

Abstract

Spatial and temporal variability in patterns and processes are poorly understood for most assemblages, including intertidal assemblages on rocky shores. Such habitats represent gradients of increasing physical harshness towards the top of the shore. Influences of environmental harshness on numbers and recruitment of animals lead to the hypotheses that there should be greater small-scale spatial variance in more extreme parts of the environment (higher on the shore) and greater short-term temporal variance in less extreme (lower) parts of the shore. These predictions were tested using two barnacles (*Tesseropora rosea* and *Chamaesipho tasmanica*) and two limpets (*Patelloida latistrigata* and *Notoacmea petterdi*) in up to five locations on wave-exposed rocky shores in New South Wales (Australia). Only *T. rosea* showed different spatial variation at higher and lower levels – but greater at lower than higher levels, contradicting the hypothesis. *T. rosea* and *P. latistrigata* had greater temporal variation at lower levels, supporting the hypothesis. *C. tasmanica*, however, showed the opposite pattern, contradicting the hypothesis. Despite close associations between *T. rosea* and *P. latistrigata*, patterns of relationship of spatial variances were inconsistent. The study demonstrates the need for tests in numerous sites and that ecological patterns can differ markedly even for closely related species. It also demonstrates that much more widespread sampling and experimentation are needed to retain predictive and explanatory understanding of environmental gradients.

Introduction

Variations in numbers of many animals and plants on intertidal rocky shores are large, often unpredictable. Usually, temporal changes are not in phase from one place to another. So, there is often great interaction in the patterns of spatial variation from one time to another and in patterns of temporal change from one place to another (e.g., Chesson, 1978). This, in turn, leads to complex spatial and temporal variation in the intensities and sometimes the directions of interactions among species, for example competition (Dayton, 1971), predation (Fairweather, 1988a) and interactions between the two (Underwood et al., 1983). It also leads to great difficulties in making predictions about events in such habitats, decreasing the operational value of ecology (Peters, 1991). In particular, it makes generalizations about patterns and processes very difficult (Foster, 1990) and makes assessments of environmental impacts complex, requiring considerable effort in experimental design (Underwood, 1994a).

Approaches to theory about variations have included Pielou's (1969) recommendations about modelling spatial patterns and temporal dynamics separately. Chesson (1985) discussed models for 'pure spatial' and 'pure temporal' environmental variation and demonstrated how these may interact. He indicated the problems in some models for processes regulating populations and demonstrated that spatial scales must be chosen carefully before interactions of spatial and temporal variation would be identifiable.

At a very large spatial scale (marine versus terrestrial habitats), Steele (1985) has described differences between marine and terrestrial environments in the temporal variation in environmental variables. The distribution of temporal variances at different scales was quite different between the two types of habitat. This sort of variation has been analysed in detail by Schneider (1994). In contrast, Taylor (1961) and

Taylor & Taylor (1977) have examined relationships between variance and mean abundance. Although the patterns and relations were examined using data from different times of sampling, the temporal order and pattern were lost.

This all results in the need for ecologists to be more concerned with the variability in abundances of populations and their interactions (see particularly Gaston & McArdle, 1993, 1994; McArdle & Gaston, 1992). The present paper is a contribution to investigating variability in space and time in abundances of intertidal populations.

It is widely described that rocky intertidal habitats are an environmental gradient of increasing harshness from low to high tidal levels. There is no doubt that regular cyclical patterns of tidal rise and fall lead to greater periods of emersion at higher levels, particularly during neap tides when upper reaches of a shore can be out of water for long periods. This has led to generalizations about the upper limits of distribution of intertidal species being set by harshness of the physical environment (Connell, 1972, 1975; Lubchenco, 1980). Alternative explanations must, however, be rejected experimentally (Underwood, 1991) before the generalization is accepted for any specific case (Underwood & Denley, 1984). There are also gradients of environmental harshness laterally along shores, due to combinations of increasing stress in wave-exposed habitats and shelter from desiccation during low tide provided by algal cover, which is often more abundant in sheltered areas (Menge, 1978a, b).

Lower on a shore, there are often large influences on abundances of species due to predation by animals which are more numerous, active or voracious at lower levels (Dayton, 1971; Edwards, 1969; Paine, 1974). Recruitment of animals is often more dense at lower levels (Connell, 1961a,b; Denley & Underwood, 1979; Wethey, 1984). This varies enormously from time to time in many habitats (Connell, 1985; Dayton, 1971; Underwood & Denley, 1984). As a result, processes of competition (Underwood, 1984a), predation (Fairweather, 1985) and provision of habitat for one species by another (Lohse, 1993) will all be variable from time to time, but the variation should be different at higher from that at lower levels.

If rocky shores represent a gradient of increasing harshness, the higher levels must be considered more extreme – the physical harshness is greater than lower down. This leads to specific models about temporal change. Under many circumstances, physical harshness at higher levels, will continuously reduce survivorship of animals, so that abundances and temporal variability are relatively small. Occasionally, however, there will be periods of relief from harshness, so animals, if they settle in the higher areas, will be able to persist in large numbers (Connell, 1975). At lower levels on a shore, environmental conditions are less harsh and numbers of animals will generally be very variable through time because of variations in processes of recruitment, predation, competition, etc., which are set by the often unpredictable recruitment of species.

From these models, it must be predicted that, in general, temporal variability in mean abundances over short periods (within a year or so) will be smaller at higher than at lower levels.

At the same time, behaviour, including behaviour during settlement, will cause animals to be more able to survive in certain microhabitats by avoiding physical harshness (Fretter & Manly, 1977; Wolcott, 1973). With other small-scale processes (reviewed by Underwood, 1979), this will contribute to small-scale spatial variance in abundances. Presumably, the outcome of these behaviours or the consequences for differential mortality in or out of the most suitable microhabitats lead to greater differences in local abundance where conditions are harsh. If environmental harshness is important at higher-levels on the shore, small-scale spatial variability should be greater than at lower levels.

This study is a preliminary test of these hypotheses (short-term temporal variability will be greater lower than higher on the shore; small-scale spatial variability will be greater higher than lower on the shore). It is preliminary in the sense that it considers each component separately and does not examine the spatio-temporal interactions that are probably very important.

Data were collected to test these specific hypotheses and to examine the relationships between spatial dispersion of the limpet *Patelloida latistrigata* and the barnacle *Tesseropora rosea*, with which it is often associated. Often, the numbers of the limpet are positively correlated with the numbers of barnacles (Creese, 1982; Jernakoff, 1985; Underwood et al., 1983) which provide resources of shelter from waves and reduced competition amongst barnacles. Therefore, the hypothesis can be proposed that small-scale variations in numbers of the barnacle will be correlated with the small-scale spatial variances in numbers of the limpets. Data were collected to test this hypothesis.

The choice of spatial scale for examining patterns is critical (Greig-Smith, 1964; Pielou, 1969; Schneider, 1994; Underwood, 1996; Underwood & Chapman, 1996). In this study, the spatial scale is at distances of tens of centimetres. This was chosen because known natural history of animals on local rocky shores (reviewed by Underwood, 1994b; Underwood & Chapman, 1995) demonstrates the consequences of behaviour during settlement, while moving to feed and in relation to predatory foraging (Fairweather, 1988b) which all create small-scale variation. In previous analyses (Underwood, 1996; Underwood & Chapman, 1996), spatial variation was greater at the smallest scales examined than at scales of metres, 10s, 100s of metres (although there was considerable variation for the barnacle *Chamaesipho tasmanica* among shores kilometres apart).

The species chosen were common (in many areas) and are ones for which considerable experimental ecological information is available. Because it is important to generate general understanding, attempts were made to choose two types of animals (barnacles and limpets) and two species of each. The barnacle *Tesseropora rosea* is a dominant component of midshore regions of semi-exposed and exposed shores (Underwood & Denley, 1984). Accounts of aspects of its recruitment, competitive interactions, roles as prey for whelks, interactions with algae and their molluscan grazers can be found in papers by Denley & Underwood (1979), Fairweather (1988b, c), Fairweather & Underwood (1983), Jernakoff (1985a), Moran (1985a). The second barnacle *Chamaesipho tasmanica* is a smaller (3–6 mm diameter is typical) species. Individuals can fuse to form dense mats (giving it the common name of 'honeycomb' barnacle). Aspects of its ecology have been described by Fairweather et al. (1984) and Underwood (1999), Underwood & Denley (1984). Previous studies of spatial variation in these two species are in Underwood (1996).

The limpet *Patelloida latistrigata* is small (maximal length 14 mm, but most about 10 mm). It is often found associated with barnacles, particularly *T. rosea*, at midshore levels on exposed shores, where it grazes on algae amongst their tests (Jernakoff, 1985b). In open areas it is out-competed for food by larger groups (Creese, 1982). It is a major prey of the whelk *Morula marginalba* (Fairweather & Underwood, 1983; Moran, 1985a; Underwood et al., 1983). The second limpet, *Notoacmea petterdi*, lives at higher levels on the shore, usually alone on vertical surfaces. In most places, it is the only macroscopic invertebrate at these heights on steep wave-exposed rocks. Nearly all individuals have home-scars to which they return after every feeding excursion (Creese, 1980). Both limpets have short-term lecithotrophic larvae. General accounts of the shores and species are in Dakin et al. (1987), Underwood (1994b) and Underwood & Chapman (1995).

Materials and methods

Study sites

This study was done on five shores, approximately equally spaced along 800 km of the coast of New South Wales, Australia (details in Chapman & Underwood, 1998). Each shore was relatively exposed to waves and had assemblages at different levels that were typical of other shores in the area. The shores were, from south to north, Tura Head, Flat Rock Island (Flat Rock), the Cape Banks Scientific Marine Research Area (Cape Banks), Bluey's Head and Scott's Head (see map in Chapman & Underwood, 1998). The different assemblages were sampled in each of two sites on each shore four times per year (at approximately 3-monthly intervals) from 1980 to 1982. Sites were all about 2.5 m from seaward to landward and about 5.5 m alongshore and were separated by 30–100 m on each shore. Details of the shores and sizes and spacing of the sites are given in Chapman & Underwood (1998) and Underwood & Chapman (1998).

Scott's Head was steeply sloping, composed of very smooth basaltic rock. Bluey's Head lies approximately 200 km south of Scott's Head; the shores were also very steep, but the substratum was pitted and corrugated by irregular erosion. Lowshore regions were subject to occasional periods of scour by sand. Cape Banks is approximately 230 km south of Bluey's Head. Its shores are sandstone platforms with steep gradients at the top and bottom of the shore. Flat Rock Island is approximately 230 km south of Cape Banks; its shores are moderately sloping and very similar in shape and aspect to those at Cape Banks. Tura Head is approximately 150 km south of Flat Rock Island; the sandstone shores at Tura Head were very variable in topography and aspect. The shores at Cape Banks have been described in detail elsewhere (Fairweather & Underwood, 1991; Underwood, 1975). The other shores were quite similar to Cape Banks in overall composition of species.

Three to four assemblages were sampled on each shore. The lowest assemblage on each shore was composed primarily of macro-algae (Underwood & Chapman, 1998). Above this were assemblages of animals, dominated by different species of barnacles, grazing limpets and gastropods. These assemblages were different at different heights on the shore. The highest levels on the shore were dominated by the small littorinid *Littorina unifasciata* and the limpet *Notoacmea petterdi*. There were few larger gastropods or barnacles. Lower on the shore, the different assemblages were distinguished primarily according to the relative abundances of the barnacles *Tesseropora rosea* and *Chamaesipho tasmanica* and a suite of grazing limpets and gastropods. Some species, e.g., *T. rosea* and the small limpet *Patelloida latistrigata,* were found in two different assemblages at different levels on each shore. Other species, e.g. *C. tasmanica* and *N. petterdi*, were only found in one assemblage on some shores. In this study, therefore only those shores on which they were found at two different levels are included. This allowed comparable data for *T. rosea* and *P. latistrigata* from all five locations, for *C. tasmanica* from one site at Bluey's Head and Cape Banks and from two sites from Flat Rock Island and Tura Head and for *N. petterdi* from each site at Bluey's Head.

Sampling design and analysis

At each time, each site was sampled using five randomly placed (50 × 50 cm) quadrats. The percentage covers of bare rock, sessile animals and primary and canopy cover of algae were sampled from 100 evenly spaced points in each quadrat. All animals which could be identified as individuals were then counted. Large and sparse species were counted in the entire quadrat. The species discussed in this manuscript were counted in five random subquadrats (20 cm^2) per quadrat. The counts in subquadrats are independent of each other at each time of sampling (e.g., Underwood & Chapman, 1996) and the data are independent from one time of sampling to the next (Underwood, 1996).

The mean and variance among the five subquadrats were calculated for each quadrat in each place (i.e., site at each level) at each time of sampling. Comparisons of variance are more difficult if the variances are related to the means of the distributions being sampled and the means differ among the places or times being compared. Here, the variances were directly related to the means (see also Underwood, 1996). It is therefore appropriate to compare variances between habitats without clarifying the influences of mean densities. The appropriate procedure under these circumstances is to examine the relationship of variances to means. The regressions of variances on means can be compared, 'removing' or controlling the influence of mean numbers.

The variance: mean relationship for all quadrats in each site (i.e., n = 60 from five quadrats at each of 12 times of sampling) was then calculated using regression on the untransformed data. These relationships were compared across levels on the shore in each site separately using the test for heterogeneity of regressions (Winer et al., 1991). Locations and sites were compared by binomial tests, i.e., the probability that slope $_{higher}$ > slope $_{lower}$ = probability that slope $_{higher}$ < slope $_{lower}$ = 0.5. Finally, temporal variance was calculated from the appropriate components of variation (explained in detail in Table 1, as in Kennelly & Underwood, 1992; Stewart-Oaten et al., 1995; Underwood, 1997).

The tests used here were tests for homogeneity of regressions, the precursor test for analysis of covariance (Huitema, 1980). In no case were tests of any aspects of individual regressions made; these were irrelevant and would have been inappropriate because the variance of the dependent variable (the sampled variances) tended to increase with the independent variable (the sampled means). Also note that no comparisons of adjusted means (analyses of covariance) were done. These were irrelevant and would have violated assumptions of homogeneity of slopes. These facts about irrelevant relationships are mentioned at the behest of referees.

The statistical assumptions underlying the analyses were as follow, but were all met by the data. First, because we used linear regressions, it must be assured that regressions are linear. This was examined by the use of higher-order polynomials. No formal tests could be done because of heterogeneity of variances within each regression, but decreases in residual sums of squares were always very small when fitting non-linear polynomial regressions instead of a linear one. Second, variances of residuals around the linear regression are assumed to be equal in all populations compared. This was tested by Cochran's tests (Underwood, 1997; see also Huitema, 1980).

For components of analysis of covariance, there has been considerable work on consequences of departures from assumptions. Unlike single regressions, using a fixed model of regression when the dependent variable (in this case the sampled means) is a random

Table 1. Calculation of temporal variance from analysis of variance of mean number per subquadrat; $n = 5$ subquadrats in each of $q = 5$ quadrats 12 times of sampling. Times of sampling and quadrats are random factors; quadrats are independently, randomly sampled at each time (and therefore nested in times; see Underwood (1997)). $\sigma_R^2, \sigma_{Q(T)}^2, \sigma_T^2$ are variances among subquadrats, among quadrats and among times, respectively.

Source of variation	Degrees of Freedom	Mean Square estimates
Among times of sampling = T	11	$\sigma_R^2 + n\sigma_{Q(T)}^2 + nq\sigma_T^2$
Among quadrats at each time = $Q(T)$	48	$\sigma_R^2 + n\sigma_{Q(T)}^2$
Residual	240	σ_R^2
Total	299	

Estimated variance among times = $\sigma_T^2 = \frac{\text{Mean Square}_T - \text{Mean Square}_{Q(T)}}{25}$

variable, measured with error causes no problems for our analyses. This is because there is little to no problem for estimates of slopes, nor for probabilities in tests when samples are balanced (all were) and where size of samples are >12 (the ones here had $n = 60$) (Huitema, 1980; Scheffé, 1959; Winer et al., 1991). Where the independent variable is random and measured with error (as here), there can be problems for interpretations of differences in analysis of covariance; such analyses of adjusted means were not done here.

Some scientists (e.g., anonymous referees for this paper) perversely prefer variances to be analysed as coefficients of variation. This is not an appropriate option because any interpretation of differences in the ratio of standard deviation to mean is confounded. Differences may be due to difference in variances (as hypothesised) or differences in means.

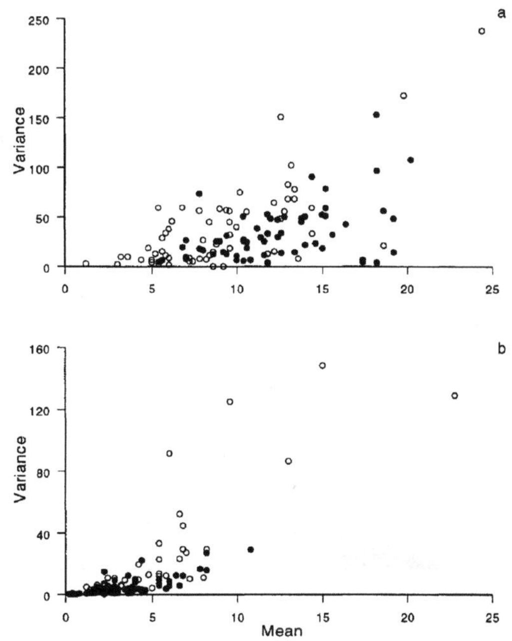

Figure 1. Relationships between variance (among subquadrats, per quadrat) and mean number per subquadrat for five quadrats from 12 times of sampling ($n = 60$) for *Tesseropora rosea* at: (a) Scott's Head, Site 1; (b) Flat Rock, Site 1. (○) Lower level; (●) upper level.

Results

The barnacle Tesseropora rosea

In nine of the 10 places examined (two sites in five locations), the slope of relationships between variances and means was steeper in the lower than upper height examined (Table 2). The only exception was at Scott's Head, Site 2, which was one of the few (four out of 10) where there was no significant difference between the slopes of the relationships in the two levels of the shore (Table 2). The probability of there being nine out of 10 cases, by chance, where differences were in the same direction is $P < 0.01$ (binomial test).

Thus, spatial variation was larger at larger mean densities and increased more for a given increase in mean density at the lower levels. The patterns are illustrated for two sites (Figure 1; Scott's Head Site 1 and Flat Rock Site 1). When mean density is small, variance is similar in lower and upper levels of the shore. At larger densities, variances are larger in lower levels.

The barnacle Chamaesipho tasmanica

Comparisons were only possible in six sites (over the 5 locations; Table 3). In all six places, there were greater abundances of the barnacles at the lower level

Table 2. Relationships between variances and means (numbers per sub-quadrat) for *Tesseropora rosea* in two sites in each of five locations. Slopes of regressions (n = 60) for a higher and lower level on each shore are compared. (F-ratio; 1,116 df; *indicates significant $P < 0.05$).

	Tura Head		Flat Rock		Cape Banks		Bluey's Head		Scott's Head	
Site:-	1	2	1	2	1	2	1	2	1	2
Upper level	3.78	2.12	2.16	2.44	1.26	1.67	3.39	3.85	3.69	10.46
Lower level	9.16	19.16	7.33	8.72	2.29	2.39	6.03	10.50	7.35	7.13
F-ratio	4.82*	0.59	32.67*	7.48*	3.34	2.60	5.07*	8.16*	7.75*	1.30

Table 3. Relationships between variances and means (numbers per sub-quadrat) for *Chamaesipho tasmanica* in a total of six sites over four locations. Slopes of regressions (n = 60) for a higher and lower level on each shore are compared. (F-ratio; 1,116 df; $*$ indicates significant $P < 0.05$).

	Tura Head		Flat Rock		Cape Banks	Bluey's Head
Site:-	1	2	1	2	1	2
Upper level	7.79	7.48	13.81	11.48	10.75	19.62
Lower level	6.18	9.10	8.36	16.52	9.52	18.88
F-ratio	0.01	0.11	0.07	0.79	0.38	0.05

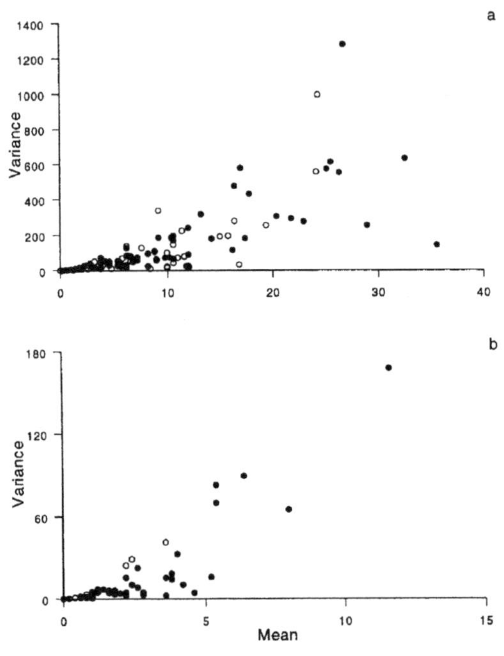

Figure 2. Relationships between variance (among subquadrats, per quadrat) and mean number per subquadrat for five quadrats from 12 times of sampling (n = 60) for *Chamaesipho tasmanica* at: (a) Bluey's Head, Site 2; (b) Cape Banks, Site 1. (○) Lower level; (●) upper level.

examined. This is illustrated for two sites in Figure 2. The relationships between variances and means were not different between the two levels on any shore examined (Table 3). Four of the six places, however, had greater slopes in the upper levels, but this is not inconsistent with chance (probability of four or more cases in one direction is $P > 0.30$, binomial test). *C. tasmanica*, therefore, showed no pattern of difference between the two levels.

The limpets Patelloida latistrigata *and* Notoacmea petterdi

In six out of the 10 areas sampled, variances of numbers of *P. latistrigata* increased more with an increase in mean number on the lower level. This is entirely consistent with chance (Table 4; $P = 0.10$, binomial test). In only two out of the 10 areas were the slopes significantly different. So, there is no evidence for any pattern of difference in spatial variance between the two levels (Figure 3).

Data on *N. petterdi* were only available for the two sites at Bluey's Head. In both cases, relationships between variances and mean numbers were similar (neither was significantly different; Table 5, Figure 4).

Thus, for neither species of limpet was there any evidence for differences in dispersion in the upper and lower levels of distribution on the shore.

Table 4. Relationships between variances and means (numbers per sub-quadrat) for *Patelloida latistrigata* in two sites in each of 5 locations. Slopes of regressions ($n = 60$) for a higher and lower level on each shore are compared. (F-ratio; 1,116 df; *indicates significant $P < 0.05$.)

	Tura Head		Flat Rock		Cape Banks		Bluey's Head		Scott's Head	
Site:-	1	2	1	2	1	2	1	2	1	2
Upper level	0.97	2.21	1.11	2.26	0.97	1.52	2.07	7.07	2.53	3.80
Lower level	2.41	4.46	1.87	1.56	2.45	2.11	3.17	7.12	2.52	2.50
F-ratio	2.55	11.88*	3.08	2.09	21.97*	2.67	1.62	0.001	0.0003	2.20

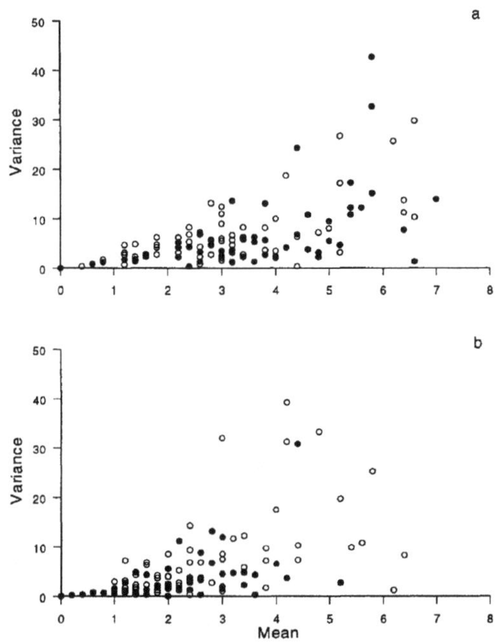

Figure 3. Relationships between variance (among subquadrats, per quadrat) and mean number per subquadrat for five quadrats from 12 times of sampling ($n = 60$) for *Patelloida latistrigata* at: (a) Scott's Head, Site 1; (b) Bluey's Head, Site 1. (○) Lower level; (●) upper level.

Table 5. Relationships between variances and means (numbers per sub-quadrat) for *Notoacmea petterdi* in a total of two sites at one location. Slopes of regressions ($n = 60$) for a higher and lower level on the shore are compared. (F-ratio; 1,116 df; * indicates significant $P < 0.05$).

	Bluey's Head	
Site:-	1	2
Upper level	2.46	3.64
Lower level	1.92	3.10
F-ratio	0.59	1.73

Relationships between P. latistrigata *and* T. rosea

The spatial patterns of dispersion of *P. latistrigata* and *T. rosea* were examined in the two levels at the two sites at Cape Banks (Figure 5). In three of the four cases, there were significant positive correlations between the means of the two species ($r = 0.25, 0.35, 0.43$; $n = 60$, all $P < 0.05$ for Site 1 Upper, Site 1 Lower and Site 2 Upper, respectively). There were also significant positive correlations between the variances of the two species ($r = 0.30, 0.27, 0.39$; $n = 60$, all $P < 0.05$, sites and levels as before). These patterns are illustrated in Figure 5a,b for Cape Banks, Upper level Site 1.

In the fourth area (Site 2, lower level), neither correlation was significant ($r = 0.11, 0.17$; $n = 60$, $P > 0.05$, for means and variances, respectively; see Figure 5c, d).

On all other shores, there was no pattern of variability between *P. lastistrigata* and *T. rosea*. In some areas, there were significant relationships between mean abundances of the limpets and the barnacles, but no correlations between their variances.

Temporal variation

The final comparisons were of the estimated temporal variance (see Table 1 for method of calculation) for upper and lower levels of the shore. For *T. rosea*, there was greater temporal variation in most lower areas than in the corresponding upper areas (Table 6). Most of this variation was attributable to sudden and dramatic increases in numbers due to episodes of recruitment of juveniles, illustrated for two areas in Figure 6. For nine of the 10 areas examined, tem-

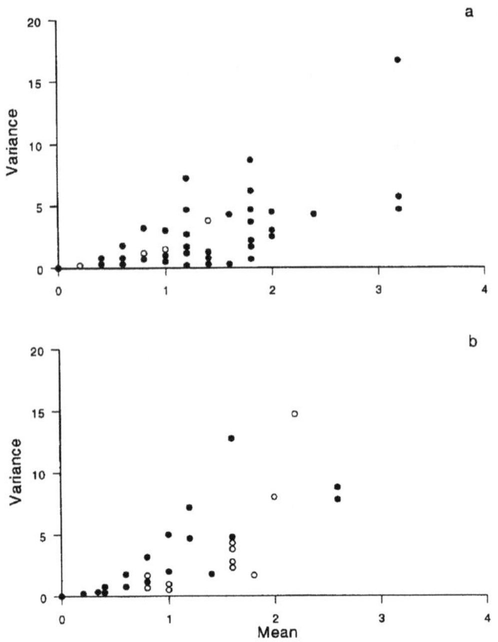

Figure 4. Relationships between variance (among subquadrats, per quadrat) and mean number per subquadrat for five quadrats from 12 times of sampling ($n = 60$) for *Notoacmea petterdi* at (a) Bluey's Head, Site1; (b) Bluey's Head, Site 2. (○) Lower level; (●) upper level.

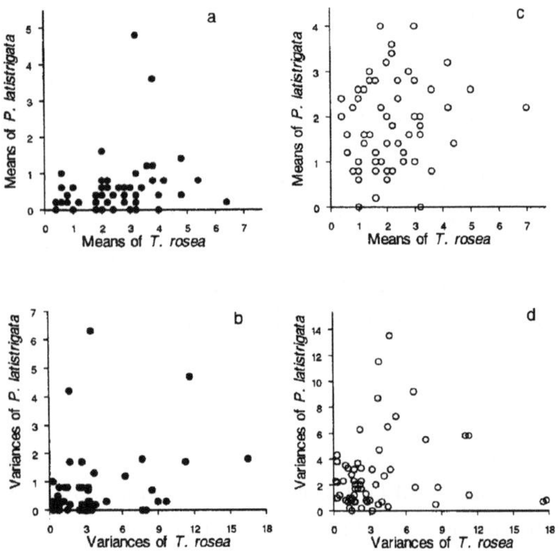

Figure 5. Relationships between spatial patterns of abundances of the limpet *Patelloida latistrigata* and the barnacle *Tesseropora rosea*: (a, c) Mean numbers of the two species; (b, d) variances of the two species. (a, b) At Cape Banks, Site 1, upper level; (c), (d) at Cape Banks, Site 2, lower level.

Table 6. Estimates of temporal variation (see Table 1).

	Upper level	Lower level
(a) *Tesseropora rosea* (10 areas)		
Range	0.28–4.76	0–178.21
Mean (S.E.)	1.37 (0.47)	38.28 (21.14)
No. of areas Lower > Upper	9/10, $P < 0.02$	
(b) *Chamaesipho tasmanica* (six areas)		
Range	0–215.39	0–6.43
Mean (S.E.)	58.10 (32.93)	1.49 (1.02)
No. of areas Lower > Upper	1/6, $P < 0.02$	
(c) *Patelloida latistrigata* (10 areas)		
Range	0–0.60	0.11–2.18
Mean (S.E.)	0.28 (0.06)	0.72 (0.19)
No. of areas Lower > Upper	8/10, $P < 0.02$	

poral variation was greater in the lower level, which is unlikely to be due to chance ($P < 0.02$, binomial test).

The limpet *P. latistrigata* showed the same pattern of greater temporal variation in the lower level (in eight out of 10 areas; $P < 0.02$; Table 6). Again, temporal changes were largely attributable to periods of recruitment, which were more frequent and there were greater numbers of juveniles (unpublished data) in the lower levels (Figure 7). Following increases due to recruitment, numbers declined again over 6–9 months.

In contrast, *C. tasmanica* showed greater temporal variation at higher levels (Table 6). Mean abundances are not shown here, but there are no simple explanations for this trend (see Underwood, 1996, for details).

Discussion

Environmental gradients

The barnacle *Tesseropora rosea* showed a pattern of greater increases in variances with increased mean at lower than at higher levels on the shore. This was completely *opposite* to that predicted from models of increasing harshness of the physical environment at higher levels. For the other three species, there was no trend with height and absolutely no support for the proposed hypothesis, nor any other that might predict different dispersion at different heights. This

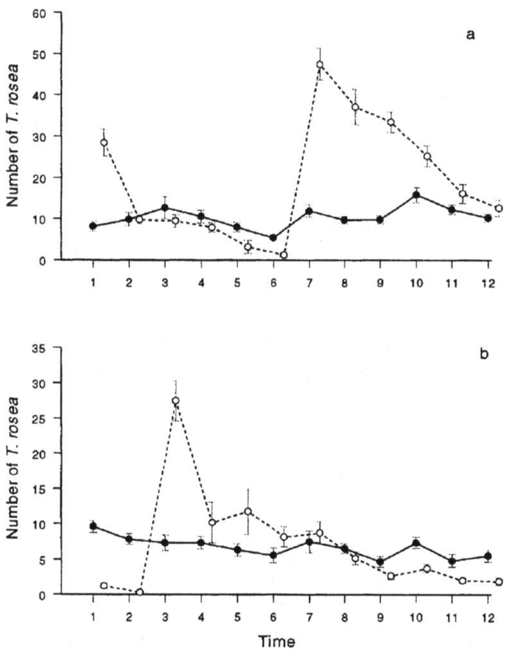

Figure 6. Temporal variation in numbers of *Tesseropora rosea*. Mean (± S.E.; n = 25 subquadrats) in (a) Scott's Head, Site 2 and (b) Tura Head, Site 1. Times of sampling are at 3-monthly intervals from early 1980. (○) Lower level; (●) upper level.

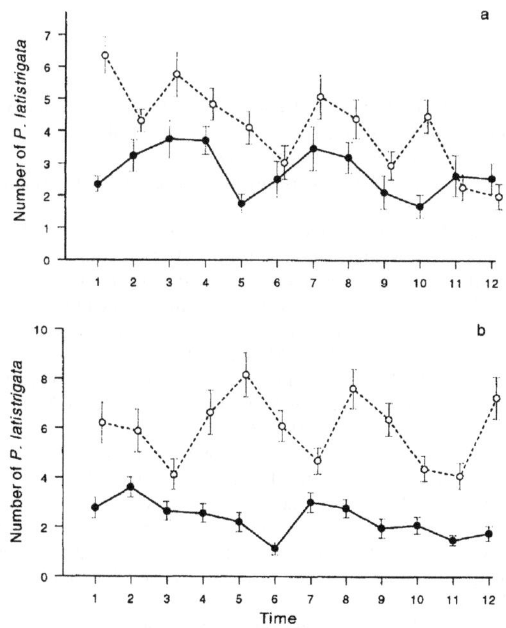

Figure 7. Temporal variation in numbers of *Patelloida latistrigata*. Mean (± S.E.; n = 25 subquadrats) in (a) Flat Rock, Site 1 and (b) Tura Head, Site 1. Times of sampling are at 3-monthly intervals from early 1980. (○) Lower level; (●) upper level.

is surprising, given the frequency with which physical factors are invoked to explain other aspects of distribution and abundance (Lewis, 1964; Stephenson & Stephenson, 1972). It is also surprising given that physical harshness certainly decreases availability of micro-algal food for grazers such as *Notocmea petterdi* and *Patelloida latistrigata* (Creese, 1980; MacLulich, 1987; Underwood, 1984b). Other interactions among the species examined here or on these shores are also affected by seasonal differences (competition amongst gastropods; Underwood, 1984a), differences in wave-exposure (predation by whelks and interactions among space-users, algae and grazers; Fairweather & Underwood, 1991; Underwood et al., 1983) and differences associated with time of low tide in relation to solar cycles (Moran, 1985b).

Despite these findings, it is clear that for several species, there is no translation of influences of the physical environment on resultant spatial patterns. For *Tesseropora rosea*, explanations must be sought elsewhere and will presumably involve choices of sites at the time of settlement and subsequent patterns of foraging by predators (notably whelks) and how these may affect the dispersion of the barnacles at small spatial scales. Such models will not be pursued further here.

Temporal patterns of variation

The temporal patterns shown by the three species for which there were sufficient comparative data all indicated striking differences between higher and lower levels. For *T. rosea* and *P. latistrigata*, this was undoubtedly attributable to greater variation in arrivals due to recruitment and subsequent mortality. In most locations, most of the mortality can be attributed to direct effects of waves and abrasion by sand during storms on these exposed coasts (unpub. data). At Cape Banks and Scott's Head, some of the mortality of barnacles and limpets was due to predation by whelks (*Morula marginalba* and a few *Thais orbita*).

Experimental removals of whelks made a significant difference to survival of *T. rosea* and *P. latistrigata* at Cape Banks (unpub. data, but see Denley & Underwood, 1979; Fairweather, 1988b; Fairweather & Underwood, 1991; Moran, 1985b; Underwood et al., 1983). In some areas, loss of *P. latistrigata* is a direct consequence of loss of tests of dead *T. rosea* when these are removed by storms (pers. obs.).

So, environmental harshness appears to 'iron out' short-term (seasonal or 3-monthly) variability in supply-side processes (Underwood & Fairweather, 1989). This is important for structuring the assemblages at different levels on these exposed shores.

For *Chamaesipho tasmanica*, however, the story is different. There was greater temporal variation at the higher levels. The reasons for this are not clear and need more work. What this does show, however, is that it is always going to be difficult to attempt generalities about sets of species, even, in this case, when they are closely related taxonomically. Functional groups are often used where diversity is complex and analyses of every species impractical (e.g. Menge et al., 1986). Here, however, is a demonstration of some of the problems in lumping things together that happen to have the same sorts of morphology or evolutionary background. The other result was that there were (in only three of the 20 cases examined) relationships between the mean abundances of the limpet *P. latistrigata* and the barnacle *T. rosea* and also between their variances. This result was consistent with the hypothesis proposed from original observations and their association and published accounts of the processes operating in their interactions. At Cape Banks, the positive relationship was found in three of four sites or levels (see Results and Figure 5). It is not usual to have data on such complex patterns from more than one site (see below). Finding a consistent pattern of relationships in three of the four places examined allows a claim of generality. This rather fades when it is realized that the general pattern is, in fact, only found in three of twenty cases! Attempting to find generality requires examination of enough cases for the generality to be convincing. This is not a new message (Foster, 1990), but is still very important.

The final comment is that most studies of relationships between variation and average abundance have been done with only one set of data from which to understand the relationships. This has led to some elegant theory (e.g., Gaston & McArdle, 1993, 1994; Taylor, 1971). None of this will progress if we cannot understand that the relationships between means and variances are themselves a variable in ecological patterns. They vary from place to place (here) and time to time (Underwood, 1996) and, in complex ways from one scale of measurement to another (Underwood, 1996). Much of the variability on rocky shores is at very small scales (Underwood & Chapman, 1996). This is a feature that makes collection of data in this habitat relatively straightforward. What is clear, however, is that the detection of relationships between variances and means, let alone discovering how they relate to physical and biological features of the environment is going to depend on more than just underlying mathematical relationships and statistical properties (e.g., the Poisson distribution). They are also going to depend on empirical, experimental tests of hypotheses derived from such theories.

This study is one attempt to do this in relation to concepts of environmental gradients and the influences of such gradients on the distribution and dispersion of intertidal invertebrates. That it failed to find any strikingly obvious patterns is a testament to why such studies are rather urgently needed in many different habitats and in relation to many different hypotheses.

Acknowledgements

This study was supported by grants from the Australian Research Council to both authors and from the ARC Special Research Centres Programme. We are grateful to various colleagues and former colleagues for assistance with field-work and to K. Astles and T. Davies for assistance with analyses; V. Mathews kindly produced the figures. Anonymous referees shed considerable dark on the topic.

References

Chapman, M. G. & A. J. Underwood, 1998. Inconsistency and variation in the development of rocky intertidal algal assemblages. J. exp. mar. Biol. Ecol. 224: 265–289.

Chesson, P. L., 1978. Predator-prey theory and variability. Ann. Rev. Ecol. Syst, 9: 323–347.

Chesson, P. L., 1985. Coexistence of competitors in spatially and temporally varying environments: a look at the combined effects of different sorts of variability. Theor. Pop. Biol. 28: 263–287.

Connell, J. H., 1961a. Effects of competition, predation by *Thais lapillus* and other factors on natural populations of the barnacle *Balanus balanoides*. Ecol. Monogr. 40: 49–78.

Connell, J. H., 1961b. The influence of interspecific competition and other factors on the distribution of the barnacle *Chthamalus stellatus*. Ecology 42: 710–723.

Connell, J. H., 1972. Community interactions on marine rocky intertidal shores. Ann. Rev. Ecol. Syst. 3: 169–192.

Connell, J. H., 1975. Some mechanisms producing structure in natural communities: a model and evidence from field experiments. In Cody, M. S. & J. M. Diamond (eds), Ecology and Evolution of Communities. Harvard University Press, Cambridge, MA: 460–490.

Connell, J. H., 1985. The consequences of variation in initial settlement versus post-settlement mortality in rocky intertidal communities. J. exp. mar. Biol. Ecol. 93: 11–46.

Creese, R. G., 1980. An analysis of distribution and abundance of populations of the high-shore limpet *Notoacmea petterdi* (Tenison-Woods). Oecologia (Berl.) 45: 252–260.

Creese, R. G., 1982. Distribution and abundance of the acmaeid limpet *Patelloida latistrigata*, and its interaction with barnacles. Oecologia (Berl.) 52: 85–96.

Dakin, W. J., 1987. Australian Seashores. Angus & Robertson, Sydney, 411 pp.

Dayton, P. K., 1971. Competition, disturbance and community organization: the provision and subsequent utilization of space in a rocky intertidal community. Ecol. Monogr. 41: 351–389.

Denley, E. J. & A. J. Underwood, 1979. Experiments on factors influencing settlement, survival and growth of two species of barnacles in New South Wales. J. exp. mar. Biol. Ecol. 36: 269–293.

Edwards, D. C., 1969. Zonation by size as an adaptation for intertidal life in *Olivella biplicata*. Am. Zool. 9: 399–417.

Fairweather, P. G., 1985. Differential predation on alternative prey and the survival of rocky intertidal organisms in New South Wales. J. exp. mar. Biol. Ecol. 109: 135–156.

Fairweather, P. G., 1988a. Correlations of predatory whelks with intertidal prey at several scales of space and time. Mar. Ecol. Progr. Ser. 45: 237–243.

Fairweather, P. G., 1988b. Predation can increase variability in the abundance of prey on seashores. Oikos 53: 87–92.

Fairweather, P. G., 1988c. Consequences of supply-side ecology: manipulating the recruitment of intertidal barnacles affects the intensity of predation upon them. Biol. Bull. mar. biol. Lab., Woods Hole 175: 349–354.

Fairweather, P. G., 1988d. Predation creates haloes of bare space among prey on rocky shores in New South Wales. Aust. J. Ecol. 13: 401–409.

Fairweather, P. G. & A. J. Underwood, 1983. The apparent diet of predators and biases due to different handling times of their prey. Oecologia (Berl.) 56: 169–179.

Fairweather, P. G. & A. J. Underwood, 1991. Experimental removals of a rocky intertidal predator: variations within two habitats in the effects on prey. J. exp. mar. Biol. Ecol. 154: 29–75.

Fairweather, P. G., A. J. Underwood & M. J. Moran, 1984. Preliminary investigations of predation by the whelk *Morula marginalba*. Mar. Ecol. Progr. Ser. 17: 143–156.

Foster, M. S., 1990. Organization of macroalgal assemblages in the Northeast Pacific: the assumption of homogeneity and the illusion of generality. Hydrobiologia 192: 21–34.

Fretter, V. & R. Manly, 1977. Settlement and early benthic life of *Littorina neritoides* (L.) at Wembury, South Devon. J. moll. Stud. 43: 255–262.

Gaston, K. J. & B. H. McArdle, 1993. Measurement of variation in the size of populations in space and time: some points of clarification. Oikos 68: 357–360.

Gaston, K. J. & B. H. Mcardle, 1994. The temporal variability of animal abundances: measures, methods and patterns. Phil. Trans. R. Soc. B 345: 335–358.

Greig-Smith, P., 1964. Quantitative Plant Ecology, 2nd edn. Butterworth's, London, 256 pp.

Huitema, B.E., 1980. The Analysis of Covariance and Alternatives. Wiley, New York, 455 pp.

Jernakoff, P., 1985a. Interactions between the limpet *Patelloida latistrigata* and algae on an intertidal rock-platform. Mar. Ecol. Progr. Ser. 23: 71–78.

Jernakoff, P., 1985b. An experimental evaluation of the influence of barnacles, crevices and seasonal patterns of grazing on the algal diversity and cover in an intertidal barnacle zone. J. exp. mar. Biol. Ecol. 88: 287–302.

Kennelly, S. J. & A. J. Underwood, 1992. Fluctuations in the distributions and abundances of species in sublittoral kelp forests in New South Wales. Aust. J. Ecol. 17: 367–382.

Lewis, J. R., 1964. The Ecology of Rocky Shores. The English Universities Press, London, 323 pp.

Lohse, D. P., 1993. The importance of secondary substratum in a rocky intertidal community. J. exp. mar. Biol. Ecol. 166: 1–17.

Lubchenco, J., 1980. Algal zonation in the New England rocky intertidal community: an experimental analysis. Ecology 61: 333–344.

MacLulich, J. H., 1987. Variations in the density and variety of intertidal epilithic microflora. Mar. Ecol. Progr. Ser. 40: 285–293.

McArdle, B. H. & K. J. Gaston, 1992. Comparing population variabilities. Oikos 64: 610–612.

Menge, B. A., 1978a. Predation intensity in a rocky intertidal community: relation between predator foraging activity and environmental harshness. Oecologia (Berl.) 34: 1–16.

Menge, B. A., 1978b. Predation intensity in a rocky intertidal community: effect of an algal canopy, wave action and desiccation on predator feeding rates. Oecologia (Berl.) 34: 17–35.

Menge, B. A., J. Lubchenco, L. R. Ashkenas & F. Ramsey, 1986. Experimental separation of effects of consumers on sessile prey in the low zone of a rocky shore in the Bay of Panama: direct and indirect consequences of food web complexity. J. exp. mar. Biol. Ecol. 100: 225–270.

Moran, M. J., 1985a. Effects of prey density, prey size and predator size on rates of feeding by an intertidal predatory gastropod *Morula marginalba* Blainville (Muricidae) on several species of prey. J. exp. mar. Biol. Ecol. 90: 97–105.

Moran, M. J., 1985b. The timing and significance of sheltering and foraging behaviour of the predatory intertidal gastropod *Morula marginalba* Blainville (Muricidae). J. exp. mar. Biol. Ecol. 93: 103–114.

Paine, R. T., 1974. Intertidal community structure: experimental studies on the relationship between a dominant competitor and its principal predator. Oecologia (Berl.) 15: 93–120.

Peters, R. H., 1991. A Critique for Ecology. Cambridge University Press, Cambridge, 366 pp.

Pielou, E. C., 1969. An Introduction to Mathematical Ecology. Wiley Interscience, New York, 286 pp.

Scheffé, H., 1959. The Analysis of Variance. Wiley, New York, 477 pp.

Schneider, D. C., 1994. Quantitative Ecology: Spatial and Temporal Scaling. Academic Press, San Diego, 395 pp.

Steele, J. H., 1985. A comparison of terrestrial and marine ecological systems. Science 313: 355–358.

Stephenson, T. A. & A. Stephenson, 1972. Life Between Tidemarks on Rocky Shores. W.H. Freeman and Co., San Francisco, 423 pp.

Stewart-Oaten, A., W. W. Murdoch & S. J. Walde, 1995. Estimation of temporal variability in populations. Am. Nat. 146: 519–535.

Taylor, L. R., 1961. Aggregation, variance and the mean. Nature 189: 732–735.

Taylor, L. R. & R. A. J. Taylor, 1977. Aggregation, migration and population mechanics. Nature 265: 415–421.

Underwood, A. J., 1975. Intertidal zonation of prosobranch gastropods: analysis of densities of four coexisting species. J. exp. mar. Biol. Ecol. 19: 197–216.

Underwood, A. J., 1979. The ecology of intertidal gastropods. Adv. mar. Biol. 16: 111–210.

Underwood, A. J., 1984a. Vertical and seasonal patterns in competition for microalgae between intertidal gastropods. Oecologia (Berl.) 64: 211–222.

Underwood, A. J., 1984b. The vertical distribution and seasonal abundance of intertidal microalgae on a rocky shore in New South Wales. J. exp. mar. Biol. Ecol. 78: 199–220.

Underwood, A. J., 1991. The logic of ecological experiments: a case history from studies of the distribution of macro-algae on rocky intertidal shores. J. mar. biol. Ass. U.K. 71: 841–866.

Underwood, A. J., 1994a. On beyond BACI: sampling designs that might reliably detect environmental disturbances. Ecol. Appl. 4: 3–15.

Underwood, A. J., 1994b. Rocky intertidal shores. In Hammond L. S. & R. Synnot (eds), Marine Biology. Longman-Cheshire, Melbourne: 273–296.

Underwood, A. J., 1996. Spatial patterns of variance in densities of intertidal populations. In Floyd, R. B., A. W. Sheppard & P. J. De Barro (eds), Frontiers of Population Ecology. CSIRO Publishing, Melbourne: 369–389.

Underwood, A. J., 1997. Experiments in Ecology: Their Logical Design and Interpretation Using Analysis of Variance. Cambridge University Press, Cambridge, 504 pp.

Underwood, A. J., 1999. Physical disturbances and their direct effect on an indirect effect: responses of an intertidal assemblage to a severe storm. J. exp. mar. Biol. Ecol. 232: 125–140.

Underwood, A. J. & M. G. Chapman, 1995. Rocky shores. In Underwood, A. J. & M. G. Chapman (eds), Coastal Marine Ecology of Temperate Australia. New South Wales University Press, Sydney: 55–82.

Underwood, A. J. & M. G. Chapman, 1996. Scales of spatial patterns of distribution of intertidal snails. Oecologia (Berl.) 107: 212–224.

Underwood, A. J. & M. G. Chapman, 1998. Variation in algal assemblages on wave-exposed rocky shores in New South Wales. Mar. Freshwat. Res. 49: 241–254.

Underwood, A. J. & E. J. Denley, 1984. Paradigms, explanations and generalizations in models for the structure of intertidal communities on rocky shores. In Strong, D. R., D. Simberloff, L. G. Abele & A. Thistle (eds), Ecological Communities: Conceptual Issues and the Evidence. Princeton University Press, New Jersey: 151–180.

Underwood, A. J. & P. G. Fairweather, 1989. Supply-side ecology and benthic marine assemblages. TREE 4: 16–20.

Underwood, A. J., E. J. Denley & M. J. Moran, 1983. Experimental analyses of the structure and dynamics of mid-shore rocky intertidal communities in New South Wales. Oecologia (Berl.) 56: 202–219.

Wethey, D. S., 1984. Spatial pattern in barnacle settlement: day to day changes during the settlement season. J. mar. biol. Ass. U.K. 64: 687–698.

Winer, B. J., D. R. Brown & K. M. Michels, 1991. Statistical Principles in Experimental Design, 3rd edn. McGraw-Hill, New York, 1057 pp.

Wolcott, T. G., 1973. Physiological ecology and intertidal zonation in limpets (*Acmaea*): a critical look at 'limiting factors'. Bull.mar. biol. Lab. Woods Hole 145: 389–422.

Hydrobiologia **426:** 37–42, 2000.
G. Liebezeit, S. Dittmann & I. Kröncke (eds), Life at Interfaces and Under Extreme Conditions.

The influence of *Corophium volutator* abundance on resuspension

E. M. G. T. de Deckere*, J. van de Koppel & C. H. R. Heip
Netherlands Institute of Ecology, Centre for Estuarine and Coastal Ecology, P.O. Box 140, 4400 AC Yerseke, The Netherlands
*E-mail: deckere@cemo.nioo.knaw.nl (*author for correspondence)*

Key words: *Corophium*, resuspension, turbidity, sediment stability

Abstract

Two experiments were performed to test the hypothesis that *Corophium volutator* affects the turbidity of water in estuaries through active resuspension of sediment. One experiment was done in a flume system under different flow velocities, and one in aquaria. A diatom film developed at the sediment surface in both experiments before *Corophium* was added. This diatom film was supposed to have a stabilising effect on the sediment. In both experiments, the concentration of suspended solids in the water column increased with the density of *Corophium* individuals. No effect of flow velocity on suspended solids concentration was found. This indicates that, in our flume experiment, active resuspension by *Corophium* was more important than physical resuspension, at least at low flow velocity (<20 cm s^{-1}) and in the presence of a diatom film. The critical erosion threshold decreased with increasing *Corophium* density in the aquarium experiment, indicating that indirect effects of *Corophium* grazing may become more important at high levels of bottom shear stress. The implications of our findings for suspended solids concentration in estuarine systems are discussed.

Introduction

Suspended solids concentration in estuaries is strongly affected by sedimentation and erosion processes occurring over intertidal flats. Sedimentation of particles from the water column is predominant when turbulence at the sediment–water interface is low. On the other hand, resuspension of particles from mudflats is a continuous process once intertidal flats are submersed. The total rate of resuspension is the sum of physical resuspension, direct biological resuspension and indirect resuspension resulting from the activity of benthic organism (Davis, 1993). Physical resuspension is dependent on bottom shear stress, resulting from water current and wave action. Biological resuspension is related to the activity of organisms, and may also occur at low bottom shear stress (Rhoads, 1974). Finally, benthic organisms may affect physical resuspension by bioturbating the sediment, by increasing the sediment microtopography (Eckman et al., 1981), by excreting mucus (Meadows & Tait, 1989) or by affecting sediment properties in other ways.

A dominant macrobenthic species on many intertidal mudflats is *Corophium volutator* (Pallas), which may reach densities up to 100 000 ind. m^{-2} during the summer period (Flach, 1992; de Deckere, unpublished results). *Corophium volutator* predominately occur at high tidal flats with high mud content, low current velocity and the presence of diatoms (Beukema & Flach, 1995). Several studies have shown that *Corophium* can affect resuspension of sediment by consuming mucus-producing diatoms (Gerdol & Hughes, 1994a; Grant & Daborn, 1994) and by binding sediment particles into their burrows (Meadows & Tait, 1989; Grant & Daborn, 1994). *Corophium* may furthermore influence the sediment properties by changing porosity, water content (Meadows & Tait, 1989; Jones & Jago, 1993; Gerdol & Hughes, 1994a; Limia & Raffaelli, 1997) and the microtopography. The openings of their U-shaped burrow protrude about 1–1.5 mm above the sediment surface (Meadows & Reid, 1966). All these activities of *Corophium* have significant effect on the susceptibility of sediment to erosion, as was found for instance by Daborn et al. (1993).

Most studies focus on the effect of grazing on diatoms by *Corophium* and the indirect effects on the physical resuspension of particles. The consequences of direct resuspension of particles by *Corophium*, however, have not yet been investigated. Continuous pumping of water through *Corophium* tubes induces active resuspension of sediment. The water removes particles released by excavating the tube, faeces and unwanted particles collected by feeding (Meadows & Reid, 1966). Because of frequent high densities of *Corophium*, active biological resuspension may significantly affect sediment characteristics and sediment transportation. In this study, we report on experiments performed to investigate the resuspension of sediments by *Corophium* and the consequences of *Corophium* grazing on sediment properties. The relation between *Corophium* density and the concentration of suspended solids within the water column under different flow regimes was investigated. Furthermore, the effect of increasing density of *Corophium* on chlorophyll *a* concentrations within the sediment and on the sediment stability was studied. Consequences of biological resuspension of sediment for sediment exchange on intertidal flats are discussed.

Materials and methods

The first experiment to study the active resuspension of *Corophium volutator* was carried out in small flumes, measuring 130*12*25 cm. The flumes were constructed from stainless steal, following the specifications of Kamermans & Huitema (1994). The flumes were placed in a 15 °C room with continuous illumination (Osram TL-tubes). The test section consisted of a 50*12*10 cm compartment in the centre of the flume that was filled with sediment (median grain size = 120 μm). Sediment was collected at a tidal flat in the eastern part of the Westerschelde estuary, the Netherlands. It was sieved over a 500 μm sieve and frozen before the start of the experiment, to exclude all macrobenthic organisms. The sediment was inoculated with diatoms, which were collected on a tidal flat in the middle part of the Westerschelde estuary. The flumes were subsequently illuminated for 7 days at a day-night rhythm of 16 h light and 8 h dark. The flow velocity of the overlying water, Oosterschelde water with a nutrient solution according to Kester et al. (1967), was adjusted to about 10 cm s^{-1}. During this period, a diatom film formed on the sediment surface. In the meantime, *Corophium* were collected at the same site as the sediment, sieved over a 500 μm sieve and adults of at least 1 cm length were selected. The *Corophium* were added to two series of five flumes with densities of, respectively, 0, 1083, 2083, 4167 and 8333 ind. m^{-2}, while the flow velocity was reduced to 0 cm s^{-1} for 1 day to allow *Corophium* individuals to construct their tubes in the sediment. The two series were then exposed for 2 weeks to an average flow speed of 5 cm s^{-1} and 20 cm s^{-1}, respectively. At the end, 3 l of water were sampled and filtered over glass fibre filters. The filters were freeze-dried, weighed and the carbon and nitrogen content was analysed with a Carlo Erba Elemental Analyser NA-1500 (Nieuwenhuize et al., 1994). The remaining water was removed and 10 cores were taken from each flume to sample chlorophyll, carbon and nitrogen in the upper 5 mm of the sediment. Chlorophyll *a* concentration was analysed following Lorenzen (1967), using 90% methanol for extraction, and carbon and nitrogen concentrations were analysed with the Carlo Erba Analyser NA-1500. The rest of the sediment was sieved to count the remaining number of *Corophium*. This experiment was repeated twice. The results were analysed statistically with analysis of covariance (ANCOVA).

A second experiment, using sediment from the same site, was done in aquaria (30 * 30 cm). The aquaria were placed in a 15 °C room with a day–night light rhythm of 12 h light, 12 hours dark. The organic fraction of the sediment was removed by adding 5 cm^3 hydrogen peroxide twice. This was removed the next day by cooking, demineralised-water was added and subsequently the sediment was freeze-dried. The sediment was homogenised with filtered water from the Oosterschelde (30 ‰), transferred into the aquaria and allowed to settle and consolidate for 4 weeks. This resulted in a sediment layer of 10 cm. Diatoms, from a *Cylindrotheca* culture, were added in the fourth week and a diatom film formed on the sediment surface. At the end of this period, the water level was filled up to a water column of 10 cm. The water was aerated, which created a little turbulence. *Corophium* was collected in the field during July, sieved over a 500 μm sieve and a mixed population of adults and juveniles was added to the aquaria with increasing densities of 0, 1000, 10 000, 25 000, 50 000 and 100 000 ind. m^{-2}, respectively. The water was removed after 2 weeks and suspended solids were measured. Sediment stability was measured with a cohesive strength meter (CSM)(Paterson, 1989), which measures the sediment strength by exposing the sediment bed to an increasing vertical water current. This method gives relative values for the sediment stability expressed as a critical water velocity, which can be related to a horizontal bed shear stress (Tolhurst et al., 1999). Porosity was measured with a resistivity probe according to Andrews & Bennett (1981). The relation between resistivity and porosity for the sediment was $1/F = 1.074\ \phi^{3.636}$ (F is resistivity and ϕ porosity). Finally, three cores were taken in each aquarium and sliced into sections of four times half a centimetre and three times 1 cm from top to bottom. Samples were freeze-dried and analysed for grain size, chlorophyll *a* and carbohydrates. Grain size was measured with a Malvern Particle Sizer 3600 EC. Chlorophyll *a* was analysed by reverse phase HPLC (Mantoura & Llewellyn, 1973). Carbohydrates were extracted from the sediment in two steps, separating a loosely bound fraction extracted with water from a more tightly bound fraction extracted with EDTA (Underwood et al., 1995). The carbohydrate concentration was then measured with the phenol-sulphuric acid assay (Dubois et al., 1956). The sediment characteristics were tested statistically using linear regression analysis.

Results

At the end of the flume experiments, 50% of the *Corophium* was found alive in the sediment, indicating a decrease of *Corophium* density during the experimental period of 12 days. Starvation was probably due to handling or the pumping system. *Corophium* losses in the aquarium experiment were smaller. About 90% of the *Corophium* were recovered from the sediment after the experiment, independent of the initial densities. Since the actual densities during the experiments were unknown and most likely variable, the initial densities of *Corophium* were used in the statistical analysis.

The amount of suspended solids in the flume experiment increased significantly with increasing densities of *Corophium volutator* in the first series (Figure 1a; ANCOVA: $F_{1,12} = 33.6$; $p < 0.001$). Significantly higher concentrations of suspended solids were found during the second series, for both treatments with low and high flow velocity (Figure 1b; ANCOVA: $F_{1,12} = 11.8$; $p < 0.01$). This was possibly caused by a higher concentration of suspended solids in the Oosterschelde water that was used for the experiment. No significant effect of the flow velocity in the flumes on suspended solids concentration was found (ANCOVA: $F_{1,12} = 1.6$; n.s.).

The suspended solids consisted mostly of inorganic material. Organic carbon content was 4.60 ± 1.23% and total nitrogen content was 0.57 ± 0.17%

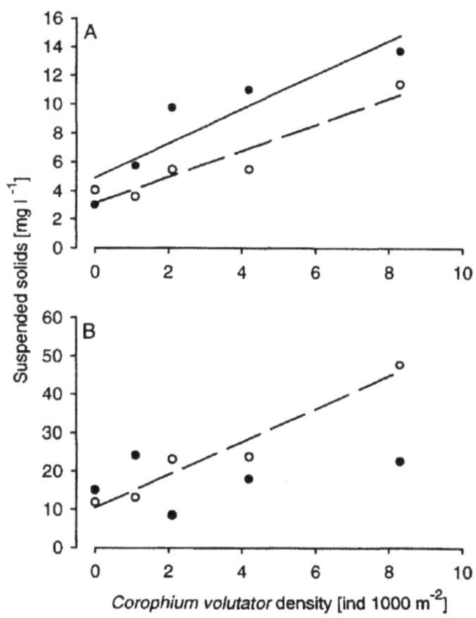

Figure 1. Relation between the suspended solids concentration and *Corophium* densities in the flume experiment, under flow velocities of 5 cm s^{-1} (●) and 20 cm s^{-1} (○), for (A): first series and (B): second series. The solid and dashed lines represent linear regression at 5 cm s^{-1} ((A) $r^2 = 0.84$, (B) not significant) and 20 cm s^{-1} ((A) $r^2 = 0.90$, (B) $r^2 = 0.94$).

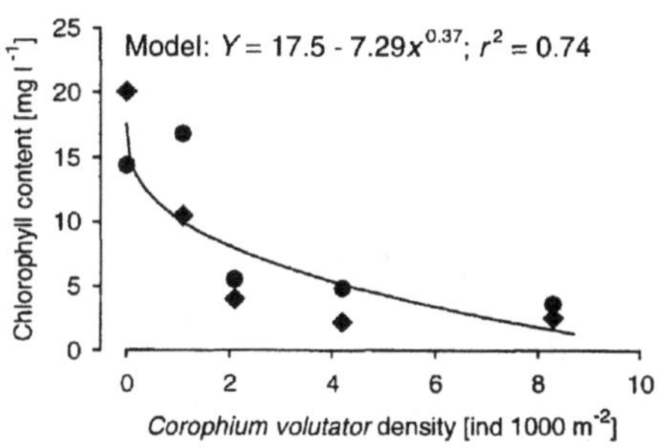

Figure 2. Chlorophyll *a* concentration in the first flume-experiment versus the *Corophium* density under two flow velocities, 5 cm s^{-1} (●) and 20 cm s^{-1} (♦). Line is based on a least square model fit.

Table 1. The organic carbon content of the sediment and of the suspended solids in the miniflumes

Corophium [ind. m^{-2}]	Org. C. in sediment [%]	Org. C. in water [%]
0	0.73 ± 0.07	3.72 ± 1.32
1083	0.61 ± 0.11	4.17 ± 1.56
2083	0.62 ± 0.07	4.58 ± 1.49
4167	0.59 ± 0.05	4.63 ± 1.00
8333	0.62 ± 0.06	4.67 ± 0.92

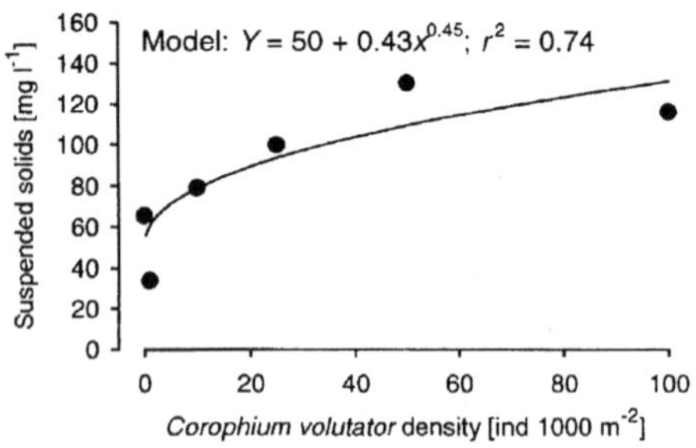

Figure 3. The relation of suspended solids concentration and *Corophium* density in the aquarium experiment. Line is based on a least square model fit of the data.

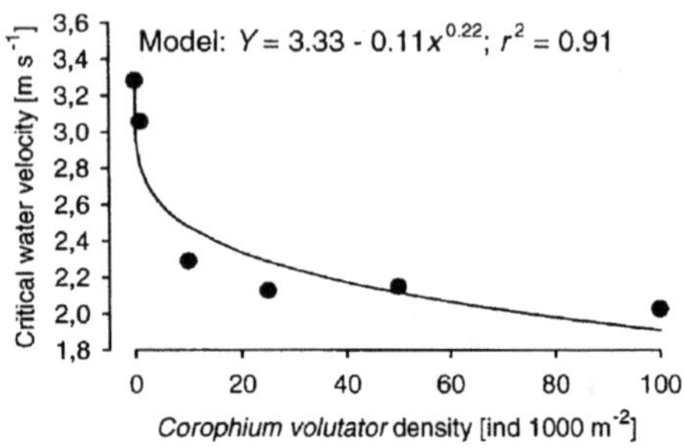

Figure 4. Sediment stability in the aquaria, measured with a CSM, versus *Corophium* density. Line is based on a least square model fit of the data.

when *Corophium* was abundant, while these values were 3.72 ± 1.32% and 0.45 ± 0.08%, respectively, in treatments without *Corophium* (as shown in Table 1). The organic carbon content in the sediment was slightly higher when *Corophium* was absent. The chlorophyll *a* concentration showed a significant decrease with increasing density of the *Corophium* (Figure 2; $F_{3,7} = 23.0$; $p < 0.001$). Both the decreases in carbon content and in chlorophyll *a* most likely resulted from the feeding activities of *Corophium.*

An increase of suspended solids in the water column was found with increasing *Corophium* density in the aquaria (Figure 3: $F_{3,3} = 31.0$; $p < 0.01$). The high concentration of suspended solids in the aquarium without *Corophium* may be caused by resuspension of sediment while water was siphoned out. The resuspended material probably originated from the diatom mat. The surface of the sediment was completely covered with a diatom film in aquaria with *Corophium* densities of 0 and 1000 ind. m^{-2} at the end of the experiment. A patchy distribution of diatom spots was visible in the aquaria with densities of, respectively, 10 000 and 25 000 ind. m^{-2}. No diatoms were observed at the end of the experiment in the aquaria with the highest densities of *Corophium*. The decrease of surface area covered by diatoms was not reflected by the chlorophyll *a* concentration in the upper half-centimetre of the sediment (as shown in Table 2). No relation was found between the *Corophium* density and the chlorophyll *a* concentration, probably because the concentrations were very low in all treatments. The chlorophyll *a* concentration in the deeper layers was constant in all aquaria with an average of 3.1 ± 0.8 μg g^{-1} dry weight. Both water and EDTA extractable carbohydrates were unaffected by the density of *Corophium*. The concentration of carbohydrates was low (as shown in Table 2) and did not change with depth. The average porosities within the upper half centimetre and within the upper two centimetres are shown in Table 2. *Corophium* did not affect the porosity in these sediment layers. Although *Corophium* density had little influence on sediment characteristics, a significant change in sediment stability was found (Figure 4; $F_{3,3} = 279.9$; $p < 0.001$). Sediment stability decreased rapidly with increasing density of *Corophium* (Figure 3) and was highest in sediments covered with a layer of diatoms.

Discussion

A large body of literature exists on the grazing of diatoms by the amphipod *Corophium volutator* (Pallas), and it's effects on the erosion and resuspension of sediment. The effects of *Corophium* on sediment stability are mainly thought to be indirect. *Corophium* grazing reduces diatom densities, which in turn increases sediment erosion and resuspension. Our experiments show that *Corophium* also directly affects sediment resuspension and hence the stability of intertidal mudflats. The concentration of suspended solids in the water overlying experimental sediments was found to increase with the density of *Corophium* individuals. At the low flow velocities used in the experiments (<20 cm s^{-1}), no effect was found of flow velocity on the concentration of suspended solids, which indicates that biological resuspension is more important than physical resuspension under these conditions.

Corophium may affect the resuspension of sediment by changing sediment properties. The burrowing and deepening of tubes may result in an increase of porosity as described by Gerdol & Hughes (1994a). Sediment stability is negatively related to the porosity (Meadows & Tait, 1989; Jones & Jago, 1993) and to the number of protruding tubes (Eckman et al., 1981). A positive relation, on the other hand,

Table 2. The chlorophyll *a* and carbohydrate concentration in the surface half centimetre of the sediment in the aquaria and the porosity in the top half and top 2 cm. n.s.: not significant

Corophium [ind. m^{-2}]	Chlorophyll *a* [μg g^{-1}d.w.]	H_2O-EPS [μg g^{-1}d.w.]	EDTA-EPS [μg g^{-1}d.w.]	Porosity [%] (0–0.5 cm)	Porosity [%] (0–2 cm)
0	5.9	153	144	85	76
1000	9.7	190	135	87	77
10 000	3.9	76	174	88	75
25 000	7.9	121	211	84	75
50 000	6.9	204	236	89	77
100 000	5.6	206	163	86	76
$F_{1,4}; p$	0.2; n.s.	1.6; n.s.	0.4; n.s.	0.1; n.s.	0.1; n.s.

is found between sediment stability and secretion of mucus (Meadows et al., 1990). The secretion of mucus networks by *Corophium* sticks sediment grains together. In this study, no effect of *Corophium* density was found either on sediment porosity or on extracellular carbohydrate concentrations. Similar to our study, Jones & Jago (1993) did not find any effect of *Corophium arenarium* on porosity in the field with densities of about 4000 ind. m^{-2}. The increase in the number of tubes protruding the sediment surface may have affected resuspension in the flume experiment. In the aquarium experiment, however, no horizontal water flow was present. Hence, resuspension due to an increased microtopography is likely to be neglegible.

Corophium may also affect sediment resuspension indirectly by reducing diatom densities. Diatoms protect the sediment against erosion by incorporating sediment particles into a matrix of mucus (Paterson, 1989). We found a strong negative correlation between chlorophyll *a* concentration and *Corophium* numbers in the flume experiment, but no correlation in the aquarium-experiment. The much higher density of *Corophium* in this last experiment, where diatom densities were most likely depleted, may explain this.

Despite the absence of effects of *Corophium* densities on sediment properties, we found a strong negative correlation between the 'critical water velocity' of the sediment and *Corophium* densities in the aquarium-experiment. Although observation of diatom cover indicated a higher density of diatoms in aquaria with low *Corophium* numbers, no significant effect of *Corophium* on chlorophyll content was found in this experiment. We are, therefore, unable to present conclusive evidence of a mechanism explaining the differences in critical erosion thresholds. Further studies on the effects of *Corophium* on sediment properties are required to gain insight in their effect on sediment stability.

Active resuspension of particles by *Corophium* does not only take place during burrowing, but may also result from feeding. *Corophium* ingests organic and inorganic particles <60 μm by either deposit feeding or suspension feeding (Meadows & Reid, 1966; Fenchel et al., 1975; Hawkins, 1985; Icely & Nott, 1985). Despite the high organic carbon fraction in this size range, still a considerable part of the gut content consists of inorganic material (Murdoch et al., 1986). Another way of feeding is epipsammic browsing. The organic film, build up of micro-organisms and organic matrix, is scraped of the larger grains with a teeth-like structure on the first pair of gnathopods (Nielsen & Kofoed, 1982; Stuart, 1985). Feeding reduces the organic carbon content and the diatom population in the sediment (this study; Gerdol & Hughes, 1994b). Unwanted particles and faeces will be removed from the tube with the respiratory current (Meadows & Reid, 1966), which is also found for *Pagurus*, *Macoma*, *Pectinaria* and *Yaldia* (Davis, 1993). The respiratory current in tubes of *Corophium* occurs almost continuously at rates of 25–100 ml h^{-1} (Reise, 1985). The current will also transport nutrients such as nitrate and ammonium out of the burrow (Pelegri et al., 1994), which may promote growth of diatoms under nutrient limited conditions. Hence, grazing and burrowing of *Corophium* has multiple effects on stability and functioning of sediment.

Our results indicate that active resuspension of sediment by *Corophium* may be an important factor influencing the suspended solid concentration of water in estuarine systems, especially when current velocity is low. At high current velocities, direct physical resuspension is probably the most important determinant of sediment resuspension. The relative importance of active and physical resuspension most likely depends on the stress upon the sediment by water currents and wave action. Since *Corophium* is most abundant in calm areas where physical stress on the sediment is minimal (Beukema & Flach, 1995), active resuspension may be an important factor determining the turbidity of water overlying intertidal mudflats.

Acknowledgements

The authors wish to thank Jurgen Dietvorst for his help by the experiment and Jan Sinke and Peter van Breugel for the analysis in the lab. We thank Jan Beukema and an anonymous reviewer for their comments on the manuscript. This work was supported by the Dutch organisation NWO-GOA as a part of the BOA research theme on tidal areas and by the European Community grants MAS3-CT95-0022-INTRMUD and ENV4-CT96-0216 ECOFLAT. This is publication no. 2528 of Netherlands Institute of Ecology, Centre for Estuarine and Coastal Ecology, Yerseke.

References

Andrews, D. & A. Bennett, 1981. Measurements of diffusivity near the sediment-water interface with a fine- scale resistivity probe. Geochim. Cosmochim. Acta 45: 2169–2175.

Beukema, J. J. & E. C. Flach, 1995. Factors controlling the upper and lower limits of the intertidal distribution of two *Corophium* species in the Wadden Sea. Mar. Ecol. Prog. Ser. 125: 117–126.

Daborn, G. R., C. L. Amos, M. Brylinsky, H. Christian, G. Drapeau, R. W. Faas, J. Grant, B. Long & D. M. Paterson, 1993. An ecological cascade effect: Migratory birds affect stability of intertidal sediments. Limnol. Oceanogr. 38: 225–231.

Davis, W. R., 1993. The role of bioturbation in sediment resuspension and its interaction with physical shearing. J. exp. mar. Biol. Ecol. 171: 187–200.

Dubois, M., K. A. Gilles, J. K. Hamilton, P. A. Rebers & F. Smith, 1956. Colorimetric method for determination of sugars and related substances. Anal. Chem. 28: 350–356.

Eckman, J. E., A. R. M. Nowell & P. A. Jumars, 1981. Sediment destabilization by animal tubes. J. mar. Res. 39: 361–374.

Fenchel, T., L. H. Kofoed & A. Lappalainen, 1975. Particle size-selection of two deposit feeders: the amphipod *Corophium volutator* and the prosobranch *Hydrobia ulvae*. mar. Biol. 30: 119–128.

Flach, E. C., 1992. The influence of four macrozoobenthic species on the abundance of the amphipod *Corophium volutator* on tidal flats of the Wadden Sea. Neth. J. Sea Res. 29: 379–394.

Gerdol, V. & R. G. Hughes, 1994a. Effect of *Corophium volutator* on the abundance of benthic diatoms, bacteria and sediment stability in two estuaries in southeastern England. Mar. Ecol. Prog. Ser. 114: 109–115.

Gerdol, V. & R. G. Hughes, 1994b. Feeding behaviour and diet of *Corophium volutator* in an estuary in southeastern England. Mar. Ecol. Prog. Ser. 114: 103–108.

Grant, J. & G. Daborn, 1994. The effects of bioturbation on sediment transport on an intertidal mudflat. Neth. J. Sea Res. 32: 63–72.

Hawkins, C. M., 1985. Population carbon budgets and the importance of the amphipod *Corophium volutator* in the carbon transfer on a cumberland basin mudflat, upper bay of Fundy, Canada. Neth. J. Sea Res. 19: 165–176.

Icely, J. D. & J. A. Nott, 1985. Feeding and digestion in *Corophium volutator* (Crustacea: amphipoda). Mar. Biol. 89: 183–195.

Jones, S. E. & C. F. Jago, 1993. *In situ* assessment of modification of sediment properties by burrowing invertebrates. Mar. Biol. 115: 133–142.

Kamermans, P. & H. J. Huitema, 1994. Shrimp (*Crangon crangon* L.) browsing upon siphon tips inhibits feeding and growth in the bivalve *Macoma balthica* (L.). J. exp. mar. Biol. Ecol. 175: 59–75.

Kester, D. R., I. W. Duedall, N. Connors & R. M. Pytkowicz, 1967. Preparation of artificial seawater. Limnol. Oceanogr. 12: 176–179.

Limia, J. & D. Raffaelli, 1997. The effects of burrowing by the amphipod *Corophium volutator* on the ecology of intertidal sediments. J. mar. biol. Ass. U.K. 77: 409–423.

Lorenzen, C. J., 1967. Determination of chlorophyll and pheopigments: spectrophotometric equations. Limnol. Oceanogr. 12: 343–346.

Mantoura, R. F. C. & C. A. Llewellyn, 1973. The rapid determination of algal chlorophyll and carotenoid pigments and their breakdown products in natural waters by reverse-phase high-performance liquid chromatography. Anal. chim. Acta. 151: 297–314.

Meadows, P. S. & A. Reid, 1966. The behaviour of *Corophium volutator* (Crustacea: Amphipoda). J. Zool. 150: 387–399.

Meadows, P. S. & J. Tait, 1989. Modification of sediment permeability and shear strength by two burrowing invertebrates. Mar. Biol. 101: 75–82.

Meadows, P. S., J. Tait & S. A. Hussain, 1990. Effects of estuarine infauna on sediment stability and particle sedimentation. Hydrobiologia 190: 263–266.

Murdoch, M. H., F. Bärlocher & M. L. Laltoo, 1986. Population dynamics and nutrition of *Corophium volutator* (Pallas) in the Cumberland Basin (Bay of Fundy). J. exp. mar. Biol. Ecol. 103: 235–249.

Nielsen, M. V. & L. H. Kofoed, 1982. Selective feeding and epipsammic browsing by the deposit-feeding amphipod *Corophium volutator*. Mar. Ecol. Prog. Ser. 10: 81–88.

Nieuwenhuize, J., Y. E. M. Maas & J. J. Middelburg, 1994. Rapid analysis of organic carbon and nitrogen in particulate materials. Mar. Chem. 45: 217–224.

Paterson, D. M., 1989. Short-term changes in the erodibility of intertidal cohesive sediments related to the migratory behavior of epipelic diatoms. Limnol. Oceanogr. 34: 223–234.

Pelegri, S. P., L. P. Nielsen & T. H. Blackburn, 1994. Denitrification in estuarine sediment stimulated by the irrigation activity of the amphipod *Corophium volutator*. Mar. Ecol. Prog. Ser. 105: 285–290.

Reise, K., 1985. Tidal Flat Ecology. Springer-Verlag, Heidelberg: 191 pp.

Rhoads, D. C., 1974. Organism-sediment relations on the muddy sea floor. Oceanogr. Mar. Biol. ann. Rev. 12: 263–300.

Stuart, V., 1985. Seasonal changes in the digestive enzyme levels of the amphipod *Corophium volutator* (Pallas) in relation to diet. J. exp. mar. Biol. Ecol. 88: 243–256.

Tolhurst, T. J., K. S. Black, S. A. Shayler, S. Mather, I. Black, K. Baker & D. M. Paterson, 1999. Measuring the *in situ* erosion shear stress of intertidal sediments with the Cohesive Strength Meter (CSM). Estuar. coast. shelf. Sci. 49: 281–294.

Underwood, G. J. C., D. M. Paterson & R. J. Parkes, 1995. The measurement of microbial carbohydrate exopolymers from intertidal sediments. Limnol. Oceanogr. 40: 1243–1253.

Hydrobiologia **426:** 43–55, 2000.
G. Liebezeit, S. Dittmann & I. Kröncke (eds), Life at Interfaces and Under Extreme Conditions.

Preying at the edge of the sea: the nemertine *Tetrastemma melanocephalum* and its amphipod prey on high intertidal sandflats

Inken Kruse[1,*] & Frank Buhs[2]
[1] *Wadden Sea Station Sylt, Alfred Wegener Institute Foundation for Polar and Marine Research, Hafenstr. 43, D-25992 List/Sylt, Germany*
[2] *Universität Kiel, Zoologisches Institut, Arbeitsgruppe Marine Ökologie und Systematik, Olshausenstr. 40–60, D-24118 Kiel, Germany*
*E-mail: ikruse@awi-bremerhaven.de (*author for correspondence)*

Key words: predation, *Corophium arenarium*, prey escape, drift, zonation, *Nereis diversicolor*

Abstract

In the European Wadden Sea, the nemertine *Tetrastemma melanocephalum* occurs together with its prey, the amphipod *Corophium arenarium*, in the upper intertidal zone. *T. melanocephalum* leaves the sediment when the tide has receded and captures *C. arenarium* in its U-shaped burrow. Highest abundances of *T. melanocephalum* on the sediment surface were found on summer evenings, 2–4 h after high tide, when just a thin film of water was left on the flats. Laboratory Y-maze experiments indicated that gradients of substances produced by *C. arenarium* in this film of water play a role in tracking the prey. In the field, *T. melanocephalum* appeared in significantly higher numbers on experimental high density patches of *C. arenarium.* The amphipod in turn is able to recognize the nemertine. In aquarium experiments, significantly more amphipods escaped from the sediment into the water column when the predator was present. In the field, both predator and prey showed a high mobility by drifting in tidal waters. Benthic abundance maxima of *T. melanocephalum* and *C. arenarium* usually did not coincide spatially. It is assumed that the nemertines avoid tidal flats that dry out quickly leaving too little time for prey capture. *T. melanocephalum* is not able to dig into the sediment, but lives in burrows of *Nereis diversicolor.* The abundance of this polychaete was inversely related to *C. arenarium*, presenting a dilemma for *T. melanocephalum*: the spatial overlap of food and accommodation was rather small.

Introduction

The amphipods *Corophium arenarium* (Pallas) and *Corophium volutator* Crawford are dominant species in upper zones of many intertidal flats of the North Atlantic Ocean (Hart, 1930; Muus, 1967; Bousfield, 1973; Larsen & Doggett, 1991). They appear in high abundances, *C. volutator* up to 118 000 ind. * m^{-2} (Andres, 1970), and are important food for migratory shorebirds (Bengtsson & Svensson, 1968; Goss-Custard, 1977; Boates & Smith, 1979; Hicklin & Smith, 1981; Peer et al., 1986; Raffaelli & Milne, 1987; Wilson, 1989, 1990), fishes (Smidt, 1951; Imrie & Daborn, 1981; Dadswell et al., 1984; Mattila & Bonsdorff, 1989), crabs (Hart, 1930; Linke, 1939), shrimps (Plagmann, 1939) and polychaetes (Muus, 1967; Commito, 1982; Ambrose, 1984a; Commito & Schrader, 1985).

Distributional segregation of both species seems to be mainly determined by sediment composition: *C. volutator* lives in muddy, *C. arenarium* in more sandy sediments (Watkin, 1941; Meadows, 1964; Flach, 1993, 1996). Populations of *Corophium* spp. have long been known to exhibit largely unexplained spatial and temporal fluctuations in density (Watkin, 1941; Muus, 1967; Reise, 1978; Ólafsson & Persson, 1986; Wilson, 1988; Beukema & Flach, 1995, Wilson & Parker, 1996). Some studies highlight the significance of parasites (Jensen & Mouritsen, 1992; Jensen et al., 1998), others explain intertidal zonation patterns with interactions of *Corophium* spp. with associated species. According to Beukema & Flach (1995), the upper *Corophium* limit is set by abiotic factors and the lower limit by biotic interrelations. On their transects, the lower *Corophium* limit coincided with the upper level of high densities of adult lugworms *Arenicola marina*. Sediment reworking by this polychaete is likely to have a strong negative impact on *Corophium* spp. (Cadée, 1976; Jensen & Kristensen, 1990; Flach & de Bruin, 1992, 1994; Flach, 1993; Beukema & Flach, 1995). Similar disturbance effects are caused by dense aggregations of cockles *Cerastoderma edule* (Jensen, 1985; Flach, 1992, 1993; Flach & de Bruin, 1994) and by the polychaete *Nereis diversicolor* (Rönn et al., 1988; Jensen & André, 1993). Both species increased migration rates of *Corophium* spp. in laboratory experiments. Earlier findings of *N. diversicolor*, being also an important predator of *Corophium* spp. (Muus, 1967; Goerke, 1971; Rönn et al., 1988), could not be corroborated (Jensen & André, 1993). Instead, it was found that *N. diversicolor* is mainly a suspension and deposit feeder (Reise, 1979; Esselink & Zwarts, 1989; Riisgaard, 1991).

In contrast, the nemertine *Tetrastemma melanocephalum* (Johnston) is an obligatory predator, which does feed on *Corophium volutator* (Bartsch, 1973, 1975, 1977). *T. melanocephalum* attacks the amphipod with its long proboscis, which is everted within fractions of a second. A stylet at the end of the proboscis pierces the exosceleton of the amphipod and toxins and enzymes are injected – the toxins immobilize *C. volutator* within seconds and the enzymes predigest prey tissue (Gibson, 1972). The nemertine then sucks out the amphipod (Bartsch, 1973, 1975). Prey capture normally happens inside the U-shaped burrow of *C. volutator*. *T. melanocephalum* enters the burrows from the sediment surface when the tide has receded and a thin film of water is left on the flats (Bartsch, 1973, 1975). Although infaunal predators are known as an important structuring force in benthic communities (Commito & Ambrose, 1985a; Ambrose, 1991), the role of *T. melanocephalum* in the *Corophium*-community has not yet been elucidated. Reports on distribution of *T. melanocephalum* on intertidal flats suggest that *Corophium* spp. are specific prey for *T. melanocephalum* because an association with either *Corophium volutator* (Bartsch, 1973, 1975, 1977; Reise, 1978, 1985) or *C. arenarium* (Gibson, 1995) is always mentioned. However, abundance of *T. melanocephalum* has rarely been quantified, and the intertidal zonation has not been determined yet. Likewise, the complex interactions between this nemertine and its prey exceeding pure predation have not been studied. Predatory interactions are subject to a considerable number of variables (Murdoch & Oaten, 1975) relating to characteristics of predator, prey and their special environment.

Nemertines are often neglected in interactions between species (McDermott, 1984, 1993), although their predatory effects can lead to a substantial decrease in prey density (Ambrose, 1991). The nemertines *Amphiporus lactifloreus* and *Lineus viridis* may force escape behaviour of their prey (Thiel, 1992; Thiel & Reise, 1993). Predator avoidance behaviour in the marine benthos is also mentioned by many authors (Witte & De Wilde, 1979; Commito, 1982; Peterson et al., 1982; Ambrose, 1984b; Commito & Ambrose, 1985a,b; Ólafsson & Persson, 1986; Rönn et al., 1988). It may have strong effects on prey distribution while the actual consumption of prey may be relatively modest (Thiel & Reise, 1993). If the predators depend on special features of the habitat, predatory effects can be concentrated in special zones or structures. For instance, *A. lactifloreus* and *L. viridis* occur preferentially in clusters of blue mussels (*Mytilus edulis*). Both predators and their respective prey show inverse relationships in these clusters along the tidal slope, which is explained by refuge behaviour of the prey (Thiel & Reise, 1993). For *T. melanocephalum*, a dependence on the burrows of *N. diversicolor* has been described (Bartsch, 1977).

The purpose of this study was to examine whether *T. melanocephalum* contributes to the zonation pattern in an upper intertidal sandflat inhabited by *C. arenarium* and *N. diversicolor*. On the one hand, we investigated whether abundance maxima of *T. melanocephalum* and *C. arenarium* coincided. Furthermore, direct interactions between predator and prey were elucidated: laboratory experiments were carried out to test whether *T. melanocephalum* tracks its prey by perceiving cues in the water produced by *C. arenarium* and whether *C. arenarium* in turn reacts to the presence of *T. melanocephalum* by escape. In a field experiment, we examined if *T. melanocephalum* invades experimental high density patches of *C. arenarium*.

Materials and methods

Description of sites studied

Sampling and field experiments were carried out on the tidal flats near the peninsula of Eiderstedt in the Northern Wadden Sea (Figure 1). The flats lie sheltered between the mainland and a sandbank, which

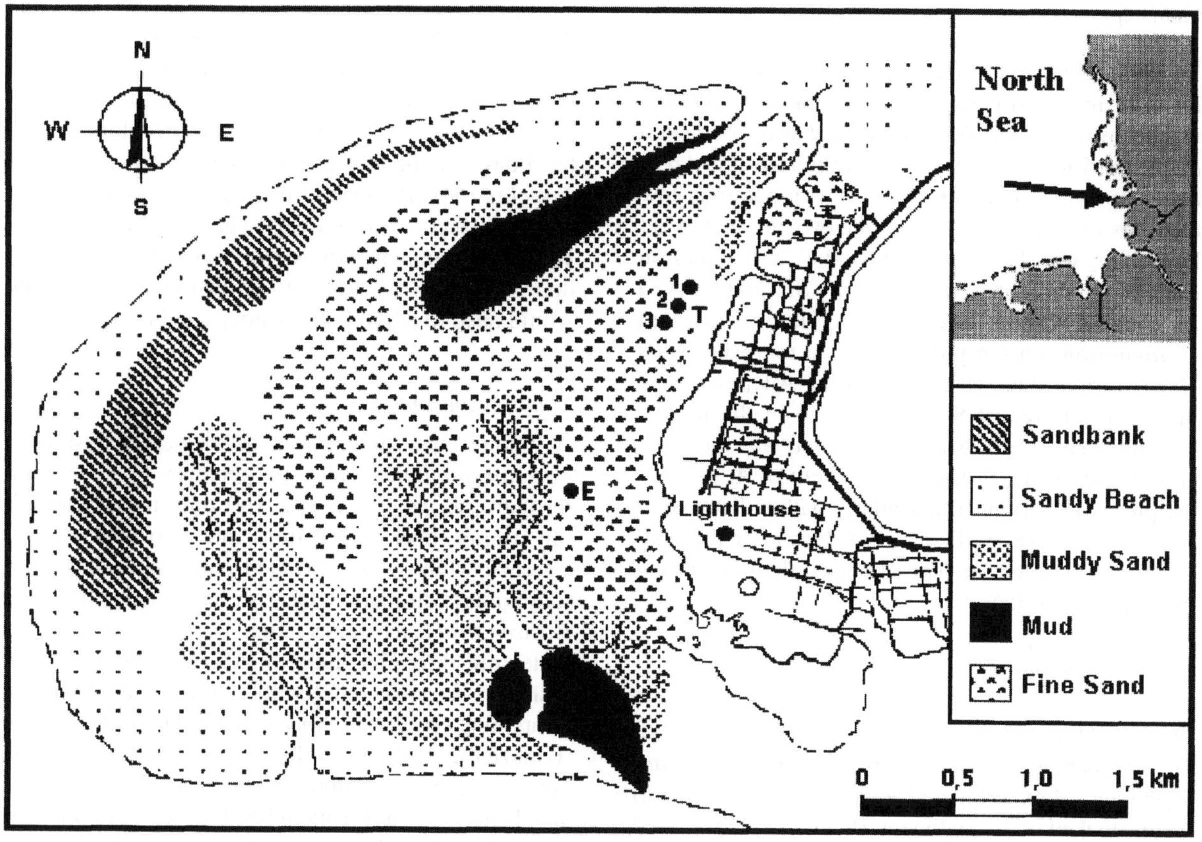

Figure 1. Study area with sediment zonation pattern of the flats. 'E' experimental site; 'T' transect through the upper intertidal with three sites, site 1 as the lowest.

is not submerged during regular high tides. Tidal range is about 3 m. Southwesterly winds prevail and blow fine grained sand from the sandbank onto the flats. Mean grain size of the sediment is 0.125–0.25 mm (Brey, 1991).

Tidal heights of sampling and experimental sites were estimated over several high waters using field observations. These observations for each site were later calibrated using the measured high water levels and tidal curves obtained by the gauge 'Kolumbusloch' near the Westerhever Sandbank.

Sampling

Abundances of *Tetrastemma melanocephalum* were estimated via surface activity, abundances of *Corophium arenarium* and *Nereis diversicolor* via sediment samples and abundances of *Arenicola marina* via counting the lugworm castings on 100 m^2 plots. These different methods were used because abundances of *T. melanocephalum* and *A. marina* were too low to be determined by sediment samples. All investigations were carried out along a transect with three sites through the upper intertidal (Figure 1). In addition, the surface activity of *T. melanocephalum* in the lower intertidal was determined occasionally (on 10 days, between 10 July and 25 August 1994, at two sites below the transect).

T. melanocephalum was quantified during periods favourable for surface activity (i.e. in the evening, 7:00–10:30 pm, on exposed flats with a water film). Nemertines crawling on the sediment surface of plots of 100 m^2 (in a few cases 25 and 50 m^2) were counted from mid June until the end of August in 1994. To avoid disturbing the same plot by footprints, counts alternated between four plots of 100 m^2 placed side by side at each transect site.

Sediment samples of 110 cm^2 surface area were taken with a plastic tube corer to a depth of 15 cm and sieved through a 500 μm mesh in the laboratory.

Field experiment

In 1995, we added *C. arenarium* to experimental plots in order to investigate whether *T. melanocephalum* would invade them. This experiment was conducted in the transition zone between *C. arenarium* and *Arenicola marina* beds (Figure 1 ('E')) with naturally low densities of *C. arenarium.* On one evening in July, at low water, we placed about 20 000 *C. arenarium* on each of six circular plots of 0.5 m^2. Six plots served as controls without prey addition. Plastic rims kept *C. arenarium* on the plots and were removed when *C. arenarium* had settled after 1 h. All plots were placed in an area of 8 * 10 m, each in the middle of 1 m^2 areas where *Arenicola marina* was naturally absent. According to the given pattern of *A. marina*, the plots were arranged irregularly with a distance of at least 1 m between them. Four days after prey addition, we counted the nemertines on the flats during their activity period.

Laboratory experiments

Aquarium experiment

To test whether *C. arenarium* escapes from *T. melanocephalum*, 12 experimental aquaria (20 * 20 * 20 cm) were stocked with sediment (8 cm depth; passed through a 300 μm sieve, from *C. arenarium*-habitat) and about 300 *C. arenarium* each. The sediment was covered with a thin film of water, simulating low-tide conditions. After 1 day when *C. arenarium* had settled, *T. melanocephalum* were inserted into six aquaria as predator addition. The other six aquaria served as controls without predators. All replicates were kept along a row of windows and exposed to a natural day/night-rhythm with the two treatments in alternate order. Then, during daytime, all replicates were flooded with 2 dm^3 of seawater, one after the other. To avoid disturbance of the sediment surface, a plastic foil was placed on the sediment surface before flooding and removed immediately afterwards. Directly after this in the same order, all swimming *C. arenarium* were removed with a pipette, which took about ten minutes for each replicate. All replicates contained standing seawater and had no outflow through which the amphipods could escape. Those *C. arenarium* remaining in the sediment were sieved out and both were counted. For each replicate, the percentage of swimming *C. arenarium* was calculated from the sum of swimming *C. arenarium* and those remaining in the sediment. The whole experiment was repeated four times. In the first three experiments, 10 nemertines were added into each predator-replicate and the amphipods were exposed to predatory impact for 1–2 days. In the 4th experiment, only one nemertine was inserted and the amphipods were exposed to predatory impact for 5 days. In the four experiments, *C. arenarium* and *T. melanocephalum* were re-used, only the dead ones were replaced. The animals were set back into the same aquarium where they had been before. This was done because we did not know how long *C. arenarium* would need to adapt to laboratory conditions.

Y-maze experiment

A Y-maze experiment was performed to test if *T. melanocephalum* is able to perceive *C. arenarium* through the water. From a 100 dm^3 tank, seawater ran trough tubes (ϕ 12 mm) into two dishes (13 * 21 cm), which were arranged parallel, each dish in longitudinal direction. One dish contained about 2000 *C. arenarium*, the other none. Out of each dish, the water ran through a tube of 20 mm ϕ into a single dish (9 * 21 cm), filled with 4 mm of sediment and containing one *T. melanocephalum.* The water left the Y-maze out of this dish through 20 mm-tubes. A 200 μm gauze was installed to keep the animals within the dishes, except for the tubes leading water into the nemertine dish. *T. melanocephalum* was placed in the middle of the nemertine-dish and could now decide to move into one of these tubes. When it had crawled through one, it was removed and another nemertine was inserted. Prior to the experiment, we tested if a non-turbulent flow would establish a gradient in the nemertine dish that *T. melanocephalum* could orientate on during crawling back and forth by putting ink into one *C. arenarium* dish. In the first run, the reaction of 28 nemertines was tested, in a second run for artefact control *C. arenarium* was put into the other dish and 20 of the same nemertines had to make their choice again. A final test with no *C. arenarium* in either dish was conducted with 9 nemertines. Before each run, the 100 dm^3 tank was filled with 200 μm- filtered seawater and between runs the experimental apparatus was rinsed with new seawater. The outflow rate was adjusted to about 1 dm^3 min^{-1} and the water level in the dishes was kept between 2 and 2.5 cm. The whole experiment was conducted in a dark room under red light.

Observations on the behaviour of the nemertines and amphipods

In the field, we investigated the behaviour of *T. melanocephalum* and *C. arenarium* drifting in tidal waters. In the laboratory, we offered alternative prey (two amphipods, one isopod and one harpacticoid copepod) to *T. melanocephalum* and observed prey capture behaviour.

Results

Surface activity of T. melanocephalum

T. melanocephalum was active on the sediment surface in summer, we could not find it there in winter. In 1994, only single individuals were present in May, activity increased in June and reached a maximum in July (8 ind. m^{-2}; Figure 2). In August and September, densities decreased again.

T. melanocephalum left the sediment when the tide had receded and reached highest surface activity on summer evenings, between 2 and 4 h after high tide. For best estimates of abundance, surface activity of *T. melanocephalum* was determined during these periods. During high tide and at night, only very few individuals were found.

Spatial distribution of the predators and the prey

Looking at the whole tidal slope, *T. melanocephalum* preferred the upper intertidal like its prey *C. arenarium* (Beukema & Flach, 1995). In the lower intertidal, surface abundances of *T. melanocephalum* were on average 5% of overall abundance, which was counted in the whole intertidal.

Within the upper intertidal, *C. arenarium* was most abundant at the uppermost site of the transect, while *T. melanocephalum* showed highest surface activity at the intermediate site (see Figure 3). *Nereis diversicolor* was also abundant at the intermediate site and reached highest densities at the lowest site. *Arenicola marina* was only found at the lowest transect site with 1–2 ind. m^{-2} (Table 1).

Mobility of the predators and the prey

T. melanocephalum and *C. arenarium* both drifted in the water column. *C. arenarium* actively held its position in the water. *T. melanocephalum* adopted a long, slender shape – nearly twice as long compared to when it is crawling. Drifting *T. melanocephalum* were mainly found during flood and in water depths between 5 and 30 cm. By walking through the water, we counted up to one nemertine m^{-1}. Several times we observed that *T. melanocephalum* coiled up and sank to the sediment. By producing mucus, they tried to attach to the sediment, presumably in order to search for a burrow to vanish in. Observations in aquaria indicate that *T. melanocephalum* can only hold its position in the water in the presence of currents. In the field, *T. melanocephalum* sank to the ground when they were isolated from the currents by lowering perspex cylinders around them.

Table 1. Tidal height and submersion of sampling (transect: 1, low – 3, upper) and experimental (E exp.) sites and abundances *A. marina*; submersion data of the transect for June–August 1994, for the experiment June–August 1995. (MHW: 138 cm above MTL)

Site	Tidal height below MHW [cm]	Mean submersion time / tide [h:min]	% of non-flooding tides	*A. marina* [ind m^{-2}]
1 low	34	2:53	4	2
2 mid	16	1:57	17	0.02
3 upper	8	1:33	34	0
E exp.	27		12	1–2

Direct interactions between T. melanocephalum *and its prey*

In order to test whether *T. melanocephalum* would accept prey organisms other than *C. arenarium*, four crustacean species were offered in plastic jars (ϕ 8.7 cm) with different water levels. Several individuals of *Talorchestia brito*, a semi-terrestrial amphipod (7–14 mm in length), were attacked and sucked out in jars with few water, when they were unable to swim. In jars filled with 1 cm of water, *T. melanocephalum* did not capture them. The same was found for *Bathyporeia* spp., a very mobile species in the field. *C. arenarium*, on the other hand, could be captured in jars with 1 cm of water. One 10 mm long *Idotea balthica* was killed and sucked out by a 10 mm long *T. melanocephalum.*

For juvenile *T. melanocephalum,* harpacticoids may play an important role. In the laboratory, one 4 mm long and 0.3 mm broad *T. melanocephalum,* which had starved for one month, fed on 5 *Mesochra lilljeborgi* (0.3–0.5 mm) and afterwards one juvenile *C. arenarium* (1.5 mm in length).

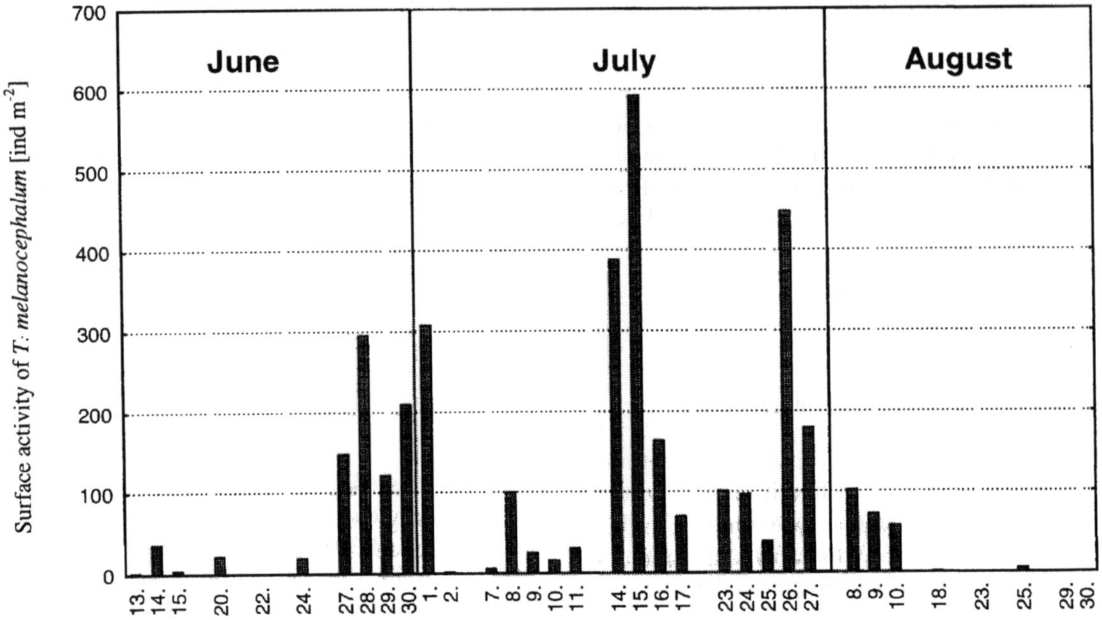

Figure 2. Surface activity of *T. melanocephalum* in summer 1994, mean values of the lowest and intermediate sites on a transect within the upper intertidal for single days, except for: 13–24 June lowest site only; 24 July, 18, 29–30 August intermediate site only. Sites were only counted at the given dates, spaces between the dates on the x-axis indicate an interruption of counting for one or more days.

Perception of prey by T. melanocephalum

In the laboratory Y-maze experiment, *T. melanocephalum* preferred the tube originating from the *C. arenarium* dish over the other tube carrying neutral seawater. In the first run, 26 individuals of *T. melanocephalum* decided for the '*C. arenarium*-tube' and two for the other. In the cross check, when *C. arenarium* was put into the other dish, we found the same preference of the '*C. arenarium*-tube'. In the blank check, when there was no *C. arenarium* in either dish, *T. melanocephalum* was undecided. The nemertine took twice as long to decide for one tube compared to when *C. arenarium* was in one of the dishes. In the first two runs, we observed several *T. melanocephalum* attacking the tweezers when we removed *T. melanocephalum* from the experiment after it had crawled through the *C. arenarium*-tube.

Tracking of prey by T. melanocephalum

In the field, *T. melanocephalum* reacted to experimentally increased density of *C. arenarium*. Four days after prey addition, significantly more *T. melanocephalum* were counted on the *C. arenarium* plots (29 ± 21 ind. 0.5 m^{-2}; mean ± SD) compared to controls (3 ± 2; Mann-Whitney-*U*-test, $p < 0.05$). Prior to the experiment, a natural density of 15 ± 37 *C. arenarium* m^{-2} was determined. Abundance of *Arenicola marina* was 1–2 ind. m^{-2} and of *Nereis diversicolor* 503 ± 230 (ind. m^{-2}; mean ± SD).

Escape from the predator by C. arenarium

C. arenarium was able to detect *T. melanocephalum* and escaped into the water column when the predators appeared in high abundances. However, this result was obtained only in the 2nd and 3rd aquarium experiment. The three experiments were carried out one after the other, re-using the animals. In the first experiment, swimming activity of *C. arenarium* did not depend on the presence of *T. melanocephalum*, but in the following two experiments it did (Table 2). In the first experiment, *C. arenarium* showed a very high swimming activity (mean 63%) in both treatments, in the following two experiments a much lower one (2nd experiment: mean 28%, 3rd experiment: mean 24% in the presence of the predator and 6% resp. 3% in the absence of predators, Mann-Whitney-*U*-test, $p < 0.05$). In a 4th experiment with only one instead of 10 *T. melanocephalum*, there was no difference in *C. arenarium* swimming activity (Mann-Whitney-*U*-test, $p > 0.05$).

Table 2. Mean percentage of swimming *C. arenarium* in aquaria with and without *T. melanocephalum*; six replicates for each treatment; *U*-test: n.s.: not significant, *: $p < 0.05$

Experiment	Date of flooding	Number of nemertines per predator replicate	Percentage of swimming *C. arenarium* without nemertines	Percentage of swimming *C. arenarium* with nemertines	*U*-test	Days of predatory impact
1	24 Aug.	10	63 ± 16	63 ± 14	n.s.	2
2	10 Sept.	10	6 ± 5	28 ± 6	*	1
3	23 Sept.	10	3 ± 3	24 ± 7	*	2
4	1 Oct.	1	8 ± 2	10 ± 2	n.s.	5

Table 3. Absolute numbers of swimming *C. arenarium* and those remaining in the sediment after flooding of aquaria with and without predators (see Table 2); $\sum$: number of surviving amphipods per replicate

Aquarium experiment 1			Aquarium experiment 2			Aquarium experiment 3		
Amphipods swimming	Amphipods sediment	$\sum$	Amphipods swimming	Amphipods sediment	$\sum$	Amphipods swimming	Amphipods sediment	$\sum$
With 10 nemertines								
189	89	278	51	195	246	30	224	258
148	82	230	58	209	267	68	250	272
125	85	210	111	213	324	89	248	274
95	140	235	119	230	349	60	234	265
144	90	234	88	249	337	70	225	248
180	33	213	80	195	275	70	218	249
Without nemertines								
123	84	207	5	249	254	2	260	281
136	74	210	12	246	258	6	251	269
136	160	296	15	267	282	5	252	281
129	154	283	6	273	279	5	263	275
185	94	218	23	268	218	25	240	218
133	39	172	42	262	304	4	236	257

At the end of the 2nd and 3rd experiment, methodological problems occurred because we found more than 300 *C. arenarium* in some replicates although we inserted only 300 (Table 3). Maybe we introduced small *C. arenarium* with the sediment, which grew to sizes remaining in the sieve during the experiments – *C. arenarium* was in the experimental aquaria for 5 days in the 2nd experiment and for four days in the 3rd experiment. Because of the different *C. arenarium* numbers between replicates, we calculated percentages of swimming *C. arenarium* from the sum of all *C. arenarium* retrieved from each aquarium. Numbers of *C. arenarium* were not different between treatments (Mann-Whitney-*U*-test, $p < 0.05$).

Apart from swimming, *C. arenarium* showed additional reactions to the presence of *T. melanocephalum*. In aquaria with *T. melanocephalum*, 2 days after addition of the nemertines and before flooding, fewer burrow openings of *C. arenarium* were found compared to those aquaria where the predator was absent. Burrow openings were closed with small rolls of cemented sediment. In absence of *T. melanocephalum, C. arenarium* accumulated mounds of sediment around the openings, giving the sediment surface quite a different appearance.

After we had flooded the aquaria and had kept them undisturbed for 2 days without aeration, another difference developed between the treatments. *C. aren-*

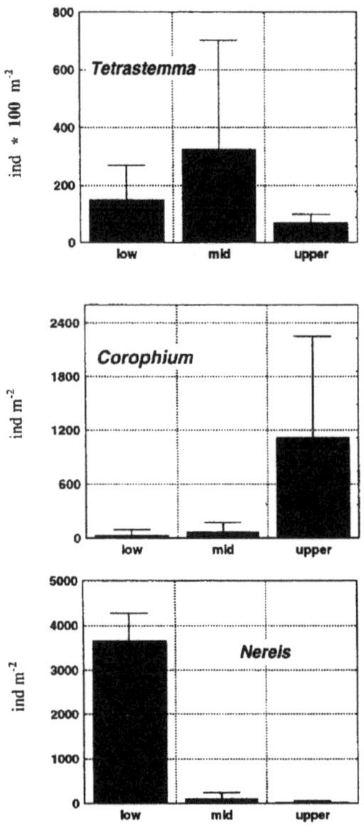

Figure 3. Surface activity of *Tetrastemma melanocephalum* * 100 m^{-2} (counts on 8 days, 14–26 July 1994) and abundance of *Corophium arenarium* and *Nereis diversicolor* [ind m^{-2}] (+SD) in the sediment along the transect (sampling dates of sites: upper site: 22 July 1994, intermediate (mid) site: 25 July 1994, lowest site: 27 July 1994).

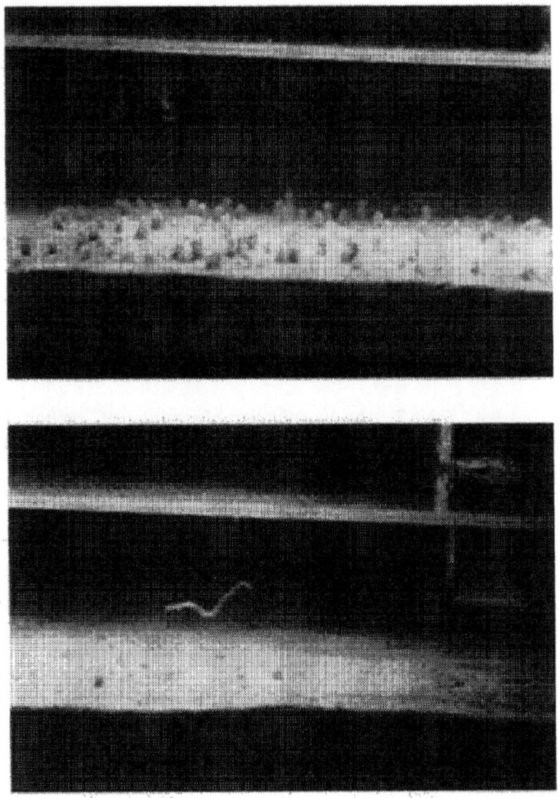

Figure 4. Corophium arenarium, change in the tube construction in the presence of *Tetrastemma melanocephalum.* Above: aquarium with *C. arenarium* only; below: aquarium with *C. arenarium* and *T. melanocephalum,* three days after insertion of the nemertines and two days after flooding of the aquaria. Above: shadows of swimming *C. arenarium* in the water column; below: *T. melanocephalum* on the aquarium wall; 0.5:2 cm.

arium built small chimneys around its burrows when *T. melanocephalum* was absent (Figure 4). In presence of the predator fewer and much shorter chimneys occurred.

Discussion

This study revealed that *Tetrastemma melanocephalum* is highly adapted to preying on *C. arenarium* which represents the only important prey for adult *T. melanocephalum* on high intertidal sandflats in the Wadden Sea. Other possible prey organisms were not abundant there or too mobile for being captured. However, amphipod prey that is probably poorly available for *T. melanocephalum* in nature was accepted in the laboratory. This indicates that *T. melanocephalum* is not dependent on a single prey item as was also described for other nemertines (McDermott, 1984).

Juvenile *T. melanocephalum* fed on harpacticoids in the laboratory (Bartsch, 1973; this study). This may be of crucial importance for small *T. melanocephalum* until they are able to capture *C. arenarium.* Harpacticoids should be available to *T. melanocephalum* because they are highly abundant both on the sediment surface and in burrows of *N. diversicolor* (Reise, 1985), where *T. melanocephalum* receives accommodation and may raise its brood. In addition, high abundances of harpacticoids in summer may coincide with the reproductive period of *T. melanocephalum.* We observed *T. melanocephalum* spawning in July/August (Kruse, 1996).

High surface activity of *T. melanocephalum* during its spawning season may be related to mate searching, but only in part. In July/August mature *T. melanocephalum* fed 2–3 *C. arenarium* d^{-1} in the lab (Kruse, 1996) and in the field, many immature *T. melanocephalum* were observed active on the sediment surface during this time.

Prey capture and foraging behaviour of T. melanocephalum

T. melanocephalum searches for prey during low tide, as was described for most other intertidal nemertines (Roe, 1971; Gibson, 1972; Nordhausen, 1988; Thiel & Reise, 1993; Thiel et al., 1995; Thiel, 1998). Since *T. melanocephalum* is a very small and light-weighted nemertine, low-tide activity may reduce the risk of being drifted into unfavourable sites by tidal currents. Furthermore, it has been suggested that prey cannot escape into the water column at low tide (Roe, 1970, 1976; McDermott, 1976; Thiel & Reise, 1993; Thiel et al., 1995). In the laboratory, *C. arenarium* left the sediment stocked with *T. melanocephalum* and swam into the water column when the aquaria were flooded one, resp. 2 days after the introduction of the nemertines. But *C. arenarium* can also escape the nemertines at low tide. *T. melanocephalum* captures *C. arenarium* inside its burrow. If an attacked *C. arenarium* succeeds to leave its U-shaped burrow at low tide, it can crawl faster over the sediment surface than *T. melanocephalum.* So we assume that the amphipods can equally well escape their predator at high or low tide. Escape from predator during low tide has also been found for the polychaete *Scolelepis squamata* who escapes from the polychaete *Eteone longa* (Behrends & Michaelis, 1977). Furthermore, some nemertine species successfully prey on tube-living amphipods in the subtidal, where escape into the water column is always possible (McDermott, 1976, 1988, 1993; McDermott & Roe, 1985).

Another explanation for *T. melanocephalum's* foraging activity during low tide may be that locating of its prey by chemoreception is facilitated. During high tide, chemical substances that could lead the predator to the prey may be diluted (Atema, 1988). Chemoreception is the main means of prey detection for nemertines (Amerongen & Chia, 1982). Our experiments suggest that *T. melanocephalum,* as well, uses cues in the water to find *C. arenarium.* The rather sedentary life-style of *C. arenarium* and its pumping activity inside the U-shaped burrow may favour the development of a gradient of chemical substances around the burrow openings in the thin film of water on the sediment surface during low tide. Searching behaviour of *T. melanocephalum* observed in the field supports the assumption that it orientates on such gradients. We saw the nemertine continuously bending its head to the side while creeping over the sediment surface. Once *T. melanocephalum* had picked up a strong scent of *C. arenarium,* a tactile stimulus of a moving object caused *T. melanocephalum* to eject its proboscis. In the laboratory, *T. melanocephalum* attacked tweezers after being exposed to water containing *C. arenarium.*

On a larger spatial scale, prey location may be facilitated by the flow regime. Thiel (1998) considers the steady unidirectional flow of water during low tide to offer the most favourable condition for locating prey. For scavenging nemertines, prey location was described for distances of 5–10 m (Thiel, 1998) or even up to 20 m (McDermott & Roe, 1985). However, long-distance chemoreception may be more important for scavenging nemertines than for those preying on live prey like *T. melanocephalum* (Roe, 1970; McDermott & Roe, 1985). Nemertines which prey on tube-dwelling amphipods do not have to crawl long distances since their prey often occurs in patches of relatively high densities (McDermott & Roe, 1985).

Concerning the diurnal activity rhythm, *T. melanocephalum* differs from many other nemertines. *T. melanocephalum* preferred the evening while many other nemertines are more active at night (Gibson, 1972; Thiel & Reise, 1993; Thiel et al., 1995). An adaption to higher temperatures may be one reason for the higher activity of *T. melanocephalum* during the evening compared to the night. The coincidence of the seasonal maximum of activity of *T. melanocephalum* and maximum temperatures in July may indicate such an optimum. For the nemertines *Amphiporus lactifloreus* and *Lineus viridis,* both common in the Wadden Sea, seasonal and diurnal maxima of activity also lie within the same temperature range, but at lower temperatures. These species were more active on the surface in late autumn and winter and at night (Thiel et al., 1995).

Impact of T. melanocephalum *on* C. arenarium

A predator may influence its prey population in different ways: by predation itself, by causing escape and avoidance behaviour and by altering of other behaviour, e.g. feeding activity. Most studies on predator-prey interactions concentrate on predation only and neglect other predatory impacts. *T. melanocephalum* may prey on 3 *C. arenarium* nemertine^{-1} d^{-1} (Bartsch, 1973; Kruse, 1996). In our laboratory experiments, *T. melanocephalum* increased the escape of *C. arenarium* into the water column 4–8-fold. This indicates that disturbance of the prey may also be an important factor determining prey distribution. However, results of our aquarium experiment are not quite clear. In the first experiment, the very high swimming activity of *C. arenarium* in the treatment with and without predator posed questions. Maybe the amphipods were not yet adapted to laboratory conditions, lunar migration rhythms or mate searching acitivities of the amphipods could have also blurred the picture. Furthermore, the four experiments conducted in sequence were not independent from each other since the animals were re-used. Consequently, the higher swimming activitiy of amphipods in replicates with than without predators in those without in the 2nd and 3rd experiment may be a consequence of the exposure to nemertines in the first experiment.

Our aquarium experiment also indicated behavioural change of the prey in the presence of *T. melanocephalum. C. arenarium* retreated into its burrow and closed the burrow openings with sediment. This may protect *C. arenarium* from being captured because *T. melanocephalum* is not able to dig into the sediment (pers. obs.; Bartsch, 1973) and may not be able to locate *C. arenarium* in a sealed burrow. *C. arenarium* in turn probably stops feeding in closed burrows. Consequences of behavioural changes and predator avoidance should be considered in future assessments of predatory impacts.

Mobility of predator and prey

Both *T. melanocephalum* and *C. arenarium* exhibited high mobility by drifting in tidal waters. For *Corophium volutator*, drift is well described (Ólafsson & Persson, 1986; Hughes, 1988; Essink et al., 1989; Hughes & Horsfall, 1990; Lawrie & Raffaelli, 1998a,b) and is reflected by shifts in upper distribution limits due to varying tide levels (Beukema & Flach, 1995). Escape from a predator may be one reason for *Corophium* spp. to enter the water column. For *Corophium arenarium*, escape from sediment disturbance by *Arenicola marina* and *Cerastoderma edule* into the water column has been shown by Flach (1993) and from *N. diversicolor* by Jensen & André (1993). Our results reveal that *T. melanocephalum* may be an additional factor to provoke emergence of *C. arenarium.*

For *T. melanocephalum,* drift in tidal waters had not been described before and scarcely for other nemertine species with an endobenthic life style (Bürger, 1897–1907; Dean, 1978). One reason for the drift of *T. melanocephalum* may be mate searching. Drift, reproduction and high surface activity of *T. melanocephalum* all coincide in July/August. In other nemertine species, mobility increases during reproduction as well. In Maine, specimens of *Cerebratulus lacteus* were found in driftnets in March (Dean, 1978) when this nemertine spawns (Bürger, 1897–1907). *Amphiporus lactifloreus* and *Lineus viridis* show higher surface activity on tidal flats during their reproductive periods (Thiel et al., 1995; Thiel & Dernedde, 1996).

Probably mate searching is not the only reason for drifting of *T. melanocephalum.* We observed many immature individuals drifting. Drift may also be an adaption to the high mobility and population dynamics of *C. arenarium* and may facilitate to reach the same sites, especially in tidal waters. Because *T. melanocephalum* is able to perceive *C. arenarium* through the water, a certain density of drifting *C. arenarium* might be a trigger for *T. melanocephalum* to enter the water column as well. On the other hand, *T. melanocephalum* may just leave sites with low prey density by emergence. This may result in a density dependent response race between the predators and the prey over a large spatial scale. Thiel (1998) proposed to scale the predators' mobilities and foraging ranges before assessing their role rather than classifying them as epibenthic or endobenthic. *T. melanocephalum* has several mobility strategies with being more or less stationary in a certain area – leaving the sediment for foraging and retreating again during low tide – or drifting in the water column. For the prey, predictablility of presence or absence of predators may be important in determining the attractivity of a locality for settling (Armonies, 1994). Thus, with respect to predator avoidance, the high intertidal is the most favourable zone for mobile *C. arenarium.*

Distribution of predator and prey

In sandflats, *T. melanocephalum* prefers the upper intertidal and lives in close association with *Corophium arenarium* (Gibson, 1995; this study). However, at a finer spatial scale within the upper intertidal, we found a maximum surface activity of *T. melanocephalum* at a transect site below that of maximum prey abundance. Limited time for capturing prey may be one factor preventing *T. melanocephalum* from exploiting the higher intertidal with highest prey densities. For searching food, *T. melanocephalum* depends on a thin film of water on the sediment surface. In summer 1994, we found an area with high *C. arenarium* density not submersed at all for about 30% of all tides. As a consequence, this upper zone may dry up quickly. Therefore, the availability of food for *T. melanocephalum* may not be highest where prey was most abundant. The upper limit of both predator and prey was controlled by abiotic factors, but restricting *C. arenarium* less than *T. melanocephalum*.

In contrast, the lower limit of *T. melanocephalum* is most probably determined by prey density. In an area with naturally low density of *C. arenarium*, we found a strong invasion of nemertines into experimental plots with increased prey densities. This area is located about 20 cm below that of high prey density and was submersed more frequently: only about 10% of all tides did not reach this zone.

Another factor controlling *T. melanocephalum* distribution may be their dependence on burrows of *Nereis diversicolor* for accommodation (Bartsch, 1973). *T. melanocephalum* is not able to dig into the sediment itself and follows chemical cues of *N. diversicolor* (Bartsch, 1977). High on the shore *N. diversicolor* was scarce and we found an inverse relation of abundance between *N. diversicolor* and *C. arenarium*. Thus, *T. melanocephalum* faces a dilemma. Spatial separation of accommodation and food may result in the choice of an area where both are sufficiently available. Furthermore, *N. diversicolor* has a negative impact on *C. arenarium*. This polychaete may cause emigration of *C. arenarium* by disturbance (Jensen & André, 1993). However, Flach (1993) and Beukema & Flach (1995) did not find negative interactions between *C. arenarium* and *N. diversicolor* and stress the significance of *Arenicola marina* in determining the lower limit of both *C. arenarium* and *C. volutator*. Our results indicate that additional factors may control the lower limit of *C. arenarium*. In our study area, *A. marina* did not occur at mid tide level, but *C. arenarium* became scarce there nevertheless. There was no sharp boundary between dense assemblages of sediment stabilizers (tube-building invertebrates like *Corophium* spp.) and sediment destabilizers (burrowing invertebrates like *Arenicola*) as described by Woodin (1983). In summary, there may exist a 'ménage à trois' between *T. melanocephalum, C. arenarium* and *N. diversicolor*. *T. melanocephalum* feeds on *C. arenarium* and depends on *N. diversicolor* as a provider of accommodation. Both drive *C. arenarium* out of the sediment, whereupon *C. arenarium* may find refuge high on the shore with an upper limit set by abiotic constraints.

Acknowledgements

We thank S. Lorenzen who drew our attention to *T. melanocephalum* and followed the study with interest. W. Armonies, U. Pieloth, M. Straßer, S. Dittmann, M. Thiel and an anonymous reviewer gave helpful comments on the manuscript. This is publication No. 1761 of the Alfred Wegener Institute Foundation for Polar and Marine Research.

References

Ambrose, W. G., Jr., 1984a. Influences of predatory polychaetes and epibenthic predators on the structure of a soft-bottom community in a Maine estuary. J. exp. mar. Biol. Ecol. 81: 115–145.

Ambrose, W. G., Jr., 1984b. Increased emigration of the amphipod *Rhepoxynius arbonius* (Barnard) and the polychaete *Nephtys caeca* (Fabricius) in the presence of invertebrate predators. J. exp. mar. Biol. Ecol. 80: 65–75.

Ambrose, W. G., Jr., 1991. Are infaunal predators important in structuring marine soft-bottom communities? Am. Zool. 31: 849–860.

Amerongen, H. M. & F.-S. Chia, 1982. Behavioural evidence for a chemoreceptive function of the cerebral organs in *Paranemertes peregrina* Coe (Hoplonemertea: Monstilifera). J. exp. mar. Biol. Ecol. 64: 11–16.

Andres, H. G., 1970. Zur Biologie und Ökologie des Amphipoden *Corophium volutator* PALLAS (Corophiidae). Unpubl. Diploma Thesis, Univ. Hamburg: 71 pp.

Armonies, W., 1994. Drifting meio- and macrobenthic invertebrates on tidal flats in Königshafen: a review. Helgoländer Meeresunters. 48: 299–920.

Atema, J., 1988. Distribution of chemical stimuli. In Atema J., R. Fay, A. N. Popper & W. N. Tavolga (eds), Sensory Biology of Aquatic Animals. Springer-Verlag, New York: 29–56.

Bartsch, I., 1973. Zur Nahrungsaufnahme von *Tetrastemma melanocephalum* (Nemertini). Helgoländer wiss. Meeresunters. 25: 326–331.

Bartsch, I., 1975. Nahrung und Nahrungsaufnahme bei zwei Schnurwurm- (Nemertinen-) Arten. Mikrokosmos 1: 16–19.

Bartsch, I., 1977. Zur Biologie des Nemertinen *Tetrastemma melanocephalum* (Johnston). Faun.-ökol. Mitt. 5: 125–128.

Behrends, G. & H. Michaelis, 1977. Zur Deutung der Lebensspuren des Polychaeten *Scolelepis squamata*. Senckenberg. marit. 9: 47–57.

Bengtson, S. A. & B. Svensson, 1968. Feeding habits of *Calidris alpina* L. and *C. minuta* Leisl. (Aves) in relation to the distribution of marine shore invertebrates. Oikos 19: 152–157.

Beukema, J. J. & E. C. Flach, 1995. Factors controlling the upper and lower limits of the intertidal distribution of two *Corophium* species in the Wadden Sea. Mar. Ecol. Progr. Ser.: 117–126.

Boates, J. S. & P. C. Smith, 1979. Length–weight-relationships, energy content and the effects of predation on *Corophium volutator* (Pallas) (Crustacea, pp. Amphipoda). Proc. NS Inst. Sci. 29: 489–499.

Bousfield, E. C., 1973. Shallow-water Gammaridean Amphipoda of New England. Comstock Publ. Assoc., Ithaca, New York: 312 pp.

Brey, T., 1991. The relative significance of biological and physical disturbances: an example from intertidal and subtidal sandy bottom communities. Estuar. coast. shelf. Sci. 33: 339–360.

Bürger, O., 1897–1907. Nemertini (Schnurwürmer). H. G. Bronn's Klassen und Ordnungen des Tierreichs 4, Suppl.: 542 pp.

Cadée, G. C., 1976. Sediment reworking by *Arenicola marina* on tidal flats in the Dutch Wadden Sea. Neth. J. Sea Res. 10: 440–460.

Commito, J. A., 1982. Importance of predation by infaunal polychaetes on controlling the structure of a soft-bottom community im Maine, U.S.A. Mar. Biol. 68: 77–81.

Commito, J. A. & W. G. Ambrose, Jr., 1985a. Predatory infauna and trophic complexity in soft-bottom communities. In Gibbs, P. E. (ed.), Proceedings of the 19th European Marine Biology Symposium. Cambridge University Press, Cambridge: 323–333.

Commito, J. A. & W. G. Ambrose, Jr., 1985b. Multiple trophic levels in soft-bottom communities. Mar. Ecol. Progr. Ser. 26: 289–293.

Commito, J. A. & P. B. Schrader, 1985. Benthic community response to experimental additions of the polychaete *Nereis virens*. Mar. Biol. 86: 101–107.

Dadswell, M. J., R. Bradford, A. H. Leim, D. J. Scarrat, D. G. Melvin & R. G. Appy, 1984. A review on research on fishes and fisheries on the Bay of Fundy between 1973 and 1983 with particular reference to its upper reaches. Can. Tech. Rep. Fish aquat. Sci. 1256: 163–294.

Dean, D., 1978. The swimming of bloodworms (*Glycera* spp.) at night, with comments on other species. Mar. Biol. 48: 99–104.

Esselink, P. & L. Zwarts 1989. Seasonal trend in burrow depth and tidal variation in feeding activity of *Nereis diversicolor*. Mar. Ecol. Progr. Ser. 56: 243–254.

Essink, K., H. L. Kleef & W. Visser, 1989. On the pelagic occurrence and dispersal of the benthic amphipod *Corophium volutator*. J. mar. biol. Ass. U.K. 69: 11–15.

Flach, E. C., 1992. Disturbance of benthic infauna by sediment-reworking activities of the lugworm *Arenicola marina*. Neth. J. Sea Res. 30: 81–89.

Flach, E. C., 1993. The distribution of the amphipod *Corophium arenarium* in the Dutch Wadden Sea: relationships with sediment composition and the presence of cockles and lugworms. Neth. J. Sea Res. 31(3): 281–290.

Flach, E. C., 1996. Distribution of *Corophium* at different scales. Senckenberg. marit. 27: 119–127.

Flach, E. C. & W. De Bruin 1992. Effects of *Arenicola marina* and *Cerastoderma edule* on distribution, abundance and population structure of *Corophium volutator* in Gullmarsfjorden, western Sweden. Sarsia 78: 105–118.

Flach, E. C. & W. De Bruin 1994. Does the activity of cockles, *Cerastoderma edule* (L.) and lugworms, *Arenicola marina* L., make *Corophium volutator* Pallas more vulnerable to epibenthic predators: a case of interaction modification? J. exp. mar. Biol. Ecol. 182: 265–285.

Gibson, R., 1972. Nemerteans. Hutchinson University Library, London: 224 pp.

Gibson, R., 1995. A synopsis of the British nemerteans. In Synopsis of the British Fauna. New Series, 24, University Press, Cambridge: 212 pp.

Goerke, H., 1971. Die Ernährungsweise der *Nereis*-Arten, Polychaeta (Nereidae) der deutschen Küsten. Veröff. Inst. Meeresforsch. Bremerh. 13: 1–50.

Goss-Custard, J. D., 1977. Predator responses and prey mortality in redshank, *Tringa totanus* (L.), and a preferred prey, *Corophium volutator* (Pallas). J. anim. Ecol. 46: 21–35.

Hart, T. J., 1930. Preliminary notes on the bionomics of the amphipod, *Corphium volutator* PALLAS. J. mar. biol. Ass. U.K. 16: 761–789.

Hicklin, P. W. & P. C. Smith, 1981. The diets of five species of migrant shorebirds in the Bay of Fundy. Proc. NS Inst. Sci. 31: 483–488.

Hughes, R. G.,1988. Dispersal by benthic invertebrates: the *in situ* swimming behaviour of the amphipod *Corophium volutator*. J. mar. biol. Ass. U.K. 68: 565–579.

Hughes, R. G. & I. M. Horsfall, 1990. Differences in the swimming behaviour of the amphipod *Corophium volutator* from different populations. J. mar. biol. Ass. U.K. 70: 143–148.

Imrie, D. M. G. & G. R. Daborn, 1981. Food of some immature fish of Minas Basin, Bay of Fundy. Proc. NS. Inst. Sci. 31: 149–153.

Jensen, K. T., 1985. The presence of the bivalve *Cerastoderma edule* affects migration, survival and reproduction of the amphipod *Corophium volutator*. Mar. Ecol. Prog. Ser. 25: 269–277.

Jensen, K. T., 1988. Recruitment and survival of *Nereis diversicolor* O. F. Müller and *Corophium volutator* (Pallas) in an artificial saltwater lagoon in the Danish Wadden Sea. Kieler Meeresforsch. Sdhft. 6: 366–374.

Jensen, K. T. & C. André, 1993. Field and laboratory experiments on interactions among an infaunal polychaete, *Nereis diversicolor*, and two amphipods, *Corophium volutator* and *C. arenarium*: effects on survival, recruitment and migration. J. exp. mar. Biol. Ecol. 168: 259–278.

Jensen, K. T. & L. D. Kristensen, 1990. A field experiment on competition between *Corophium volutator* (Pallas) and *Corophium arenarium* Crawford, Crustacea: (Amphipoda): effects on survival, reproduction and recruitment. J. exp. mar. Biol. Ecol. 137: 1–24.

Jensen, K. T. & K. N. Mouritsen, 1992. Mass mortality in two common soft-bottom invertebrates, *Hydrobia ulvae* and *Corophium volutator* – the possible role of trematodes. Helgoländer Meeresunters. 46: 329–339.

Jensen, T., K. T. Jensen & K. N. Mouritsen, 1998. The influence of the trematode *Microphallus claviformis* on two congeneric intermediate host species (*Corophium*): infection characteristics and host survival. J. exp. mar. Biol. Ecol. 227: 35–48.

Kruse, I., 1996. Einnischung der Nemertine *Tetrastemma melanocephalum* in die Lebensgemeinschaft des '*Corophium*-Watts'. Unpubl. Diploma Thesis, University of Kiel: 74 pp.

Larsen, P. F. & L. F. Doggett, 1991. The macroinvertebrate fauna associated with the mud flats of the Gulf of Maine. Bull. mar. Sci. 7: 365–375.

Lawrie, S. M. & D. G. Raffaelli, 1998a. Activity and mobility of *Corophium volutator*: a field study. Mar. Freshwat. Behav. Physiol. 31: 39–53.

Lawrie, S. M. & D. G. Raffaelli, 1998b. *In situ* swimming behaviour of the amphipod *Corophium volutator* (Pallas). J. exp. mar. Biol. Ecol. 224: 237–251.

Linke, O., 1939. Die Biota des Jadebusenwattes. Helgoländer wiss. Meeresunters. 1: 201–348.

Mattila, J. & E. Bonsdorff, 1989. The impact of fish predation on shallow water soft bottoms in brackish waters (SW Finland), an experimental study. Neth. J. Sea Res. 23: 69–81.

McDermott, J. J., 1976. Predation of the razor clam *Ensis directus* by the nemertean worm *Cerebratulus lacteus*. Chesapeake Sci. 17: 299–301.

McDermott, J. J., 1984. The feeding biology of *Nipponemertes pulcher* (Johnston) (Hoplonemertea), with some ecological implications. Ophelia 23: 1–21.

McDermott, J. J., 1988. The role of hoplonemerteans in the ecology of seagrass communities. Hydrobiologia 156: 1–11.

McDermott, J. J., 1993. Nemertea inhabiting the *Haploops* (Amphipoda) community of the northern Øresund with special reference to the biology of *Nipponnemertes pulcher* (Hoplonemertea). Hydrobiologia 266: 15–28.

McDermott, J. J. & P. Roe, 1985. Food, feeding behavior and feeding ecology of nemerteans. Am. Zool. 25: 113–125.

Meadows, P. S., 1964. Experiments on substrate selection by *Corophium* species: films and bacteria on sand particles. J. exp. Biol. 41: 499–511.

Muus, B., 1967. The fauna of Danish estuaries and lagoons. Meddr. Danm. Fisk. Havunders. NS 5: 1–316.

Murdoch, W. W. & A. Oaten, 1975. Predation and population stability. In Macfayden, A. (ed.), Advances in Ecological Research. Academic Press, London, 9: 1–131.

Nordhausen, W., 1988. Impact of the nemertean *Lineus viridis* on its polychaete prey on an intertidal sandflat. Hydrobiologia 156: 39–46.

Ólafsson, E. B. & L. E. Persson, 1986. The interaction between *Nereis diversicolor* O. F. Müller and *Corophium volutator* Pallas as a structuring force in a shallow brackish sediment. J. exp. mar. Biol. Ecol. 103: 103–117.

Peer, D. L., L. E. Linkletter & P. W. Hicklin 1986. Life history & reproductive biology of *Corophium volutator* (Crustacea: Amphipoda) and the influence of shorebird predation on population structure in Chignecto Bay, Bay of Fundy, Canada. Neth. J. Sea Res. 20: 359–373.

Peterson, C. H., W. G. Ambrose & J. H. Hunt, 1982. A field experiment on the swimming response of the Bay Scallop (*Argopecten irradians*) to changing biological factors. Bull. mar. Sci. 32: 939–944.

Plagmann, J., 1939. Ernährungsbiologie der Garnele (*Crangon vulgaris* Fabr.). Helgoländer wiss. Meeresunters. 2: 113–162.

Raffaelli, D. & H. Milne 1987. An experimental investigation of flatfish and shorebird predation on estuarine invertebrates. Estuar. coast. shelf. Sci. 24: 1–13.

Reise, K., 1978. Experiments on epibenthic predation in the Wadden Sea. Helgoländer wiss. Meeresunters. 31: 55–101.

Reise, K., 1979. Moderate predation on meiofauna by the macrobenthos of the Wadden Sea. Helgoländer wiss. Meeresunters. 32: 453–465.

Reise, K., 1985. Tidal Flat Ecology. Ecological Studies 54, Springer-Verlag, Berlin: 191 pp.

Riisgard, H. U., 1991. Suspension feeding in the polychaete *Nereis diversicolor*. Mar. Ecol. Prog. Ser. 70: 29–37.

Roe, P., 1970. The nutrition of *Paranemertes peregrina* (Rhynchocoela: Hoplonemertea). I. Studies on food and feeding behavior. Biol. Bull. 139: 80–91.

Roe, P., 1971. Life history and predator-prey interactions of the nemertean *Paranemertes peregrina* Coe. Unpubl. Ph. D. Thesis, University of Seattle: 129 pp.

Roe, P., 1976. Life history and predator-prey interactions of the nemertean *Paranemertes peregrina* Coe. Biol. Bull. 150: 80–106.

Rönn, C., E. Bonsdorff & W. G. Nelson, 1988. Predation as a mechanism of interference within infauna in shallow brackish water soft bottoms; experiments with an infauna predator, *Nereis diversicolor* O. F. Müller. J. exp. mar. Biol. Ecol. 116: 143–157.

Smidt, E. L. B., 1951. Animal production in the Danish Waddensea. Meddr. Danm. Fisk. Havunders. 11: 1–151.

Thiel, M., 1992. Zur Ökologie von Nemertinen im Wattenmeer. Unpubl. diploma thesis, University of Kiel: 69 pp.

Thiel, M., 1998. Nemertines as predators on tidal flats – high noon at low tide. Hydrobiologia 365: 241–250.

Thiel, M. & K. Reise, 1993. Interaction of nemertines and their prey on tidal flats. Neth. J. Sea Res. 31: 163–172.

Thiel, M. & T. Dernedde, 1996. Reproduction of *Amphiporus lactifloreus* (Hoplonemertini) on tidal flats: implications for studies on the population biology of nemertines. Helgoländer Meeresunters. 50: 337–351.

Thiel, M., W. Nordhausen & K. Reise, 1995. Nocturnal surface activity of endobenthic nemertines. In Eleftheriou, A., A. D Ansell & C. J. Smith (eds), Biology and Ecology of Shallow Coastal Waters. Proceedings of the 28th European Marine Biology Symposium, Iraklio, Crete, 1993. Olsen & Olsen, Fredensborg: 283–293.

Watkin, E. E., 1941. The yearly life cycle of the amphipod *Corophium volutator*. J. anim. Ecol. 10: 77–93.

Wilson, W. H., Jr., 1988. Shifting zones in a Bay of Fundy soft-sediment community, patterns and processes. Ophelia 29 (3): 227–245.

Wilson, W. H., Jr., 1989. Predation and the mediation of intraspecific competition on an infaunal community in the Bay of Fundy. J. exp. mar. Biol. Ecol. 132 (3): 221–245.

Wilson, W. H., Jr., 1990. Relationship between prey abundance and foraging site selection by semipalmated sandpipers on a Bay of Fundy mudflat. J. Field Ornith. 61: 9–19.

Wilson, W. H. & K. Parker, 1996. The life history of the amphipod, *Corophium volutator*: the effects of temperature and shorebird predation. J. exp. mar. Biol. Ecol. 196 (1–2): 239–250.

Witte, F. & P. A. W. J. De Wilde, 1979. On the ecological relation between *Nereis diversicolor* and juvenile *Arenicola marina*. Neth. J. Sea Res. 13: 394–405.

Woodin, S. A., 1983. Biotic interactions in recent marine sedimentary environments. In Tevesz, M. J. S. & P. L. McCall (eds), Biotic Interactions in Recent and Fossil Benthic Communities. Plenum Press, New York: 3–38.

Hydrobiologia **426:** 57–64, 2000.
G. Liebezeit, S. Dittmann & I. Kröncke (eds), Life at Interfaces and Under Extreme Conditions.

Meiobenthos of hypersaline tropical mangrove sediment in relation to spring tide inundation

Emil Ólafsson[1,*], Susanna Carlström[1] & Simon G. M. Ndaro[2]
[1]*Department of Zoology, University of Stockholm, S-106 91 Stockholm, Sweden*
[2]*Department of Zoology and Marine Biology, University of Dar-es-Salaam, P.O. Box 35064, Dar-es-Salaam, Tanzania*
*E-mail: emil.olafsson@zoologi.su.se (*author for correspondence)*

Key words: hypersaline, mangrove, tidal cycles, meiofauna, nematode assemblage

Abstract

Tropical intertidal sediments often contain porewater of relatively high salinity, especially in areas exposed to longer periods without seawater inundation and high evaporation. Such an area exists on the west coast of Zanzibar: a high intertidal mangrove plateau, flooded only during spring high tides, with sediment porewater salinities commonly exceeding 100 ppt. A field survey was conducted in this area to examine variations in population density of major meiofaunal taxa and the assemblage structure of free-living marine nematodes during spring-neap tidal cycles. Samples were taken on seven occasions for two months, starting from the end of the rainy season. Porewater salinity remained high throughout the sampling period, ranging from 89 to 160 ppt. Neither spring tide inundation nor heavy rains lowered the salinity markedly. The meiofauna consisted only of four taxa, present on all sampling occasions: nematodes, harpacticoid copepods, plathyhelminthes and chironomids. Densities in surface sediments (0–5 cm) were low compared to other mangrove areas, ranging from 271 to 656 animals 10 cm^{-2} with nematodes dominant on all sampling occasions (58–87%). Density fluctuations could not be explained by the effects of spring tide inundation, but the meiofauna showed significant correlations with grain size and organic material. Despite the wide range of salinity, only the numbers of chironomids were negatively correlated with increased salinity. Nematode species diversity was low in all samples, although altogether 28 species were recorded in the samples. Four species occurred in more than 50% of the samples (*Microlaimus* sp. (100%), *Metalinhomoeus* sp. (76%), *Daptonema* sp.1 (56%), *Chromadorina* sp. (56%)) while 12 species were found only in one or two samples. Multidimensional scaling ordination (MDS) of the nematode species abundance data indicated little effects of spring tide inundation on the assemblage structure, but rather a successive change from wet to dry season with a reduction in species diversity and increased numbers of the dominant nematode species *Microlaimus* sp.

Introduction

In physically harsh environments, such as exposed sandy beaches or hypoxic sediments, the abundance and species diversity of benthic macrofauna is often greatly reduced and sometimes this group totally disappears. By contrast, benthic meiofauna may survive in such environments, often in relatively high abundance, although diversity is usually limited to a few specialised species (e.g. Ólafsson, 1991; Vopel et al., 1996; Modig & Ólafsson, 1998).

Mangrove forests occupy large areas of the world's tropical coastlines. These forests occur at all intertidal levels and may, therefore, exhibit large variations in temperature and salinity. Longer periods without inundation, together with characteristics of tropical areas such as high evaporation and periods of little rainfall, result in high water losses from the sediment. Thus, combined with limited ground water flow, excessive accumulation of salt in the porewater can occur. High salinity areas may be recognised by dwarfed trees and bare patches devoid of trees (Alongi, 1989; Wolanski et al., 1992).

In a survey of the mangroves of Zanzibar, Ólafsson (1995) described the meiobenthos in a hypersaline habitat where sediment salinities exceeded 100 ppt. In these extreme conditions, no marine macroinfauna were found and the meiofauna was restricted to three taxa in relatively low numbers. Apart from a brief account of meiofauna densities in extreme salinities (up to 135 ppt) (Alongi, 1988), the meiobenthos in hypersaline mangrove habitats has received no further attention.

The hypersaline area in Zanzibar is located at high intertidal level where the sediment is inundated only at spring high tides. This means that the sediment is exposed to incoming sea-water two times a day for ca. 4–6 days and then stays dry until the next spring tide after ca. 8–10 days. The meiofauna inhabiting this area may be affected by the tidal cycle in several ways. It is well known that meiofauna are rapid colonisers of intertidal sediments (e.g. Sherman & Coull, 1980; Savidge & Taghon, 1988), where hydrodynamic forces are relatively high, providing a quick dispersal via currents (see review by Palmer, 1988). Therefore, it is plausible that the area may be colonised at every spring tide by marine species that are not adapted to extreme saline conditions. This would presumably result in higher density and different species composition at, or just after, the spring tides compared to periods when the sediment is not flooded. Secondly, the flooding of the hypersaline sediment could act as a lifeline to those species present in the area by lowering the porewater salinity during critical periods in their life-cycle. Several meiofauna species, for instance, complete their life-cycle within hours or days (cf. Hicks & Coull, 1983; Heip et al., 1985). In the hypersaline area, we might therefore have species that thrive under the periods of flooding, while during the non-flooding periods only the most tolerant individuals will survive.

Materials and methods

Study area

Unguja island is the largest island of Zanzibar and supports six major mangrove stands of approximately 6000 hectares (Ngoile & Shunula, 1992; Shunula & Whittick, 1996). This study took place in the Muwanda mangrove forest (05° 55′ S, 39° 13′ E), the second largest mangrove stand, situated on the north-west coast of Unguja. The forest is up to 1 km wide and covers an area of about 520 hectares. A description of the zonation of mangrove vegetation is given in Shunula & Whittick (1996). *Avicennia marina* dominates the mid and high intertidal forest and is the only species extending landwards (Shunula & Whittick, 1996).

The field sampling was confined to a plateau in the high intertidal zone of the forest. The *Avicennia marina* grow there as small shrubs, occurring scattered over the plateau. Large patches (hundreds of m^2) occur as bare sediment without trees. Porewater salinity is generally very high here, typically over 100 ppt and the local residents use parts of the plateau for salt extraction.

Field sampling

A sampling station, ca. 25 m^2, was located at a site approximately 3.7 m above chart datum, 80 m from land in a part of the plateau, where the sediment appeared smooth, without trees, pneumatophores or visible crab holes. Sampling was carried out on seven occasions from the end of May to middle of July 1996. This included four spring tide samplings; June 4 (S2), June 18 (S4), July 2 (S5) and July 16 (S7) and three sampling occasions during neap tides; May 24 (N1), June 11 (N3) and July 9 (N6). No sampling was done during the neap tide between S4 and S5. Sampling was carried out at low tides during daytime. The first sampling date was 1 day after the last heavy rains from the rainy period. The station was inundated at least at four high tides prior to sampling at spring tide while it had not been inundated for at least eight high tides prior to sampling at neap tides.

At each sampling occasion, a square frame of 1 m^2 was randomly located and samples retrieved as follows: one 5 cm deep core (8.5 cm^2) was pushed into the sediment at a fixed position within the frame for meiofauna sample and left in position. Only meiofauna cores were taken on the first sampling occasion. Adjacent to the meiofauna core, two smaller cores (4 cm^2) were taken for chlorophyll *a* extraction. These were immediately combined and mixed with an equal volume of 90% acetone in a vial and wrapped with aluminium foil. Cores (8.5 cm^2) for organic content and grain size determination were taken down to 5 cm depth and finally the core for meiofauna was retrieved and fixed with 5% formalin. When possible, porewater was retrieved from the holes left by the sediment sampling and salinity measured with a refractometer. On some occasions, one had to dig deeper to get sufficient water for salinity measurements. The whole procedure was repeated 5 times, yielding 5 replicates of each variable.

In the laboratory, the meiofauna samples were washed through 500 and 40 μm sieves and meiofauna extracted from the 40 μm fraction using Ludox (colloidal silica polymer) at a specific gravity of 1.15. The meiofauna was enumerated and identified to major taxa in a Petri dish under a stereo dissecting microscope. The extracted samples were transferred to glycerine and mounted on slides for nematode identification under a high-power microscope. Nematodes were sorted to species, identified to genus level using the pictorial keys of Platt & Warwick (1983) and assigned to trophic groups according to the scheme of Wieser (1953). Grain size samples were oven dried at 85 °C for 24 h and subsequently sieved through a series of sieves (2, 1, 0.5, 0.25, 0.125 and 0.063 mm) and grain size determined on the basis of the weight of each size fraction (Giere et al., 1988). Organic content was determined by oven drying samples at 85 °C for 48 h and then combusting them at 550 °C for 12 h (Giere, 1993). Chlorophyll samples were fixed with 90% ($^{v}/_{v}$) acetone and kept in the dark, refrigerated for 24 h and analysed using a spectrophotometer according to Parsons et al. (1984). Temperature was measured at 3 cm depth in the sediment with a mercury thermometer. Due to technical problems, chlorophyll samples were not processed from the second sampling date (S2).

Statistics

Differences in density were investigated by means of one-way analysis of variance. Paired a posterior comparisons of density estimates were carried out with the Tukey test, using 95% confidence limits. Prior to the analysis of variance, all data were log 10 transformed apart from organic% data which were arcsine transformed. Cochran's C test was used to check the assumption of homoscedasticity. Major taxa and nematode species abundance data were double square root transformed and subjected to non-metric multidimensional scaling ordination (MDS) using the Bray-Curtis similarity measure (Warwick et al., 1990a,b).

Results

Environmental variables

During the course of the study, the environmental factors fluctuated considerably (Figure 1). Porewater salinity was on average higher at neap tides than at spring tides (Figure 1 and Table 1). Mean grain size varied also significantly ($p < 0.001$) being higher in the beginning of the study compared to later dates.

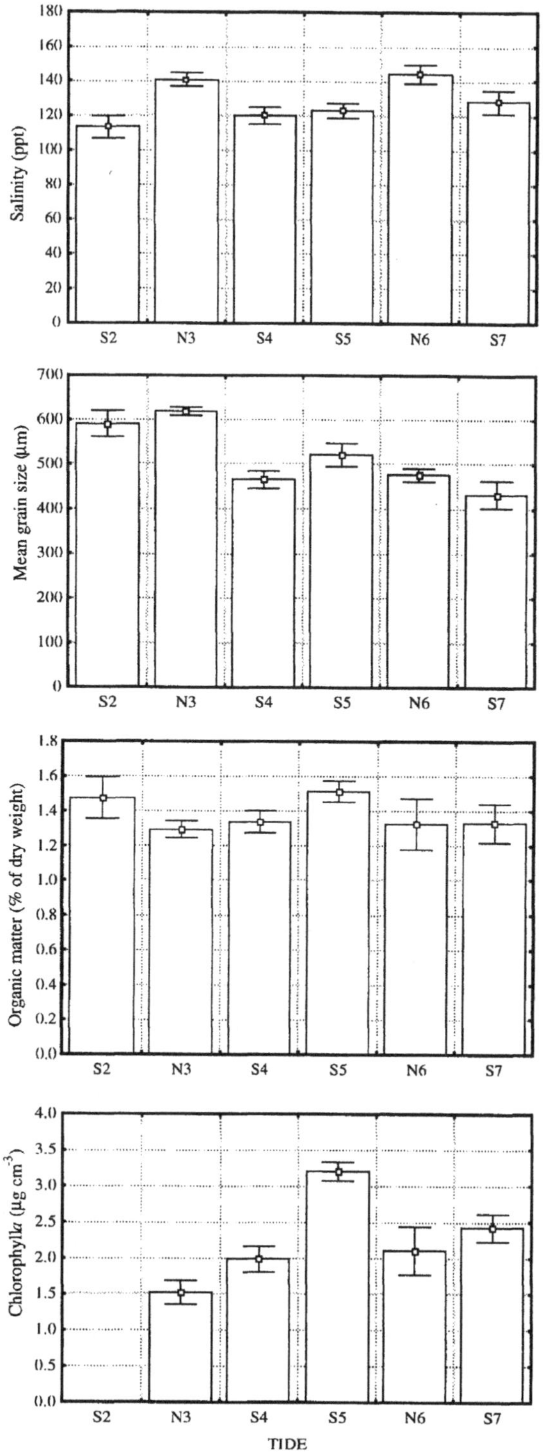

Figure 1. Average values ($n = 5$) of environmental factors from six sampling dates (S: spring tide, N: neap tide, numbers: sampling occasion). Error bars represent standard error.

Table 1. The results of one-way ANOVA and Tukey a posteriori tests on meiofauna and environmental data. NS = $p > 0.05$, * = $p < 0.05$, ** = $p < 0.01$, *** = $p < 0.001$ (S: spring tide, N: neap tide, numbers represent sampling occasions)

	ANOVA	Tukey
Meiofauna		
Nematoda	*	NS
Harpacticoida	**	N1>S4, S5, N6, S7
Chironomida	NS	
Plathyhelminthes	***	N3<S2, S4, S5, N6, S7; S2, S4<N6, S7
Environmental factors		
Salinity	**	S2<N3, N6
Grain size	***	S2>S4, N6, S7; N3>S4, N6
Chlorophyll *a*	**	N3<S5, S7; S5>S4, N6
Organic matter	NS	

There was no significant difference in organic content of the sediment among dates (Figure 1, Table 1), while chlorophyll *a* varied significantly (Figure 1, $p < 0.01$). Temperature in the sediment was similar on all sampling occasions varying from 29 to 33 °C.

Meiofauna taxa

Altogether, four taxa were recorded on all sampling occasions. Densities in surface sediments (0–5 cm) were low compared to other mangrove areas, ranging from 271 to 656 animals 10 cm^{-2}. Nematodes were most abundant in all samples, constituting between 58% and 87% of the total meiofauna. Harpacticoid copepods were usually the second most abundant group with 6–34% of total meiofauna. Chironomids (Diptera larvae) and plathyhelminthes were also found in all samples and composed between 3–9% and <1–12%, respectively, of the total fauna.

There was a significant temporal difference in abundance of nematodes, harpacticoids and plathyhelminthes (Figure 2, Table 1). These density variations did not show patterns reflecting the spring or neap tides. For the nematodes, there was an increase in overall density during the latter part of the study as was the case for the plathyhelminthes (Figure 2). The harpacticoids were, however, found in more than three times higher densities at the first sampling occasion compared to the other occasions. There was no significant difference in the density of chironomids among dates ($p > 0.05$).

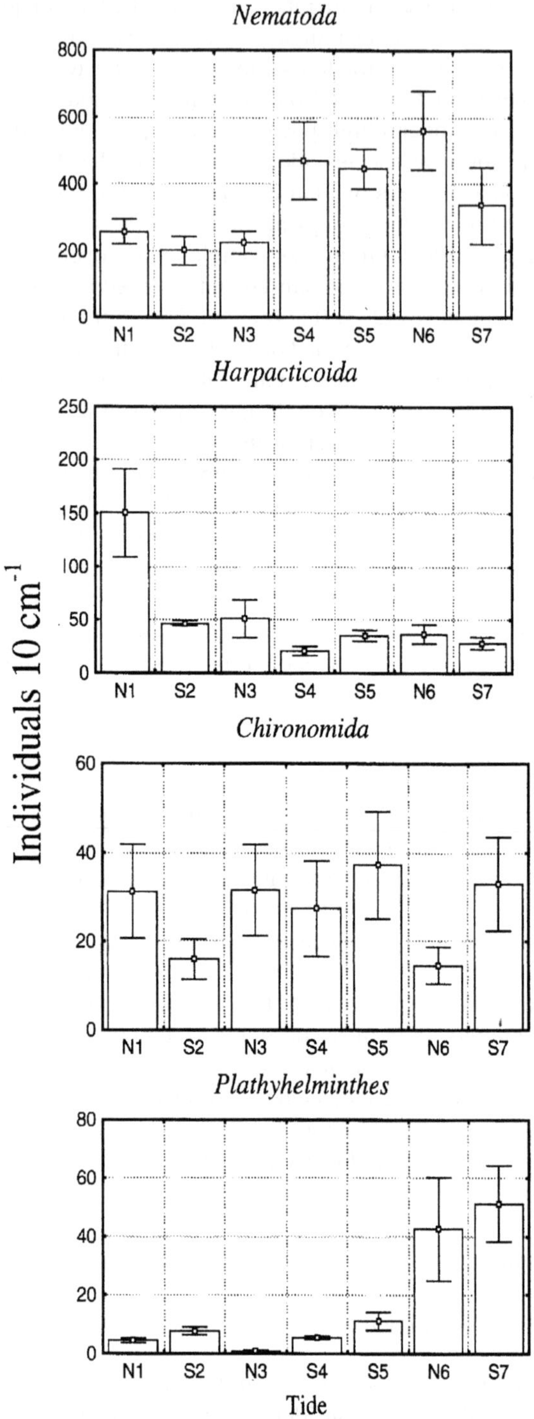

Figure 2. Average number ($n = 5$) per 10 cm^2 of individuals of the meiofauna taxa from all sampling dates (S: spring tide, N: neap tide, numbers: sampling occasion). Error bars represent standard error.

Table 2. Nematode species found in current study. Family, feeding group status (according to Wieser (1953), percentage frequency occurrence and percentage of total abundance is also presented (1a: selective deposit feeders, 1b: non-selective deposit feeders, 2a: epistrate feeders, 2b: predators/omnivores)

Family	Species	Feeding group	Frequency [%]	Abundance [%]
Oncholaimidae	*Viscosia* sp.	2b	15	1
Tripyloididae	*Bathylaimus* sp.	1b	3	<1
Trefusiidae	*Trefusia* sp.	1a	12	2
Chromadoridae	*Euchromadora* sp.	2a	6	<1
	Chromadora sp.	2a	18	3
	Chromadorina sp.	2a	56	9
	Chromadorita sp.	2a	15	1
	Ptycholaimellus sp.	2a	6	<1
Ethmolaimidae	*Gomphionema* sp.	2b	3	<1
Cyatholaimidae	*Marylynnia* sp.	2a	3	<1
Selachinematidae	*Halichoanolaimus* sp.	2b	3	<1
Desmodoridae	*Chromaspirina* sp.	2a	15	1
	Metachromadora sp.	2a	3	<1
	Spirinia sp.	2a	12	1
Microlaimidae	*Microlaimus* sp.	2a	100	50
Leptolaimidae	*Camacolaimus* sp.	2a	3	<1
	Leptolaimus sp.1	1a	24	2
	Leptolaimus sp.2	1a	3	1
Monhysteridae	*Monhystera* sp.	1b	9	1
Xyalidae	*Daptonema* sp.1	1b	56	10
	Daptonema sp.2	1b	6	<1
	Daptonema sp.3	1b	6	<1
	Paramonhystera sp.	1b	9	<1
	Promonhystera sp.	1b	9	<1
	Theristus sp.	1b	18	1
Linhomoeidae	*Linhomoeus* sp.	2a	12	1
	Metalinhomoeus sp.	1b	76	13

Several significant correlations were established between numbers of animals and environmental factors, when data from all sampling occasions were taken together. The correlations, however, were usually weak. Chironomida was the only taxon that showed significant correlation with salinity ($R = -0.41$, $p < 0.05$, $N = 30$). Significant negative correlations were detected between mean grain size and the number of nematodes ($R = -0.53$, $p < 0.01$, $N = 30$) and the number of plathyhelminthes ($R = -0.38$, $p < 0.05$, $N = 30$). A positive correlation was detected between grain size and harpacticoids ($R = 0.48$, $p < 0.01$, $N = 30$). Despite very small differences in the amount of organic material, there were significant correlations with the meiofauna. Nematodes were negatively correlated with organic material ($R = -0.55$, $p < 0.01$, $N = 30$), while there was a positive correlation for harpacticoids ($R = 0.38$, $p < 0.05$, $N = 30$).

Nematode assemblage

Altogether, 28 putative species of nematodes were recorded (Table 2). Of these, four species occurred in more than 50% of the samples (*Microlaimus* sp. (100%), *Metalinhomoeus* sp. (76%), *Daptonema* sp. (56%), *Chromadorina* sp. (56%)), while 12 were found only in one or two samples. The first three sampling dates *Microlaimus* sp. comprised on average 35% of the nematode population, but 60% for the last four sampling occasions. Multidimensional scaling ordination (MDS) of the nematode species abundance data indicated little effects of spring tide inundation

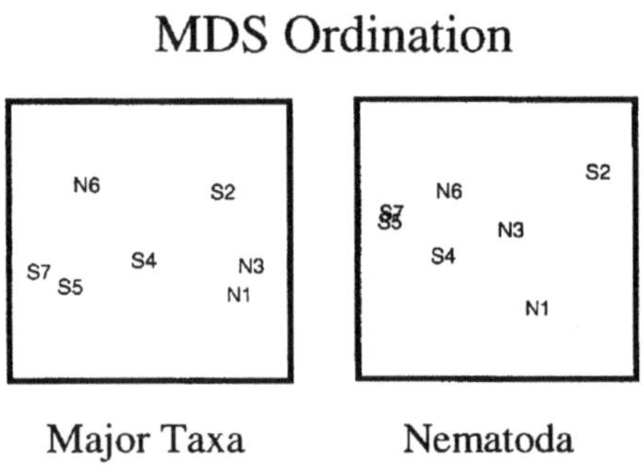

Figure 3. Non-metric MDS ordination of major taxa and nematode species abundance data based on average values (n = 5). (S: spring tide, N: neap tide, numbers: sampling occasion).

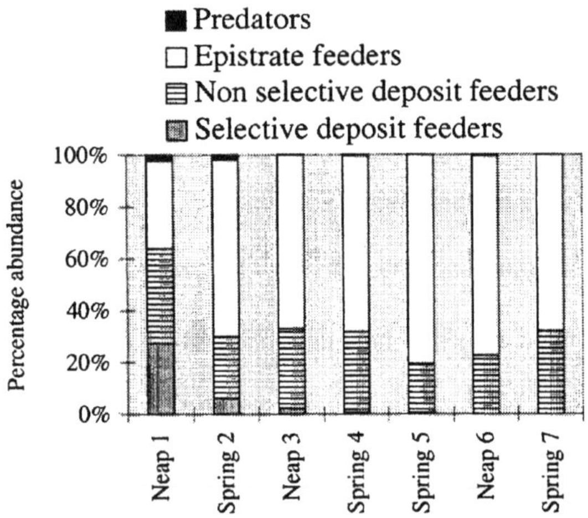

Figure 4. Percentage abundance of nematode trophic groups from each sampling occasion.

on the assemblage structure (Figure 3), but rather a successive change from wet to dry season. The number of species was far the highest on the first sampling date or 18 species (N1), but then gradually became lower i.e. 12 (S2), 12 (N3), 11 (S4), 6 (S5), 9 (N6) and 7 (S7). There was also a change in the trophic structure with time. Three species of selective deposit feeders were relatively abundant the first sampling dates but then this group totally disappeared by the end of the study (Figure 4).

Discussion

Despite the existence of an abundant and species rich meiofauna assemblage lower on the shore, approximately 200 m away from our station (Ólafsson, 1995), this field survey strongly suggests that there is not a significant immigration of meiofauna from the lower shore into the hypersaline area during spring tide flooding. There was neither an increase in numbers for any of the taxa just after flooding, nor did we find any substantial change in species composition associated with the tidal cycle. Like most mangrove forests, this one is in a sheltered area and presumably the tidal currents are not strong enough to carry significant amount of meiobenthic animals from the lower shore.

To our surprise, the porewater salinity remained very high after flooding. This was also evident in a preliminary study carried out during the last days of the heavy rains where porewater salinity was well above 100 ppt, while freshwater pools were still on the sediment surface. This suggests that hypersalinity is a permanent condition and, therefore, it was not strange to note that none of the species appeared to be dependent on the incoming seawater. While the porewater salinity was higher during the neap tide sampling, the difference was relatively small from when the area had just been flooded, and perhaps not of a physiological importance for the species adapted to already high salinities. It is clear that the fauna living in or drifting to this habitat must be able to cope with extremely high porewater salinities. Apart from chironomids, the meiofauna seems to be unaffected by salinity variations (89–160 ppt) with respect to major taxa composition and densities. Other environmental factors may, therefore, be more important in regulating the meiofauna numbers within this habitat. For example, nematodes tend to be more abundant in finer sediments (Coull, 1988) and we found a negative correlation between nematode numbers and grain size. Though there was a significant difference in the mean grain size, the difference was little on average ranging from ca. 430 to 610 μm. The coarser sediment from the first sampling dates might result from flushing out of finer sediment particles by the heavy rains. We do not believe that this is an artefact as several samples were taken randomly within the same relatively small area at each sampling date.

While we could not link the tidal cycle to changes in the structure of the meiofauna assemblage, there was some indication that wet and dry seasons affected the infauna at this site. At the first sampling date, the rainy season had just ended with the last heavy rains only 1 day prior to sampling. It is obvious that the assemblage structure of the fauna was different both in terms of abundance, diversity and species composition at the first sampling dates. Input of dissolved and particulate organic matter through runoff as a result of heavy rains might have favoured, for instance, selective deposit feeders as these may utilise dissolved organic matter. This group of nematodes was abundant during the first sampling date and was present only in low numbers on subsequent sampling dates with none recorded on the last two sampling dates. The dominant species, *Microlaimus* sp., increased in numbers with time and was approximately four times higher in numbers during the last four sampling dates compared to the first three ones. This species can be classified as an epistrate feeder, where it presumably scrapes off bacteria and algae from sediment grains. It is plausible that the rain had drastic effects on the food resources this species depends on. Unfortunately, we can only offer circumstantial evidence here, as our laboratory experiments, designed to test whether salinity significantly altered meiobenthic assemblage structure, showed unacceptably high mortality of nematodes in control and treatment mesocosms. It appears that there is no distinct nematode assemblage characteristic of mangrove areas (Ólafsson, 1995; Somerfield et al., 1998). The nematode genera found in this study are commonly found in marine sediments all over the globe. Alongi (1988) recorded low numbers of nematodes during hypersaline conditions at a high intertidal mangrove station in tropical north-eastern Australia. The dominant species at that station was *Terschellingia longicaudata*, not found in the current study. In similarly high saline areas of the Red Sea, Gerdes et al. (1985) found only one nematode species, *Oncholaimus oxyuris*, present in the sediment, a genus not found in the hypersaline area of Muwanda. They also noted harpacticoids, plathyhelminthes and insect larvae as dominant groups of the benthos, which agrees with our current observations.

Conclusions

From this study, we draw the conclusion that in this hypersaline mangrove habitat, the meiobenthos is composed of four taxa, including a number of species, which are resident and able to cope with very high salinities. Immigration to the area appears to be low and the spring tide inundations do not appear to alter the community structure in terms of abundance or species numbers.

Acknowledgements

We thank J. Francis, the director of the Institute of Marine Sciences on Zanzibar for making facilities available, and all the staff for helping in one way or another. Special appreciation goes to M. Mwadini who skilfully assisted in the field and the laboratory. Salim M. Mohammed commented on the manuscript. This study was supported by SAREC (Swedish Agency for Research Co-operation with Developing Countries) grant no: SWE-94-057 and the Institution of Tropical Ecology in Uppsala under the Sida programme of Minor field studies.

References

Alongi, D. M., 1988. Microbial-meiofaunal interrelationships in some tropical intertidal sediments. J. mar. Res. 40: 349–365.

Alongi, D. M., 1989. The role of soft-bottom benthic communities in tropical mangrove and coral reef ecosystems. Rev. Aquat. Sci. 1: 243–279.

Coull, B. C., 1988. Ecology of the marine meiofauna. In Higgins, R. P. & H. Thiel (eds), Introduction to the Study of Meiofauna. Smithsonian Inst. Press, Washington D. C.: 18–38.

Gerdes, G., J. Spira & C. Dimentman, 1985. The fauna of the Gavish Sabkha and the Solar Lake – a comparative study. In Friedman, G. M. & W. E. Krumbein (eds), Hypersaline Ecosystems: the Gavish Sabkha. Springer-Verlag, Berlin: 322–345.

Giere, O., A. Eleftheriou & D. J. Murison, 1988. Abiotic factors. In Higgins, R. P. & H. Thiel (eds), Introduction to the Study of Meiofauna. Smithsonian Inst. Press, Washington D. C.: 61–78.

Giere, O., 1993. Meiobenthology: the Microscopic Fauna in Aquatic Sediments. Springer Verlag, Berlin: 328 pp.

Heip, C., M. Vincx & C. Vranken, 1985. The ecology of marine nematodes. Oceanogr. mar. Biol. Ann. Rev. 23: 399–489.

Hicks, G. R. F. & B. C. Coull, 1983. The ecology of marine meiobenthic harpacticoid copepods. Oceanogr. mar. Biol. Ann. Rev. 21: 67–175.

Modig H. & E. Ólafsson, 1998. Responses of marine benthic invertebrates to hypoxic events in the Baltic Sea. J. exp. mar. Biol. Ecol. 229: 133–148.

Ngoile, M. A. K. & J. P. Shunula, 1992. Status and exploitation of the mangroves and associated fishery resources in Zanzibar. Hydrobiologia 247: 229–234.

Ólafsson, E., 1991. Intertidal meiofauna of four sandy beaches in Iceland. Ophelia 33: 55–65.

Ólafsson, E., 1995. Meiobenthos in mangrove areas in eastern Africa with emphasis on assemblage structure of free-living marine nematodes. Hydrobiologia 312: 47–57.

Palmer, M. A., 1988. Dispersal of marine meiofauna: a review and conceptual model explaining passive transport and active emergence with implications for recruitment. Mar. Ecol. Progr. Ser. 48: 81–91.

Parsons, T. R., Y. Mait & C. M. Lalli, 1984. A Manual of Chemical and Biological Methods for Seawater Analysis. Pergamon Press, Oxford: 173 pp.

Platt, H. M. & R. M. Warwick, 1983. Free-living marine nematodes. Part. 1. British enoplids. In Kermack, D. M. & R. S. K. Barnes (eds), Synopses of the British Fauna (New Series) 38. Cambridge University Press, Cambridge: 307 pp.

Savidge, W. B. & G. L. Taghon, 1988. Passive and active components of colonization following two types of disturbance on intertidal sandflat J. exp. mar. Biol. Ecol. 115: 137–155.

Sherman, K. M. & B. C. Coull, 1980. The response of meiofauna to sediment disturbance. J. exp. mar. Biol. Ecol. 46: 59–71.

Shunula, J. P. & A. Whittick, 1996. The Mangroves of Zanzibar. Institute of Marine Sciences. University of Dar-es-Salaam: 65 pp.

Somerfield, P. J., J. M. Gee & C. Aryuthaka, 1998. Meiofaunal communities in a Malaysian mangrove forest. J. mar. biol. Ass. U. K. 78: 717–732.

Vopel, K., J. Dehmlow & G. Arlt, 1996. Vertical distribution of *Cletocamptus confluens* (Copepoda, Harpacticoida) in relation to oxygen and sulphide microprofiles of a brackish water sulphuretum. Mar. Ecol. Prog. Ser. 141: 129–137.

Warwick, R. M., K. R. Clarke & J. M. Gee, 1990a. The effect of disturbance by soldier crabs *Mictyris platycheles* H Milne Edwards on meiobenthic community structure. J. exp. mar. Biol. Ecol. 135: 19–33.

Warwick, R. M., K. R. Clarke & Suharsono, 1990b. A statistical analysis of coral community responses to the 1982–83 El-NiÔo in the Thousand Islands Indonesia. Coral Reefs 8: 171–179.

Wieser, W., 1953. Die Beziehungen zwischen Mundhöhlengestalt, Ernährungsweise und Vorkommen bei freilebenden marinen Nematoden. Ark. Zool. 4: 439–484.

Wolanski, E., Y. Mazda & P. Ridd, 1992. Mangrove hydrodynamics. In Robertson, A. I. & D. M. Alongi (eds), Tropical Mangrove Ecosystems. American Geophysical Union, Washington D.C.: 43–62.

Hydrobiologia **426:** 65–74, 2000.
G. Liebezeit, S. Dittmann & I. Kröncke (eds), Life at Interfaces and Under Extreme Conditions.

Spatial and temporal distribution of *Myrophis punctatus* (Ophichthidae) and associated fish fauna in a northern Brazilian intertidal mangrove forest

Mário Barletta[1,*], Ulrich Saint-Paul, Audrey Barletta-Bergan, Werner Ekau & Dirk Schories
Center for Tropical Marine Ecology, Fahrenheitstr. 1, D-28359 Bremen, Germany
[1]*Present address: Centro de Estudos do MAR-CEM, Universidade Federal do Paraná – UFPR, Av. Beira mar s/n, 83255-000, Pontal do sul, Paraná, Brazil*
E-mail: mbarletta@cem.ufpr.br

Key words: Myrophis punctatus, Ophichthidae, ichthyotoxics, *Ichthyotere cunabi*, mangrove forest, macrotidal region, estuarine habitat, Caeté Estuary, East Amazon

Abstract

The mangrove forest along the northern Brazilian coast is not inundated during low tide. However, many fish species stay in the mangrove forest during this time. Tidal behaviour strategies are described for fish species that linger in the mangrove forest during low tide. The samples were taken at the end of the dry season (December 1996) and at the end of the rainy season (July 1997). Fish were captured using an ichthyotoxic plant extract (*Ichthyotere cunabi*). Spatial and temporal fish density and biomass were analyzed statistically. Thirty-six samples were taken with a total density of 2.8 ind m^{-2} and a total biomass of 17.4 g m^{-2} distributed among seven families and 14 species. *Myrophis punctatus* was the most important species in number (1.66 ind m^{-2}) and weight (12.68 g m^{-2}) of all catches. The total fish densities were not significantly different among areas and between months, although, total biomass differed significantly in time and space. The densities and biomass for the three most dominant species (*M. punctatus*, *Poecilia* spp. and *Gobionellus smaragdus*) differed significantly among species. Only the biomass of these species showed significant monthly differences. The only significant main effect on variance in the densities and biomass of *M. punctatus* were encountered between months. In addition, the factor area was significantly different for the variable number of species.

Introduction

Mangrove forests are very unstable habitats in which temperature, salinity and dissolved oxygen fluctuate greatly (Morton, 1989; Lowe-McConnell, 1991; Babler, 1997). In this habitat, there is also the risk of predation by both terrestrial and aquatic predators. Fishes that utilize the intertidal zone overcome this problem in two ways. The first is to avoid the zone at low tide and make use of its resources only when it is submerged. The second strategy is to remain in the intertidal zone at low tide. This strategy requires the presence of some form of shelter to alleviate the dangers of exposure (Gibson, 1999). Many intertidal fish, that are able to withstand these environmental changes, show morphological, physiological and behavioural adaptations which enable them to survive and reproduce in a habitat subject to regular change (Lewis, 1970; Bridges, 1993; Gibson, 1996).

Along the Northern Brazilian coast, the inhabitants of mangrove forests are exposed for a longer period during low tide due to the influence of macrotides (4–5 m). During this time, the fishermen ('ribeirinhos') of this region capture *Myrophis punctatus*, locally called 'Cutuca', in small tidal drainage creeks in the mangrove forest. Fish stay in burrows and are caught by using an ichthyotoxic plant known as 'Cunabí' (*Ichthyotere cunabi* – Compositae).

Studies utilizing rotenone in mangrove creeks during high tide reported on fish species composition, abundance, biomass and food habits in South Florida

* Author for correspondence

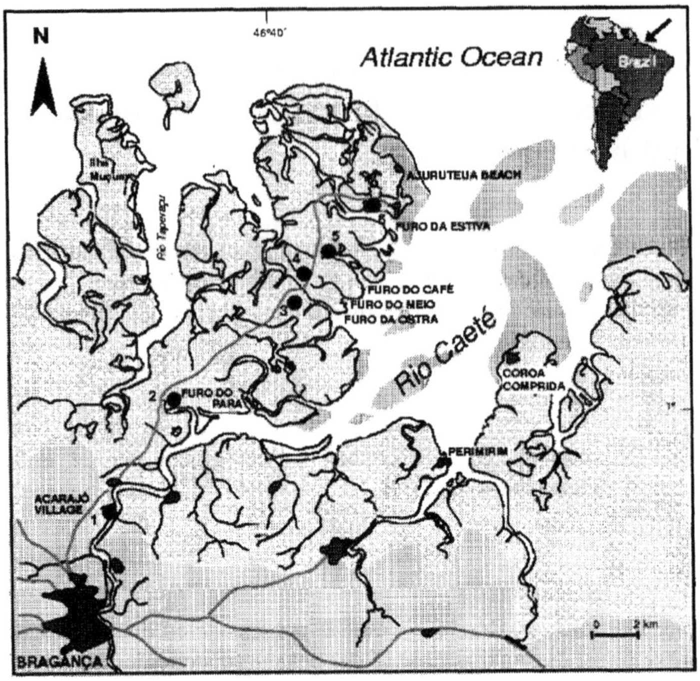

Figure 1. Study areas in Caeté Estuary: 1: Acarajó Village; 2: Furo do Pará; 3: Furo da Ostra; 4: Furo do Meio; 5: Furo do Café; 6: Furo da Estiva.

(Odum & Heald, 1972; Thayer et al., 1987), species diversity and biomass in Solomon Islands and in the Embley Estuary (Australia) (Blaber et al., 1989; Blaber & Milton, 1990). A large number of studies on fish behaviour and biology were conducted in intertidal rocky shores (Philips, 1977; Yoshiyama, 1981; Marliave, 1986), salt marches (Kneib, 1984, Cattrijsse et al., 1994), sand beach surf-zones (Harvey, 1998) and shallow sandy ecosystems (Gibson et al., 1993; Burrows et al., 1994; Gibson et al., 1998). However, little is known about which species stay in the mangrove forest during low tide, and the survival strategies of these assemblages is poorly understood.

This study investigates which species live in the intertidal mangrove forest habitat together with *M. punctatus* and describes the ecological strategies of the intertidal fish assemblage to minimize the predation and desiccation risks in the mangrove forest during low tide.

Furthermore, this study evaluates the null hypothesis that number of fish species, total density and total biomass (and components) would be equal in different mangrove areas along a salinity gradient in the Caeté estuary.

Materials and methods

Study area

The Bragança coastal plain is located in the North-East of Pará State (Figure 1). The climate is hot and humid (mean air temperature 25.7 °C), with hot dry periods from August to December and a rainy season from January to July. Annual precipitation exceeds 2,545 mm. The Caeté Bay, which connects Bragança city with the open sea, belongs to the northern part of the East Brazilian Basin. During the rainy season, freshwater run-off from the rivers Caeté and Urumajó increases and salinity drops to 6 PSU, while salinity is higher during the dry season (>35 PSU).

The coastal plain is a macrotidal (4–5 m) depositional system (Souza Filho & El-Robrini, 1997) and is characterized by sand-mud and mud sediments. In the north of the coastal plain, the mangrove forest is in contact with the littoral sand plain, and its south is in contact with the Barreiras Group of the coastal plateau (Souza Filho & El-Robrini, 1997). Between Bragança and Ajuruteua beach, the mangrove forest is drained by many tidal creeks and in this region, the mangrove forest is flooded twice daily by high tide.

Sampling methods

Samples were taken at the end of the dry season (December 1996) and at the end of the rainy season (July 1997), always during low tide, in six areas between Bragança and Ajuruteua beach (Figure 1). Water temperature (°C), dissolved oxygen (mg l^{-1}) (WTW OXI 325) and salinity (WTW LF 197) were recorded.

In each area, three small creeks in the mangrove forest were randomly sampled. Fish were captured by blocking the creeks with a hand net. About 20 m upstream, an ichthyotoxic (250 g per sample) was mixed with the water. Fish affected by the poison immediately jump out of the burrows and to the water surface. All fish were fixed with formalin (10%), weighed and measured (0.1 cm). Biomass [g m^{-2}] and density [ind m^{-2}] were calculated. The fish identification followed Böhlke (1978), Hoese (1978a, b), Figueiredo & Menezes (1978), Menezes & Figueiredo (1985), Burgess (1989), Cervigón (1991, 1993), Nelson (1994) and Eschmeyer (1998).

Statistical analysis of data

Analysis of variance was performed to determine whether significant differences in fish density, biomass and number of species occur in space and time. Two-way ANOVAs were used in order to test differences of the fish community parameters among areas (A1 – A6) and months (December 1996 and July 1997). Data were log $(x + 1)$-transformed to increase the normality of distribution. Cochranïs test was used to check the homogeneity of variances. When the variances were deemed heterogeneous at some probability level ($p < 0.05$), but not at some lower level ($0.01 < p < 0.05$), the analysis of variance was done using the lower probability level for significance tests (Underwood, 1981). A multiple comparison Tukey test ($p < 0.05$) was used whenever significant differences were detected (Day & Quinn, 1989).

Results

Area 1 (Acarajó) showed the lowest mean salinity and Area 2 (Furo do Pará) the strongest salinity variation (10–25 PSU) (Figure 2). In both areas, the salinity changed with season. The mean salinity reached the highest value (30 PSU) in areas located after Area 2 in the direction Ajuruteua beach.

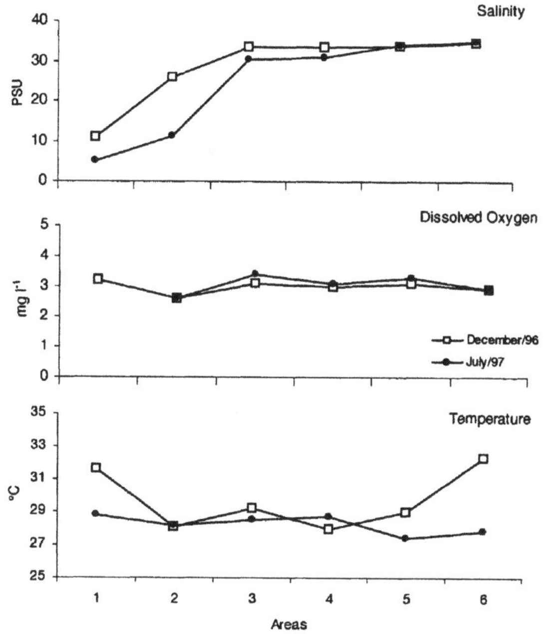

Figure 2. Mean values of water salinity [psu], dissolved oxygen [mg l^{-1}] and temperature [°C] at all sample sites during the study period.

The water temperature was highest in the Areas 1 (Acarajó) and 6 (Furo da Estiva) in December 1996 (Figure 2). In the intervening areas, the mean water temperature fluctuated between 27 and 29 °C.

The dissolved oxygen concentration was similar in all areas. However, Area 2 (Furo do Pará) showed a lower concentration (2.6 mg l^{-1}) than the other study areas (Figure 2).

Thirty-six fish samples were taken from the mangrove forest, between Bragança and Ajuruteua. A mean total density of 2.8 ind m^{-2} and a total biomass of 17.4 g m^{-2} was recorded, which was distributed among seven families and 14 species (Table 1). *M. punctatus* was the most important species, both in number (1.66 ind m^{-2}) and in weight (12.7 g m^{-2}) in all catches. Moreover, *M. punctatus* was the most frequent species (72%) in all samples (Table 1). Seven species occurred in less than 10% of all catches and are considered as rare. The family Gobiidae showed the highest species diversity, with *Gobionellus smaragdus* being the most important species in number (0.19 ind m^{-2}) and *G. oceanicus* in weight (0.96 g m^{-2}). The mean length of all species was less than 30 cm, varying among species (Table 1).

For all variables, with the exception of total fish biomass ($p = 0.024$), data were successfully trans-

Table 1. Density [ind m^{-2}] and biomass [m^{-2}] of fish species per area, and total length mean (x) and standard deviation (s.d.) per species from an intertidal mangrove forest in the Caeté Estuary

Taxonomic groups	Total length	Area 1		Area 2		Area 3		Area 4		Area 5		Area 6		Density	Biomass	Density	Biomass	Frequency
	$\bar{x}$ (s.d.)	[ind m^{-2}]	[g m^{-2}]	[ind m^{-2}]	[g m^{-2}]	[ind m^{-2}]	[g m^{-2}]	[ind m^{-2}]	[g m^{-2}]	[ind m^{-2}]	[g m^{-2}]	[ind m^{-2}]	[g m^{-2}]	$\bar{x}$ [ind m^{-2}]	$\bar{x}$ [g m^{-2}]	[%]	[%]	[%]
Ophichthidae																		
Myrophis punctatus	21.7 (5.9)	0.77	9.32	2.29	10.94	2.15	13.13	2.65	25.7	1.8	13.81	0.37	3.2	1.66	12.68	59.5	73	72
Poeciliidae																		
Poecilia spp.	3 (1.1)	0.12	0.21	1.06	0.88	0.17	0.09	0.16	0.13					0.25	0.22	9	1.3	25
Gambusia spp.	2.3 (0.8)			0.68	0.33	0.11	0.02	0.25	0.11					0.17	0.08	6.1	0.4	19
Gobiidae																		
Gobionellus smaragdus	3.9 (1.4)	0.11	0.34	0.06	0.08	0.23	0.45	0.17	0.17	0.06	0.07	0.55	0.39	0.19	0.25	7	1.4	22
Gobionellus oceanicus	8.3 (3.3)	0.06	1.11	0.59	3.48	0.11	1.19							0.13	0.96	4.5	5.5	17
Gobionellus stigmaticus	3.7 (0.7)	0.06	0.08	0.39	0.17							0.33	0.11	0.13	0.06	4.6	0.3	11
Gobiosoma hemigymnum	1.3 (0.5)					0.11	0.02	0.07	0.01			0.11	0.06	0.05	0.02	1.8	0.1	8
Coryphopterus spp.	2 (0)					0.1	0.1							0.02	0.02	0.6	0.1	6
Evortodus lyricus	5 (1)	0.07	0.14	0.07	0.14									0.02	0.05	0.8	0.3	6
Eleotridae																		
Guavina guavina	7.9 (3.8)	0.67	15.61	0.08	2.15									0.12	2.96	4.5	17.02	11
Eleotris pionis	5 (0)	0.11	0.34											0.02	0.06	0.7	0.33	3
Paralichthyidae																		
Citharichthys spp.	5 (0)	0.06	0.11											0.01	0.02	0.3	0.11	3
Loricariidae																		
Hypostomus spp.	6 (0)	0.06	0.1											0.01	0.02	0.3	0.10	3
Mugilidae																		
Mugil spp.	4.5 (0.5)	0.07	0.08											0.01	0.01	0.4	0.08	3
Total:															**2.8**	**17.4**		

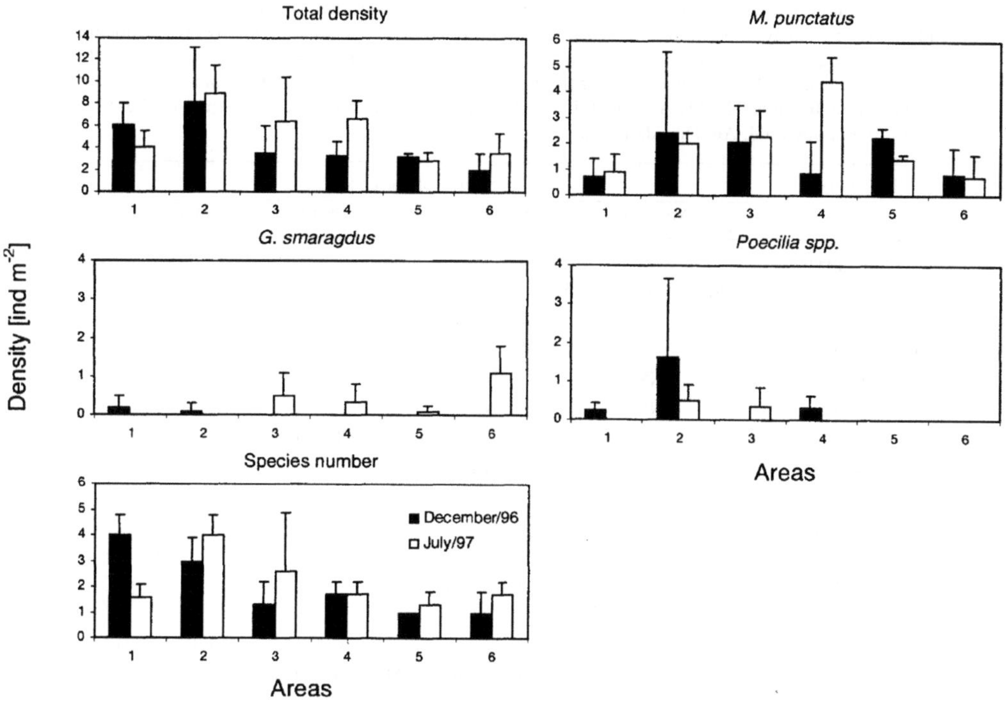

Figure 3. Mean densities (+ std.dev.) of the three numerically dominant species, total density, and number of species per sampled area.

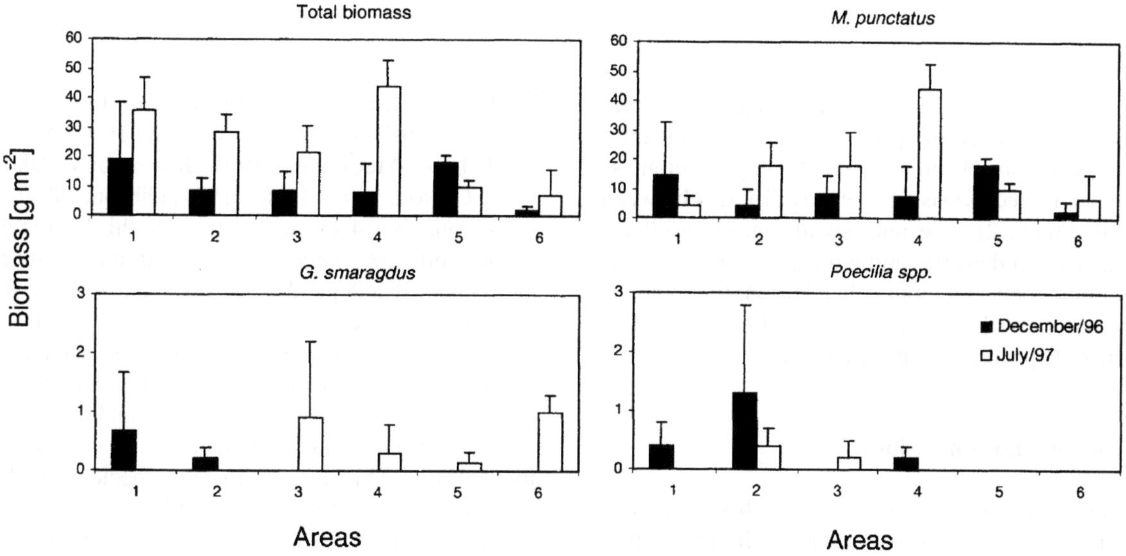

Figure 4. Mean biomasses (+ std. dev.) of the three dominant species, and total biomass.

formed to fit the requirements of the analysis of variance. In the case of the variable total fish biomass, ANOVA was performed using the lower probability level for the significance test ($p < 0.01$) (Underwood, 1981).

The total fish densities were not significantly different among areas and between months; however, for the total biomass, time and space were significantly different (Figures 3 and 4; Table 2). For the three most important species (*M. punctatus*, *Poecilia* spp. and *G. smaragdus*) densities and biomass

Table 2. Summary of the ANOVA results for number of species and total (and component) densities and biomasses. Analysis performed on log $(x+1)$-transformed data. Differences among areas, months and species were determined by Tukey *post hoc* comparisons. Where $*$, $p < 0.05$; $**$, $p < 0.01$; NS, not significant; Mp, *Myrophis punctatus*; Psp ,*Poecilia* spp.; Gs, *Gobionellus smaragdus;* Dec, December/96; Jul, July/97; A1, Acarajó Village; A2, Furo do Pará; A3, Furo da Ostra; A4, Furo do Meio; A5, Furo do Café; A6, Furo da Estiva

Parameters	Source of variance			
	Areas (1)	Months (2)	Species (3)	Interactions
No. of species	* A2 A1A4A3 A6A5	NS	-	NS
Density [ind m^{-2}]				
Total	NS	NS	-	NS
Main Spp.	NS	NS	** Mp > (Psp; Gs)	(1×3)*
M. punctatus	NS	NS	-	NS
Biomass [g m^{-2}]				
Total	** A1A2A4A5 A3 A6	** Jul > Dec	-	NS
Main Spp.	NS	* Jul > Dec	** Mp > (Psp; Gs)	(1×3)* (2×3) **
M. punctatus	NS	* Jul > Dec	- -	

differed significantly among species, and the interactions between areas and species were also significantly different (Figures 3 and 4; Table 2). Only the biomass of the main species was significantly different between months (Table 2). The only significant main effects on variance in densities and biomass of *M. punctatus* were found between months.In addition, the main effect area was significantly different for the variable number of species (Figure 3, Table 2).

Discussion and conclusions

The mangrove forest is associated with soft, muddy substrate and accretionary shorelines. Its main influence on fishes is the physical structure it provides (Day et al., 1989; Robertson & Blaber, 1992; Vose & Bell, 1994; Blaber, 1997). The pneumatophores, prop roots, trunks, fallen branches and leaves supply a complex habitat not only for fish, but also for crustaceans and mussels, both on trees and in the mud. A rich epiflora of macroalgae and diatoms is often found on the roots of mangrove trees and associated with the substrate (Day et al., 1989; Daniel & Robertson, 1990; Robertson & Blaber, 1992). In addition, abiotic factors also effect fish distribution (Gibson, 1969; Morton, 1989; Lowe-McConnell, 1991; Robertson & Blaber, 1992; Gibson, 1996; Babler, 1997; Gibson, 1999).

Among the 14 species captured in this habitat during low tide, *M. punctatus* was the dominant species and occurred independent of area and salinity fluctuations in the entire system, as predicted by the null hypothesis. However, the samples taken in the boundary area between the coastal plain and the sand plain (Area 6) showed lower biomass and density values, and *M. punctatus* no longer occurred outside the mangal zone. This suggests that this species is well adapted to survive in this very unstable habitat; however, its distribution is not restricted to the intertidal mangrove forest (Figueiredo & Menezes, 1978; Cervigón, 1991). The family Gobiidae showed the highest specific diversity, indicating that the species of this family are also well adapted to this habitat (Blaber et al., 1989; Blaber & Milton, 1990). The fact that gobiids are territorial defenders (Lowe-McConnell, 1991) may explain the low density and biomass. Furthermore, during the end of the dry season, individuals from the

Table 3. Number of species and total biomass of fish caught by ichthyotoxics in small intertidals creeks in South Florida (U.S.A.), in 10 estuaries in Solomon Islands, Embley Estuary (Northern Australia) and in Caeté Estuary (East Amazonia)

Place and estuary	Area sampled [m^2]	Number of Spp	Biomass [$g\ m^{-2}$]	References
A. South Florida - USA		64 [a]		Thayer et al., 1987
1. Whitewater Bay	-		10.8	
2. Coot Bay	-		5.3	
B. Solomon Islands				Blaber & Milton, 1990
1. Kolombangara estuaries				
Dulo	107	30	6.17	
Sambira	74	30	16.69	
Niupilesi	59	23	6.61	
Lady Lever	109	23	5.97	
Ringgi Cove One	189	22	2.83	
Ringgi Cove Two	62	30	43.79	
		Average:	13.68	
2. Rendova estuaries				
Tivu	18	16	3.56	
Poko	75	38	22.81	
Kenela	43	15	3.56	
		Average:	9.98	
3. Florida estuary				
Haïa	100	19	3,97	
C. North Australia				Blaber et al., 1989
1. Embley Estuary	-	66	8.2	
D. North Brazil				This study
1. Caeté Estuary (small intertidal creeks)	216	14	17.4	

[a] The species number in South Florida (U.S.A.) (Thayer et al., 1987) is the sum of all species captured in Whitewater Bay and Coot Bay.

gobiid *G. smaragdus* were captured only in areas 1 and 2, and in July 1997 they did not occur in these two areas anymore. This suggests that *G. smaragdus* undertakes longitudinal movements in the mangrove forest in accordance with salinity variations in the estuary. The very euryhaline freshwater fish *Poecilia* spp. was frequent but did not occur in Area 6 (Furo da Estiva), suggesting that Area 5 (Furo do Café) is the distribution limit of this species. Significant interactions between areas and the most important species (*M. punctatus, Poecilia* spp. and *G. smaragdus*) were detected. In other words, the distribution of these three species in the mangrove forest differed from each other, resulting in a rejection of the null hypothesis. In addition, the high species diversity in Area 1 and 2 could be explained by the occurrence of both fresh water and estuarine fish species in these areas.

Studies using rotenone and block net during high tide in intertidal mangrove creeks, in South Florida (U.S.A.) (Thayer et al., 1987), in 10 estuaries on the Solomon Islands (Blaber & Milton, 1990) and in the Embley Estuary (North Australia) (Blaber et al., 1989) showed very variable biomass values. The total biomass from this study is similar to those of the Whitewater Bay, Sambira and Poko Estuaries. However, this study showed a lower number of species (Table 3), perhaps because the samples were taken during low tide when the intertidal mangrove forest was not inundated. During flood tide, other species come to explore this environment, thereby increasing species diversity in the intertidal mangrove forest.

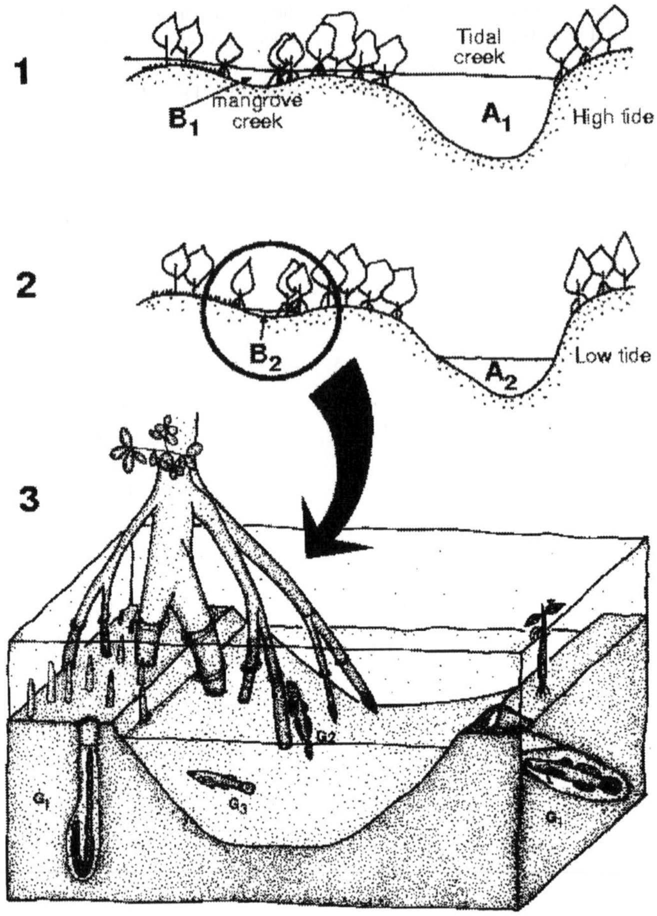

Figure 5. Idealized cross sections through a typical intertidal mangrove forest in the coastal plain in the Caeté Estuary, indicating the fish group distribution strategies during low tide. A1 and A2 represent the main tidal channel; B1 and B2 represent the very small creeks in the intertidal mangrove forest. Cross section 1: intertidal mangrove forest during high tide; Cross section 2: intertidal mangrove forest during low tide; Cross section 3: intertidal mangrove creek during low tide (B2 situation – vertical scale exaggerated). Cross section 3 represents the three different ecological strategies developed by fish to reduce interspecific competition in the intertidal mangrove forest during low tide. The first strategy (Group 1 – G1) includes fish species that stay in crab holes until the next flood tide. Group 2 (G2) represents the fish species which stay buried or attached to *Rhizophora mangle* roots. The third group (G3) represents the species which stay in the water stream (for more information see Table 4).

When we consider the species number of the present study and add those species found in a study of the intertidal creeks of the Furo do Meio and the Caeté Estuary, the number of species is higher than reported for the Solomon Islands and the Embley Estuary (North Australia).

At high tide, the conditions in the mangrove forest (intertidal zone) almost reflect those of the main tidal creeks. Factors such as pH, salinity and oxygen content tend to be uniform during flood tide. During this time, a large number of zoo- and ichthyoplankton, crustaceans and fish (juvenile and adult) inhabit the intertidal mangrove forest (Robertson & Blaber, 1992; Vose & Bell, 1994; Blaber, 1997). In the mangrove forest of tropical Australia, most fish using the forest during high tide move into shallow creeks or directly into the mainstream of the estuary at ebb tide (Robertson & Blaber, 1992).

During ebb tide in the mangrove forest of the Northern Brazilian coast (macrotidal region), the shoreline is exposed to air and this habitat changes suddenly from aquatic to virtually terrestrial. The fish

Table 4. Fish ecological strategy groups in the mangrove intertidal forest (macrotide region) during low tide (see Figure 5 for more information)

Ecological groups	Family	Species	Fish strategy in the mangrove forest during low tide
Group 1	Ophichthidae	*Myrophis punctatus*	- Stay in crab holes until the next flood tide.
	Eleotridae	*Guavina guavina*	
		Eleotris pionis	
Group 2	Gobiidae	*Gobionellus smaragdus*	- Stay buried or attached to *Rhizophora mangle* roots until the next flood tide.
		G. oceanicus	
		G. stigmaticus	
		Gobiosoma hemigymnum	
		Coryphopterus spp.	
		Evortodus lyricus	
Group 3	Poeciliidae	*Poecilia* spp.	- Stay in the water stream and use O_2 from the interface between air and water until the next flood tide.
		Gambusia spp.	

species here, during low tide, develop three different tidal behaviour strategies to alleviate the dangers of exposure, of which predation and desiccation are the most severe (Figure 5, Table 4). The first strategy (Group 1, Figure 5, G1) is to avoid tidal creeks during low tide. These fish wait for the next flood tide in crab holes situated closer to the creeks (*M. punctatus*, *Guavina guavina* and *Eleotris pionis*). The second group includes species of the family Gobiidae, which use their pelvic fins to attach to *Rhizophora mangle* roots or on/in the sediment (Figure 5, G2). Species of this family are also able to breath air by having a highly vascularized bucco-pharyngeal chamber; they are able to move into an adjacent pool during low tide (Wootton, 1998). The last behaviour strategy (Group 3, Figure 5, G3) is to stay in tidal creeks during low tide (*Poecilia* spp. and *Gambusia* sp.). These fish species are adapted to utilize the well-oxygenated water surface layer (Lewis, 1970).

All these ecological strategies reflect the interactions among individuals of the same or different species (competition and predation) during ebb and low tide in a very limited and severe habitat. These strategies can explain the high concentration (density and biomass) of *M. punctatus* (Ophichtidae) and the high diversity of the Gobiidae, and may help to reduce interspecific competition in the intertidal mangrove forest during low tide. During the spawning season, the eggs, larvae and juvenile fish of permanent inhabitants probably use the main tidal channel and river during ebb tide for dispersion. We therefore suggest that, independent of the fish species, these ecological strategies occur in other intertidal mangrove forests where macrotides also exist.

Acknowledgements

This study resulted from the cooperation between the Center for Tropical Marine Ecology (ZMT), Bremen, Germany, and the Universidade Federal do Pará (UFPA) and the Museu Paraense Emílio Goeldi (MPEG), both Belém, Brazil, under the Governmental Agreement on Cooperation in the Field of Scientific Research and Technological Development between Germany and Brazil financed by the German Ministry for Education, Science, Research and Technology (BMBF) [Project number: 03F0154A, Mangrove Management and Dynamics–MADAM], and the CNPq–Conselho Nacional de Desenvolvimento Científico e Tecnológico, of the Ministery for Science and Technology of Brazil. The first author was supported by a CNPq grant (No. 290057/95-7). We thank Dr Claudio Richter and Thorsten Dittmar (ZMT) for helpful comments on the manuscript. The manuscript benefited from the comments of anonymous referees.

References

Blaber, S. J. M., 1997. Fish and Fisheries of Tropical Estuaries. Fish and Fisheries Series 22, Chapman & Hall, London; 367 pp.

Blaber, S. J. M. & D. A., Milton, 1990. Species composition, community structure and zoogeography of fishes of mangrove estuaries in Solomon Islands. Mar. Biol. 105: 259–268.

Blaber, S. J. M., D. T. Brewer & J. P. Salini, 1989. Species composition and biomass of fishes in different habitats of a tropical Northern Australia Estuary: their occurrence in the adjoining sea and estuarine dependence. Estuar. coast. shelf Sci. 29: 509–531.

Böhlke, J. E., 1978. Ophichitidae. In Fischer, W. (ed.), FAO Species Identification Sheets for Fishery Purposes. Western Central Atlantic (Fishing Area 31). Roma: FAO.

Bridges, C. R., 1993. Ecophysiology of intertidal fish. In Rankin, J. C. & F. B. Jensen (eds), Fish Ecophysiology. Chapman & Hall, London, Fish Fisheries Ser.9; 421 pp.

Burgess,W. E., 1989. An Atlas of Freshwater and Marine Catfishes. A Preliminary Survey of the Siluriformes. T.F.H. Publications, Inc. Neptune City, N.J.; 784 pp.

Burrows, M. T., R. N. Gibson, L. Robb & C. A. Comely, 1994. Temporal patterns of movement in juvenile flatfishes and their predators: underwater television observations. J. exp. mar. Biol. Ecol. 177: 251–268.

Cattrijsse, A., E. S. Makwaia, H. R. Dankwa, O. Hamerlynck & M. A. Hemminga, 1994. Nekton communities of an intertidal creek of a European estuarine brackish march. Mar. Ecol. Prog. Ser. 109: 195–208.

Cervigón, F., 1991. Los peces marinos de Venezuela. Volume I. Fundación científica los Roques. 2a. ed. Caracas; 425 pp.

Cervigón, F., 1993. Los peces marinos de Venezuela. Volume II. Fundación científica los Roques, Caracas; 497 pp.

Daniel, P. A. & A. L. Robertson, 1990. Epibenthos of mangrove waterways and open embayments: community structure and the relationship between exported mangrove detritus and epifaunal standing stocks. Estuar. coast. Shelf. Sci. 31: 599–619.

Day, R. W. & G. P. Quinn, 1989. Comparisons of treatments after analysis of variance in ecology. Ecol. Monogr. 59: 433–463.

Day, J. W. Jr., C. A. S. Hall, W. M. Kemp & A. Yañes-Arancibia, 1989. Estuarine Ecology. John Wiley & Sons, New York; 558 pp.

Eschmeyer, W. N., 1998. A Catalog of Fishes. Calif. Acad. Sci.; 2905 pp.

Figueiredo, J. L. & N. A. Menezes, 1978. Manual de Peixes Marinhos do Sudeste do Brasil. II. Teleostei (1). Museu de Zoologia, Univ. de São Paulo, São Paulo; 109 pp.

Gibson, R. N., 1969. The biology and behaviour of littoral fish. Oceanogr. Mar. Biol. ann. Rev. 7: 367–410.

Gibson, R. N., 1996. Intertidal teleost: life in a fluctuating environment. In Pitcher, T. J. (ed.), Behaviour of teleost fishes. 2nd edn. Chapman & Hall, London. Fish and Fisheries Ser. 7: 513–533.

Gibson, R. N., 1999. Movement and homing in intertidal fishes. In Horn, M. H., K. L. M. Martin & M. A. Chotkowski (eds), Intertidal Fishes. Life in Two Worlds. Academic Press, London: 97–125.

Gibson, R. N., A. D. Ansell & L. Rob, 1993. Seasonal and annual variations in abundance and species composition of fish and macrocrustacean communities on a Scottish sandy beach. Mar. Ecol. Prog. Ser. 98: 89–105.

Gibson, R. N., L. Pihl, M. T. Burrows, J. Modin, H. Wennhage & L. A. Nickel. 1998. Diel movements of juvenile plaice *Pleuronectes platessa* in relation to predators, competitors, food availability and abiotic factors on a microtidal nursery ground. Mar. Ecol. Prog. Ser. 165: 145–159.

Harvey, C. J., 1998. Use of sandy beach habitat by *Fundulus majalis*, a surf-zone fish. Mar. Ecol. Prog. Ser. 164: 307–310.

Hoese, D. F., 1978a. Eleotridae. In Fischer, W. (ed.), FAO Species Identification Sheets for Fishery Purposes. Western Central Atlantic (Fishing Area 31). Rome: FAO.

Hoese, D. F., 1978b. Gobiidae. In Fischer, W. (ed.), FAO Species Identification Sheets for Fishery Purposes. Western Central Atlantic (Fishing Area 31). Rome: FAO.

Kneib, R. T., 1984. Patterns in the utilization of the intertidal salt marsh by larvae and juveniles of *Fundulus heteroclitus* (Linnaeus) and *Fundulus luciae* (Baird). J. exp. mar. Biol. Ecol. 83: 41–51.

Lewis, W. M., 1970. Morphological adaptation of ciprinodontids for Inhabiting oxygen deficient waters. Copeia 2: 319–326.

Lowe-McConnell, R. H., 1991. Ecological Studies in Tropical Fish Communities. Cambridge University Press, Cambridge; 382 pp.

Marliave, J. B., 1986. Lack of planktonic dispersal of rocky intertidal fish larvae. Trans. am. Fish. Soc. 115: 149–154.

Menezes, N. A. & J. L. Figueiredo, 1985. Manual de Peixes Marinhos do Sudeste do Brasil V. Teleostei (4). Museu de Zoologia, Univ. São Paulo, São Paulo, 105; pp.

Morton, R. M., 1989. Hydrology and fish fauna of Canal developments in an intensively modified Australian estuary. Estuar. coast. shelf. Sci. 28: 43–58.

Nelson, J., 1994. Fishes of the World. 3rd edn. Wiley & Sons, New York; 600 pp.

Odum, W. E. & E. J. Heald, 1972. Trophic analyses an estuarine mangrove community. Estuar. coast. shelf. Sci. 20: 637–738.

Phillips, R. R., 1977. Behavioural field study of the Hawaiian rockskipper *Istiblennius zebra* (Teleostei, Blenniidae) I. Ethogram. Z. Tierpsychol. 32: 1–22.

Robertson, A. I. & S. J. M. Blaber, 1992. Plankton, epibenthos and fish communities. In Robertson, A. I. & D. M. Alongi (eds), Tropical Mangrove Ecosystems. American Geophysical Union. Washington, DC: 173–224.

Souza filho, P. W. M. & M. El-Robrini,1997. Influência das variacões do nível do mar na sedimentação da planície costeira Bragantina Durante o Holoceno - Nordeste do Pará, Brasil. In Costa, M. L. & R. S. Angélica (eds), Contribuições à Geologia da Amazônia. Belém. FINEP/SBG. Núcleo Norte; 123 pp.

Thayer, G. W., D. R. Colby & W. F. Hetter, Jr., 1987. Utilization of the red mangrove prop root habitat by fishes in South Florida. Mar. Ecol. Prog. Ser. 35: 25–38.

Underwood, A. J., 1981. Techniques of analysis of variance in experimental marine biology and ecology. Oceanogr. Mar. ann. Rev. 19: 513–605.

Vose, F. E. & S. S. Bell, 1994. Resident fishes and macrobenthos in mangrove-rimmed habitats: evaluation of habitat restoration by hydrologic modification. Estuaries, 17: 585–596.

Wootton, R., 1998. Ecology of Teleost Fishes. Chapman & Hall, London. Fish and Fisheries Ser.1; 404 pp.

Yoshiama, R. M., 1981. Distribution and abundance patterns of rocky intertidal fishes in Central California. Env. Biol. Fishes. 6: 315–332.

Hydrobiologia **426**: 75–87, 2000.
G. Liebezeit, S. Dittmann & I. Kröncke (eds), Life at Interfaces and Under Extreme Conditions.

Variability of foraging in highshore habitats: dealing with unpredictability

M.G. Chapman
Centre for Research on Ecological Impacts of Coastal Cities, Marine Ecology Laboratories A11, University of Sydney, NSW 2006, Australia. E-mail: gee@bio.usyd.edu.au

Key words: behaviour, foraging, gastropod, highshore, intertidal, variability

Abstract

Intertidal areas are habitats at the border of two very different environments: the marine environment and the terrestrial environment. In contrast to many habitats at borders, intertidal areas are very variable in space and time. They have upshore and alongshore gradients of environmental conditions, which change through time in predictable (due to changing tides) and unpredictable (due to changing weather) ways. Because most animals and plants on rocky shores are marine in origin, extreme highshore levels are generally considered more harsh and unpredictable environments than are mid- or lowshore levels. In this study, the linear distances and directions dispersed by the littorinid *Littorina unifasciata* while foraging were compared across randomly-chosen mid- and highshore replicated sites to test the hypothesis that movement during foraging was more variable from place to place in complex midshore habitats. Experiments were repeated on different days to test the hypothesis that temporal variability in movement was greater at high- than at midshore levels because environmental conditions necessary for foraging were more variable high on the shore. Finally, the data were used to test models about differences in variability among individuals in the same patch of habitat according to their recent history of submersion/emersion. In contrast to expectations, dispersal was more variable in space and time within and among highshore sites. The implications of such variable behaviour are discussed with respect to the generalization of patterns of behaviour from sparse data and the levels of replication needed in the design of experiments to investigate behaviour of intertidal animals.

Introduction

Animals and plants in intertidal habitats provide good examples of living at borders. There is little in common between the physical features of marine and terrestrial environments. Organisms which are periodically subjected to one set of these conditions and then to the other have to be able to survive in diverse and variable environments. Some borders between adjacent environments may be relatively stable. Intertidal areas, in contrast, are very variable. They can span many degrees of latitude or longitude, encompassing different environmental conditions and containing different species (e.g., Poore, 1994). Second, environmental conditions across intertidal habitats, i.e., from the landward to the seaward edge, are also variable. Animals and plants lower on the shore are subjected to subtidal, marine conditions for longer periods and to aerial, terrestrial conditions for shorter periods than are organisms higher on the shore. Third, these two gradients of environmental conditions — alongshore and upshore — vary through time at short and long temporal scales. Some changes are fairly predictable, e.g. tidal movements. Others are unpredictable (e.g. calm seas, storms), but each can change rapidly the environmental conditions of any patch of intertidal habitat. These predictable and unpredictable environmental conditions have strong influences on the types of organisms that live in intertidal habitats. They are mostly diverse small plants and invertebrates, most of which come from marine ancestors.

Extreme highshore levels are generally considered environmentally more harsh and unpredictable than are levels lower on the shore. They may be continuously emersed for many days. When this co-incides with hot weather, there can be large, widescale mortality (Underwood, 1979; Yamaguchi, 1975). In addition, if calm seas co-incide with rain, gently-sloping intertidal habitats can become covered with warm freshwater for many days, which may cause large mortality of intertidal snails (pers. obs.). Therefore, intertidal animals and plants must be able to survive conditions that vary from marine to terrestrial to almost freshwater for varying amounts of time, or be able to compensate for large mortality. High on the shore, it is not only unpredictable when seawater will arrive, but also where it will arrive. At extreme levels, the substratum is often only washed or splashed by waves when seas are particularly high (due to combinations of tides, waves, wind, etc.). The directions from which the water arrives and where it washes under different conditions causes very variable patterns in the duration and amount of wetting from place to place and from time to time (Chapman & Underwood, 1996). Nevertheless, many organisms living at these levels are dependent on being wet by saltwater for basic activities as feeding or reproducing, although some highshore animals feed during rain (Chapman, 1994a; Ohgaki, 1988). Highshore organisms are typically opportunists (Zann, 1973), able to rapidly exploit favourable conditions to feed (Chapman, 1994a,b) or release eggs (Creese, 1980).

Foraging behaviour of mobile intertidal animals strongly influences local patterns of distribution and abundance (Chapman & Underwood, 1994), small-scale patterns of dispersion (Underwood & Chapman, 1996) and responses to micro-habitats (Chapman, 1994a, b). Where they feed depends on where food is available (which varies spatially and temporally (MacLulich, 1987; Underwood, 1984)) and where they can move over the substratum (e.g., where it is wet). This influences where they are found after foraging. Although animals from lowshore levels may move large distances during feeding (Chelazzi et al., 1983; Levings & Garrity, 1983), most highshore animals are limited in how far they can move because of the short time available to feed. For those that return to specific micro-habitats after foraging (Chapman, 1994a), how far and where they can go may also be limited by the distribution of these micro-habitats (Chapman & Underwood, 1994). One might therefore expect different patterns of dispersion and of foraging behaviour among populations of the same species when these populations live at different heights on the shore.

In New South Wales, Australia, the small littorinid, *Littorina unifasciata* Gray is very widespread and abundant from mid- to very highshore levels on intertidal rocky shores (Chapman, 1994b; Underwood, 1981). They are generally aggregated in large numbers and inactive during low tide and only forage when the substratum is wet (Chapman, 1994b, 1995; Chapman & Underwood, 1996). These aggregations are dynamic. Individuals do not home to specific sites or aggregations, but disperse gradually through local patches of habitat during subsequent periods of foraging. In general, highshore habitats on these shores are topographically less variable (Underwood & Chapman, 1989) and have smaller biotic diversity (Underwood, 1981) than midshore habitats. Each of these factors can influence foraging. Populations of *L. unifasciata* living in patches of highshore habitats must forage in habitats that are less spatially variable than do populations of snails in midshore habitats. One can therefore predict that patterns of foraging from place to place will be more variable lower on the shore than higher on the shore.

In addition, however, highshore populations are subjected to more variable environmental conditions than are midshore populations because wetting by waves and spray is unpredictable in time and space. Variable wetting may occur at the scale of centimetres to metres, providing conditions for foraging for some but not other individuals in a patch of habitat, in addition to scales of tens to hundreds of metres, only providing conditions for foraging for some patches of habitat. One can therefore predict greater temporal variability in foraging among highshore than among midshore habitats. Also, past history of submersion affects activity of *L. unifasciata* (Chapman, 1998), suggesting that there should also be greater variability among individuals at highshore levels than among those at midshore levels.

In this study, spatial and temporal variability in foraging by *L. unifasciata* was examined by testing the following specific hypotheses:

(1) there will be greater spatial variability in patterns of foraging (indicated by the distances dispersed and the directionality of movement; Underwood, 1977) among replicate sites tens of metres apart on the same shore and among shores at midshore levels than are found at highshore levels;

(2) there will be less variability in patterns of foraging among individuals in a patch of habitat at midshore levels than is found among individuals living in highshore patches of habitat;

(3) there will be less variability in patterns of foraging from one time to another in midshore sites than in highshore sites.

Materials and methods

This study was done in the Cape Banks Scientific Marine Research Area (Cape Banks) in Botany Bay, New South Wales, Australia (latitude 34° S, longitude 151.25° E). These shores vary in shape, aspect and exposure to waves (Fairweather & Underwood, 1991), but, in general, are relatively gently-sloping. High- and midshore populations of *L. unifasciata* were defined as being within 2 m (horizontal distance) of the upper and lower limit of *L. unifasciata*, respectively (*L. unifasciata* are not found in lowshore intertidal levels).

Experiments were done in three sites, 10 m apart, at each level on each of two shores (Shores 1 and 2 here = Shores 3 and 4 in Chapman, 1994a). The experiments were repeated on five different days during July and August, 1998, although, due to large seas and loss of tags, data were not available for all midshore sites on Shore 2 on two days.

Patterns of foraging were measured as the linear distances and directions dispersed by individual snails over two consecutive tidal cycles, i.e. from one low tide (when the animals are inactive and easy to find and identify) to the low tide 24 h later (Underwood, 1977). Snails were individually marked in situ with enamel paint and all dislodged animals replaced. The data are therefore for individually marked, undisturbed animals (Chapman, 1986). It is not possible to follow paths moved by individual snails when foraging because they are small, difficult to identify and only move when awash (Chapman, 1995). Therefore, larger dispersed linear distances could be caused by longer paths (perhaps due to more time spent foraging) or to less sinuous paths. Whether each or both of these influence the linear distances dispersed is not an issue in this study. These linear distances do validly represent some differences in foraging from one place or time to another. Twenty different individuals were marked in each site on each occasion.

Topographic complexity of the substratum influences foraging (Underwood & Chapman, 1989) and varies with height on the shore. Topographic complexity of each site was measured as the circular variance from a series of triangular planes measured using a 10 × 10 grid of 100 points, with 5 cm between adjacent points (described in detail in Underwood & Chapman, 1989). This index has a value of 0 when the substratum is flat, i.e., there is no topographic complexity. Values of 0.2 indicate very complex substratum (Underwood & Chapman, 1989). Two random quadrats were measured in each site.

Results

Displaced distances

Distances dispersed were transformed to natural logarithms before analysis because these measures are generally logarithmically distributed (Underwood, 1977). Heterogeneity of variances was tested using Cochran's test prior to analysis of variance and appropriate means compared using Student–Newman–Keuls (SNK) tests (Underwood, 1997). Because data were missing from some sites for two days, two different analyses of variance were done to measure spatial and temporal differences in mean distances dispersed. First, data from Shore 1 were compared across all sites and all days (Height on the shore was a fixed factor; Days was a random factor; Sites was nested within heights; $n = 13$ (random data removed from all sites with larger n to balance the data)). Then, data from the two shores were compared across all sites for those days for which there was a complete set of data (Height was a fixed factor; Shore and Day were random factors; sites was nested within Height × Shore; $n = 13$).

As predicted by the hypotheses (large spatial variability at midshore levels and large temporal variability at highshore levels), there was a significant interaction between Days and Sites (Height) when displaced distances were compared across all sites and all days on Shore 1 (Table 1a). Although transforming the data could not stabilize the variances, the F-ratios were still interpreted because analysis of variance is robust to heterogeneous variances when there are many independent estimates of variance (Underwood, 1997). Comparison of the means using SNK tests showed that the differences found were not, however, as predicted (Figure 1). There were no significant differences in the mean distances displaced during foraging among midshore sites on any day. There were, however, significant differences among highshore sites on Days 3, 4 and 5 (as shown in Figure 1c–e).

When comparisons were made across sites on the two shores for three days (Days 1, 2 and 4), spatial variability in the mean distances dispersed varied among sites from day to day and temporal variability varied from site to site (Table 1b; Figure 2). Differences on Shore 1 were as described above. On Shore 2, significant differences among highshore sites were found on Days 1 and 4 (Figure 2a, d). There were no differences among highshore sites on Day 2, nor among midshore sites on any day. Note that for those days which were not included in the analysis, there ap-

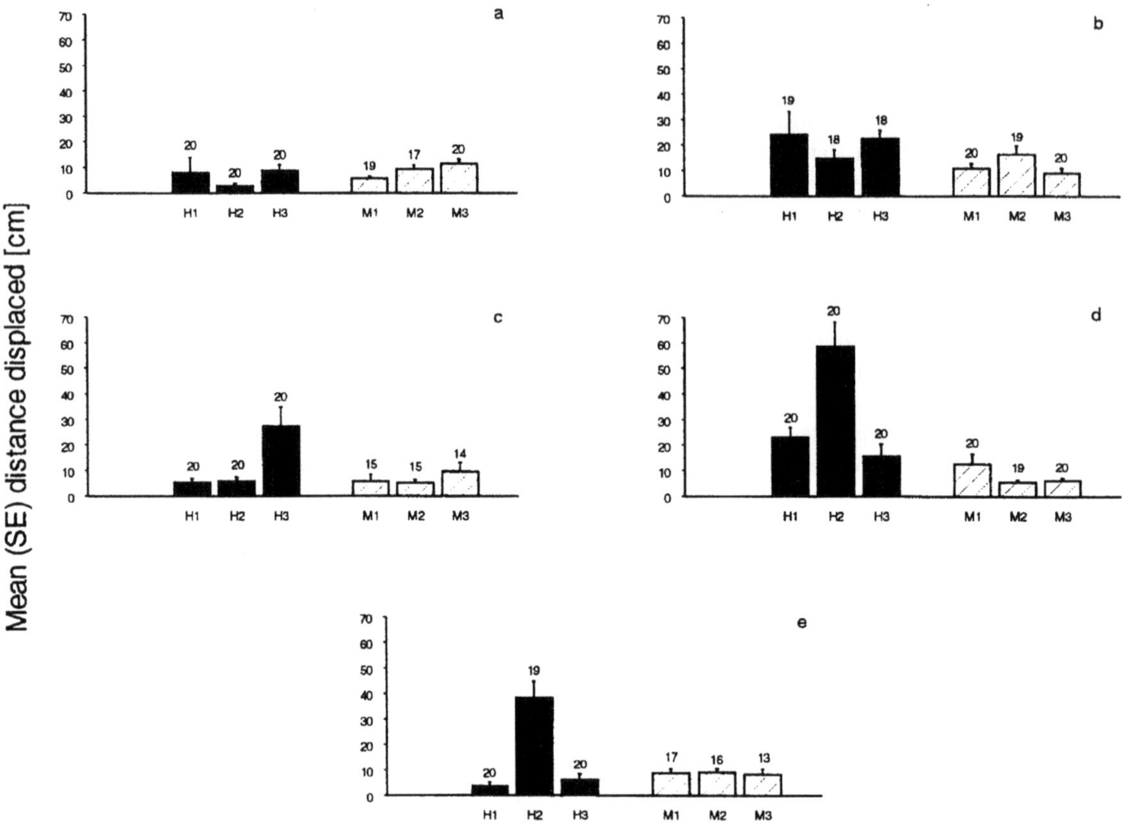

Figure 1. Mean (SE) distances displaced (cm) in three highshore (H1–H3) and three midshore (M1–M3) sites on Shore 1 on five different days (a–e); numbers above the histograms are the numbers of snails recovered in each site.

peared to be no differences among high - or midshore sites on Day 3 (Figure 2c), but highshore sites were very different from each other on Day 5 (Figure 2e).

Therefore, with respect to spatial variability in the linear distances dispersed, there was greater variation among highshore sites than among midshore sites. This was opposite to the pattern predicted by the first hypothesis. Differences in complexity of habitat at high- and midshore levels was as expected (significant differences among sites; $F = 3.27$, 8 and 12 df, $P < 0.05$).

Topography was more complex lower on the shore and, in general, more variable within and among midshore sites than within and among highshore sites (Figure 3). Data were not collected for abundances and diversity of other biota, but these midshore areas were typical of those described elsewhere (Underwood, 1981) with greater diversity than highshore areas. Therefore, movement in midshore sites was more similar among sites, despite greater variability in the complexity of these sites.

In previous studies, *L. unifasciata* dispersed smaller distances over complex topography (Underwood & Chapman, 1989). In this study, however, when the data were averaged over all days of the experiment in high and midshore sites separately, this pattern was not found (Figure 4). The mean displaced distances did not vary with topographic complexity low on the shore and were generally small. At highshore levels on each shore, the snails displaced greater distances in slightly more complex topography than in the more simple topography. Variability of topography from site to site was, however, very small at these higher levels.

Temporal variability in distances displaced varied according to height on the shore as predicted. Interpretation of variation over five days on Shore 1 (Table 1a) showed significant differences among days in mean distances displaced in Sites 1 and 2 high on the shore. In Site 3 and all midshore sites, there were no significant differences (SNK tests on transformed data; Figure 1). Patterns of temporal change in these highshore sites varied among sites, i.e., certain sites did not consistently show greater change between dif-

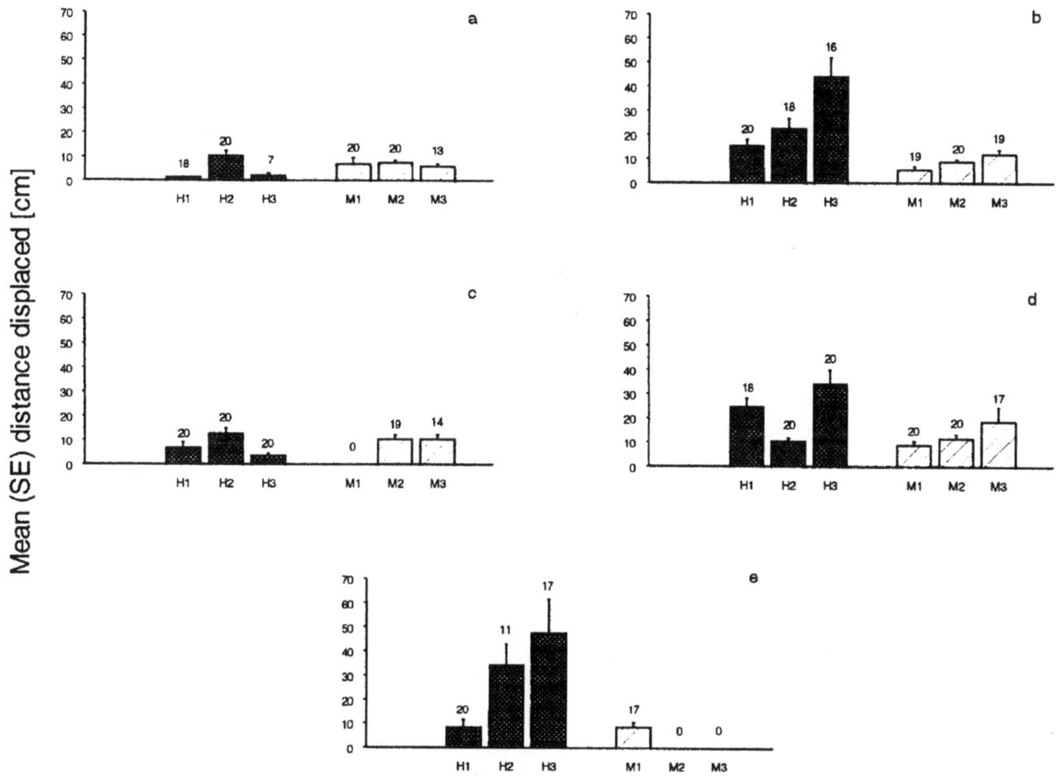

Figure 2. Mean (SE) distances displaced (cm) in three highshore (H1–H3) and three midshore (M1–M3) sites on Shore 2 on five different days (a–e); numbers above the histograms are the numbers of snails recovered in each site.

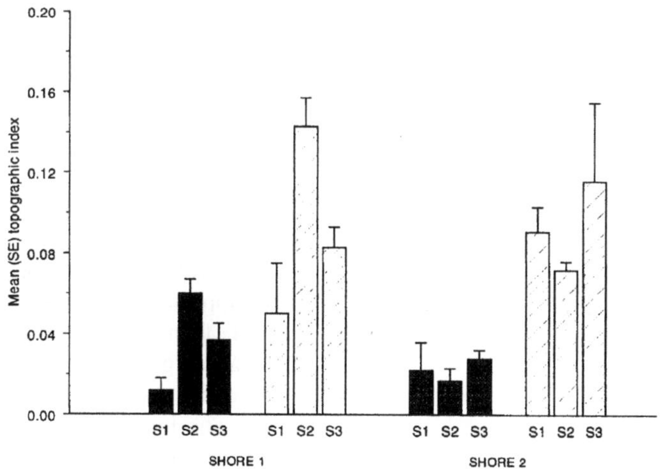

Figure 3. Mean (SE) topographic index in three highshore (H1–H3) and three midshore (M1–M3) sites on Shores 1 and 2; $n = 2$ quadrats per site (details in text).

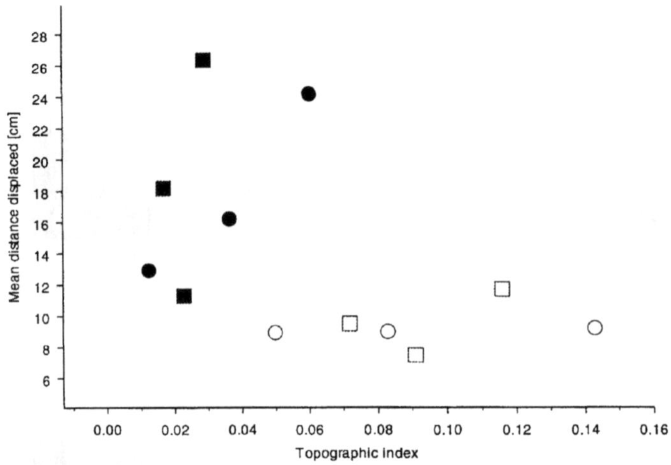

Figure 4. Relationship between topographic index and mean distance displaced (averaged over all days); ● highshore sites, Shore 1; ○ midshore sites, Shore 1; ■ highshore sites, Shore 2; □ midshore sites, Shore 2; each symbol represents a site.

Table 1. Analyses of mean distances displaced by *L. unifasciata* in each of three sites at high- and midshore levels: (a) on Shore 1 on five days; data transformed to $\log_e(X+1)$; Cochran's test $C = 0.088$, $P < 0.05$; $n = 13$; (b) on Shores 1 and 2 on three days; data transformed to $\log_e(X+1)$; Cochran's test $C = 0.085$, $P > 0.05$; $n = 13$; *F*-ratios only calculated for those terms in the analysis needed to test the hypotheses.

	df	MS	*F*	*P*
(a) Source of variation				
Days = D	4	10.25		
Height = H	1	1.30		
Site(Height) = S(H)	4	4.64		
D × H	4	11.29	2.54	>0.05
D × S(H)	16	4.45	5.77	<0.0001
Residual	360	0.77		
(b) Source of variation				
Shores = Sh	1	0.10		
Days = D	2	41.91		
Height = H	1	5.95		
Site(Sh × H) = S(Sh × H)	8	0.97		
Sh × D	2	0.77		
Sh × H	1	1.09		
D × H	2	30.55	14.47	>0.05
Sh × D × H	2	2.11	0.65	>0.05
D × S(Sh × H)	16	3.23	4.85	<0.0001
Residual	432	0.67		

Table 2. Analyses of mean variance in the distances displaced by *L. unifasciata* in each of three sites at high- and midshore levels: (a) on Shore 1 on five days; data transformed to $\log_e(X+1)$; Cochran's test $C = 0.214$, $P > 0.05$; $n = 3$ sites; (b) on Shore 2 on three days; data transformed to $\log_e(X+1)$; Cochran's test $C = 0.522$, $P > 0.05$; $n = 3$ sites.

	df	MS	*F*	*P*
(a) Source of variation				
Days = D	4	1.46	0.76	> 0.05
Height = H	1	15.56	13.26	< 0.05
D × H	4	1.17	0.61	> 0.05
Residual	20	1.93		
(b) Source of variation				
Days = D	2	8.94	3.42	> 0.05
Height = H	1	0.69	0.14	> 0.05
D × H	2	5.06	1.94	> 0.05
Residual	12	2.61		

ferent times. When comparisons were made across the two shores, similar patterns were found (Table 1b). There were significant differences in the mean displaced distances among days 1, 2 and 4 in two highshore sites (Sites 1 and 3), but not in Site 2, nor in any midshore sites (Figure 2).

Therefore, to some degree, the third hypothesis was supported. The distances dispersed were more variable from time to time in high- than in midshore sites, although the temporal patterns themselves varied among sites and shores and between different times. The temporal variables which influenced foraging were therefore patchy – they affected different sites on different days.

Variability among individual snails (Hypothesis 2) was measured as the variance among distances dispersed by the different snails in each site on each

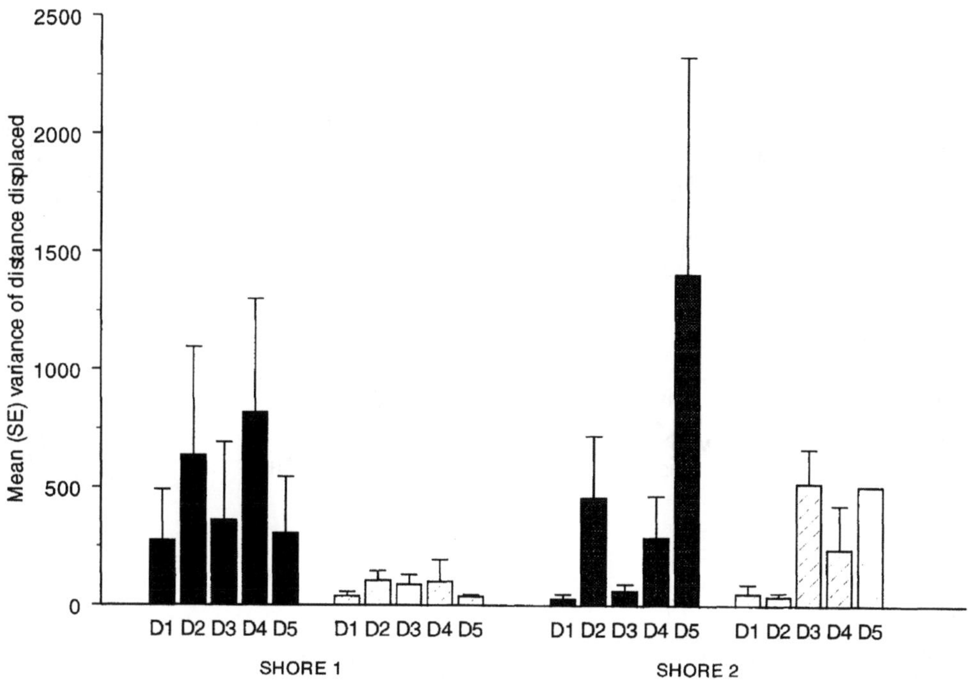

Figure 5. Mean (SE) variance of distances displaced measured among individuals in each site on each day; solid bars are highshore sites; hashed bars are midshore sites; D1 – D5 is five different days; $n = 3$ sites per day.

day (data transformed to natural logarithms before calculating variances). Each site therefore provided one measure per day. These were compared across each shore separately. On Shore 1, there was a significant effect of Height (Table 2a), but similar patterns of difference from day to day. Animals in highshore sites were more variable in the distances dispersed than were animals in midshore sites (Figure 5). They were also more variable from day to day (Figure 4), although not significantly so (Table 2a). On Shore 2, there were, however, no significant differences in individual variability between high and midshore sites (Table 2b), possibly because the two days which were most variable (Days 3 and 5) had incomplete data and could not be included in the analysis (Figure 5). Overall, however, a similar pattern was observed to that found on Shore 1; individual variability was greater at high- than at midshore levels and more variable from site to site (shown by the standard error bars in Figure 5).

Directions displaced and directionality of movement

The mean vectors displaced by snails in each site on Shore 1 on each day are shown in Figure 6. The length of each line is a measure of directionality; longer vectors indicate greater directionality (i.e., less variance among individual snails in the directions displaced) and asterisks indicate significant departure from uniform movement (Rayleigh's test; Mardia, 1972).

Patterns in the directions displaced on Shore 1 varied among sites and days (Figure 6). When movements were in random directions, the mean direction dispersed has no biological meaning. Because this was found for many sites on many days, differences in mean directions were not analysed (Underwood & Chapman, 1985). Over the five days, dispersal in highshore sites was more directional (significantly in Site 1 on 2 days, Site 2 on 2 days and Site 3 on 3 days; Rayleigh's test; Mardia, 1972) than was the case in midshore sites (significant in Site 1 on 1 day, Site 2 on 2 days and Site 3 on 1 day). Therefore, movement tended to be randomly orientated lower on the shore. When movement was directional, however, mean directions displaced varied from site to site on any day and from day to day in any site (Figure 6).

Similar patterns were found on Shore 2 (Figure 7). Movement was more directional higher on the shore (in Site 1 on 3 days, in Site 2 on 1 day and in Site 3 on 2 days) than was the case at midshore levels (only on one day in one site; Figure 7d). Therefore, as on Shore 1, the directions displaced and whether the snails tended to move randomly or in the same mean direction varied

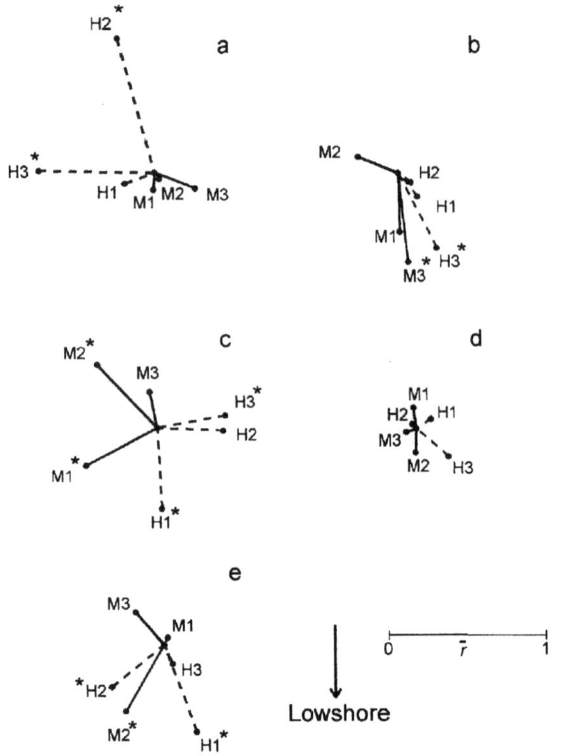

Figure 6. Mean directions displaced in three highshore (H1–H3) and three midshore (M1–M3) sites on Shore 1 on five different days (a–e); length of vectors scaled to *r* between 0 (a uniform distribution) and 1 (all directions identical); asterisks indicate significant directionality (Rayleigh's test); arrow indicates the direction downshore.

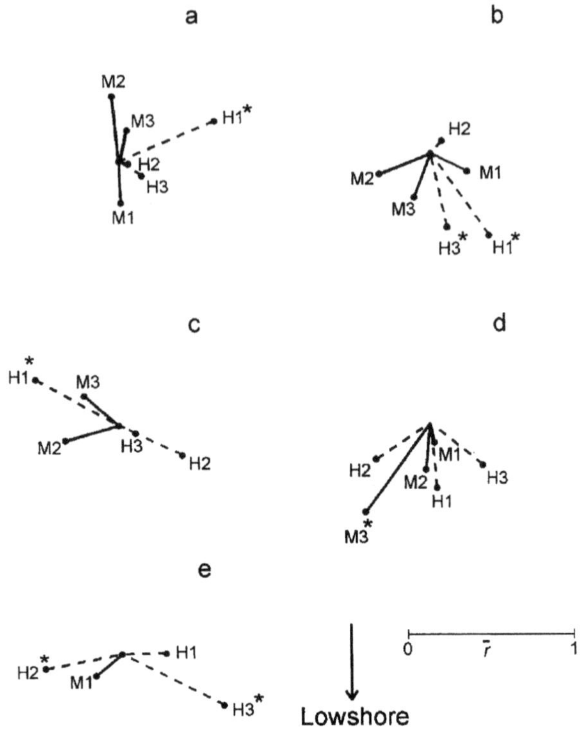

Figure 7. Mean directions displaced in three highshore (H1–H3) and three midshore (M1–M3) sites on Shore 2 on five different days (a–e); length of vectors scaled to *r* between 0 (a uniform distribution) and 1 (all directions identical); asterisks indicate significant directionality (Rayleigh's test); arrow indicates the direction downshore.

among the different sites on the same day and differently from time to time in the different sites. As was the case for the distances dispersed, there were fewer differences in directional movement between mid- and highshore levels on Shore 2 than was seen on Shore 1.

Previous work has shown correlation between distances moved and directionality of movement, with snails tending to move smaller distances and in random orientations in more complex areas (although this may be affected by experimental treatments (Chapman, 1986; Underwood & Chapman, 1989). To examine whether directional movement was correlated with the distances displaced in high- and midshore sites, the resultant direction ($\bar{R}$, which is a measure of concentration of directions) was tested against the mean distance displaced for each day in each site. Relationships varied between shores (Figure 8).

On Shore 1, mean directionality ($\bar{R}$) was not at all correlated with mean distances displaced in midshore sites ($r = -0.08$, $P > 0.50$; Figure 8a), whereas in highshore sites there was quite a strong, although non-significant negative relationship ($r = -0.46$, $P > 0.05$). Therefore, at highshore levels, snails tended to move further when moving in random directions and smaller distances when directional. On Shore 2, however, there was a positive relationship at each shore level (Figure 8b), although this was only significant in midshore sites ($r = 0.36$, $P > 0.05$; $r = 0.77$, $P < 0.01$, for highshore and midshore levels, respectively). Therefore on Shore 2, snails tended to move further when they were all tending to move in similar directions.

Discussion

Field tests of models about foraging in intertidal animals are often done by examining where and when animals move (see reviews by Chapman & Underwood, 1992a; Chelazzi et al., 1988; Hawkins & Hartnoll, 1983) because it is not usually possible to measure whether or how much they are feeding (but see Kitting, 1979; Petraitis & Sayigh, 1987). Records of the paths of foraging animals (e.g., Chelazzi et al.,

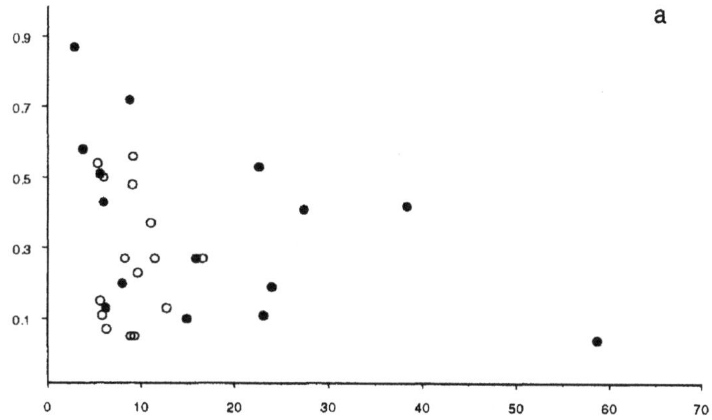

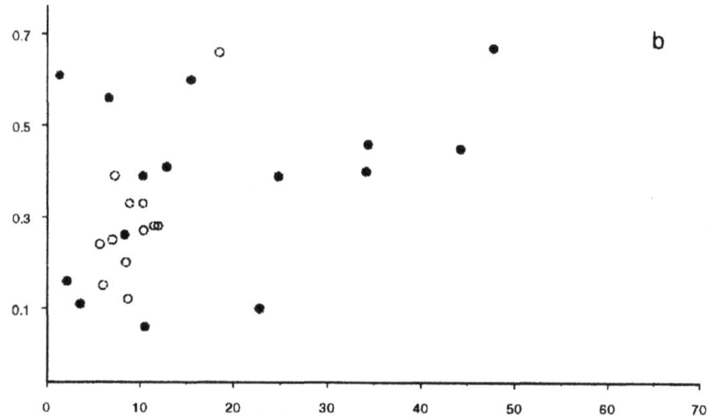

Figure 8. Relationships between directionality of movement ($\bar{R}$) and mean distance displaced (averaged over all days) on (a) Shore 1 and (b) Shore 2; ● highshore sites, ○ midshore site.

1994) can directly estimate how animals utilize habitat when active, but because of the equipment and amount of time needed to collect and analyse such data, these measurements can usually only be obtained from small numbers of individuals in one or a few places on few occasions (e.g., Chelazzi et al., 1983, 1994).

Collecting data on the positions of animals before and after movement and inferring patterns of movement from these positions (Underwood, 1977) will not directly measure certain aspects of foraging, e.g., the length or sinuosity of the path or speed of movement (Chelazzi et al., 1988; Erlandsson & Kostylev, 1995). Because these data are relatively easy to collect, however, enough data can be obtained to provide useful tests of hypotheses about spatial and temporal variability in dispersal during foraging. If spatial and/or temporal variability in the distance and/or directions dispersed when feeding is very large, many of the generalities of foraging behaviour that have been obtained from poorly replicated studies (e.g., Little et al., 1991) must be of concern. Data on subsequent positions of animals can also be used to test specific hypotheses about how aspects of habitat affect animals moving across habitat (e.g., effects of topographic complexity (Chapman & Underwood, 1994), height on the shore (Chapman, 1986), density of conspecifics (Underwood, 1988)) in properly controlled and replicated field experiments (Chapman & Underwood, 1992b; Underwood, 1988). This is not always possible using more labour-intensive or expensive techniques.

In this study, spatial and temporal variability in patterns of movement of *L. unifasciata* were measured by comparing the linear distances and directions dispersed by snails over two consecutive high tides among replicate sites at mid- and highshore levels on two shores. Because the snails do not home, these distances are an estimate of how far the snails move across patches of habitat when feeding (Underwood, 1977; Underwood & Chapman, 1989). The directions dispersed give measures of whether the snails in a patch of habitat spread out randomly, or tend to move in the same direction and of the mean direction dispersed when movement is directional. Comparisons among randomly chosen sites at each level on each shore measured variability among patches of habitat, only 10 m apart. Comparisons among five different randomly chosen days, using different samples of animals each day, allowed properly independent estimates of temporal variability in each site.

Previous work showed that *L. unifasciata* dispersed further across topographically simple substratum than across complex substratum at one level on the shore, leading to faster turnover of populations in patches of the former (Underwood & Chapman, 1989). In addition, orientation tended to be more directional in simple areas. Here, topographic complexity varied with height, with midshore areas being more topographically complex than highshore areas. More importantly, with respect to the hypothesis about spatial variability being examined here, there was more variability in complexity among mid- than among highshore sites. It was therefore predicted that, because dispersal is influenced by topographic complexity, there would be greater variability from site to site in the distances displaced and the directionality of movement at midshore than at highshore levels.

This was not, however, found. Snails dispersed very small distances in each midshore site on each day (as expected in topographically complex habitats), but there were no differences among midshore sites on any shore on any day. Movement was usually randomly orientated in all sites. Therefore, snails in midshore sites showed very similar patterns of movement from place to place and day to day.

In highshore sites, in contrast, there were very large differences in the mean distances dispersed and the directionality of movement from site to site, even though the complexity of substratum was relatively similar. Patterns of movement were not consistent from day to day in any site. In some sites on some days, snails dispersed significantly greater distances than in other sites, but, on other days, movement was similar among sites. Similarly, in some sites on some days, movement was directional, whereas at other times it was random. Therefore, despite smaller variation in complexity of habitat higher on the shore, patterns of movement varied significantly among sites in unpredictable and uncorrelated ways.

Temporal variability in opportunity to feed seemed more important than features of habitat in determining variation in patterns of movement. Midshore areas are regularly submersed; snails in these sites behaved similarly in all sites and on most days. High on the shore, wetting was unpredictable in time and extent and it was probably this that most influenced dispersal. Although no quantitative data were collected about duration of wetting during the ensuring two high tides, observations suggested that when seas were relatively calm (Days 1 and 3), sites were only wet in small patches compared to the amount of wetting that occurred when seas were rough (Days 2, 5 and 5). In general, highshore snails moved further on Days 2, 4 and 5 than on the other days. The large interactions between days and sites were probably due to the different sites getting wet by different amounts or for different lengths of time on different days.

Previous work has shown many aspects of behaviour of *L. unifasciata* to be variable from place to place and/or time to time, particularly at small spatial and temporal scales (e.g., rates of dispersal and directionality of movement: Underwood & Chapman, 1989; aggregation: Underwood & Chapman, 1992; Chapman, 1995; trail-following: Chapman, 1998). Some behaviour is influenced by the past history of submersion and emersion experienced by the snails (Chapman, 1998). In this study, midshore animals which had experienced similar and regular history of submersion and emersion dispersed in very similar ways in each patch of habitat, i.e., variability among individuals in the population was small. Highshore animals, which had more varied histories, showed greater variability in the distances displaced amongst individuals in the same patch of habitat, in addition to more variability among patches of habitat.

In contrast to what was found lower on the shore and in previous studies (Underwood & Chapman, 1989), when data were averaged over all days, the distances dispersed in highshore sites were positively correlated with topographic complexity, i.e., snails moved further in complex habitat. In fact, these highshore populations moved much further than did *L. unifasciata* moving across much more complex habitat lower on the shore. This contrasts with patterns of movement of the closely related snail, *Nodilittorina pyramidalis*, which also lives in these highshore habitats (Chapman & Underwood, 1994). Populations of a species living at different levels on the shore seem therefore to respond differently to similar features of habitat, either because of intrinsic differences in the behaviour of the snails or because the habitat varies in different ways at different levels, or at different times because of changing environmental conditions. For example, the snails may not be responding directly to observable features of habitat, i.e., the complexity that people can see and measure, but to variables that are themselves correlated with topographic complexity, e.g., the quality or variability in micro-algal food. If this relationship differs from one level on the shore to another, then snails living at different levels may show different patterns of movement in response to what are apparently similar features of habitat. Because rates of dispersal influence rates of population turnover and interchange (Underwood & Chapman, 1989), different responses by populations living at different shore-levels has large implications for models about population dynamics in this species.

Hypotheses from contrasting models can be tested only using properly designed and replicated field experiments (e.g., Underwood, 1986; Underwood & Petraitis, 1993) run for amounts of time appropriate for the hypotheses (Menge, 1997) and, wherever possible, independently repeated (Connell & Sousa, 1983; Underwood, 1997). To evaluate whether differences between rates of dispersal in response to topographic complexity at different levels on the shore is due to the characteristics of mid- or highshore animals, or differences in the habitats at the different levels, requires hypotheses about predicted patterns of movement when animals are transplanted from one habitat to another, or one shore level to another (e.g., Chapman, 1986; Chapman & Underwood, 1992b; McQuaid, 1981) or predicted patterns of movement when habitat is manipulated (e.g., Chapman & Underwood, 1994; Worthington & Fairweather, 1989). Similar hypotheses and experiments are needed to identify whether variability in behaviour among individuals in midshore and highshore populations is primarily governed by the history of the animals or features of habitat.

The design of well-replicated experiments, with all appropriate controls, investigating the movements of intertidal animals is complex (Chapman, 1986; Chapman & Underwood, 1992b; Petraitis, 1982; Underwood, 1988). Experiments are often not designed well, or replicated as often as many experiments that test hypotheses about such factors as mortality (Menge, 1978; Paine, 1974), competition (Underwood, 1986) or the effects of physical factors (Connell, 1972, 1985). Yet, behaviour of many mobile intertidal animals is very variable in space and time (e.g., Moran, 1985), is complex and influenced by many features of habitat (Chelazzi et al., 1984; Cook & Cook, 1981; Raffaelli & Hughes, 1978; Underwood, 1988) and can have widespread implications for the scale of measurement of patterns of dispersion, distribution and abundance (Underwood & Chapman, 1996; West, 1986). The data from this study show that even relatively simple measures of dispersal of this one species of intertidal snail were very variable from place to place and time to time, at small spatial and temporal scales. Variability was not predictable from features of habitat, despite earlier work suggesting that it should be. It varied according to height on the shore and, at highshore levels, varied among sites differently from one day to another. This suggests that if these extreme highshore levels are harsh environments for marine animals as they are usually perceived to be, this harshness is not determined by a simple set of physical variables that operate on a shore-wide scale. Animals living in these environments can respond rapidly to very variable conditions, remaining inactive for long periods, but then taking advantage of changing conditions to disperse and feed. How far and in which directions they move are governed by local conditions, primarily determined by where and for how long different patches of habitat remain wet. Animals living in these habitats must be extremely plastic in their behaviour to be best able to respond to such rapidly and unpredictably changing conditions.

Better documentation of spatial and temporal variability in behaviour of intertidal animals is essential if we are to understand how they respond to habitat. This is of crucial importance as restoration of habitat becomes an increasingly important option of management in the modern world. Finally, the generality of behavioural responses to habitat will never be understood unless experiments testing hypotheses about behaviour are replicated in time and space enough to incorporate these large levels of variability. Unreplicated studies, no matter how elegant, are still only applicable to those individuals used in that single experiment.

Acknowledgements

This research was supported by funds from the Australian Research Council and Centre for Research on

Ecological Impacts of Coastal Cities. J. Grayson and V. Mathews organized most of the field work; they were ably assisted by P. Barnes, M. Button, R. Forrest, S. Monteiro, A. Philips and M. Sage, all of whom also helped with entry and checking of data; J. Grayson and M. Sage assisted with analyses; J. Grayson and V. Mathews assisted with preparations of the figures. I thank A. J. Underwood, J. Grayson and two anonymous referees for helpful comments on an previous draft of this manuscript.

References

Chapman, M. G., 1986. Assessment of some controls in experimental transplants of intertidal gastropods. J. exp. mar. Biol. Ecol. 103: 181–201.

Chapman, M. G., 1992. Living at the fringe of intertidal society. Ph. D. thesis, University of Sydney, Sydney, Australia.

Chapman, M. G., 1994a. Small- and broad-scale patterns of distribution of the upper-shore littorinid *Nodilittorina pyramidalis* in New South Wales. Aust. J. Ecol. 19: 83–95.

Chapman, M. G., 1994b. Small-scale patterns of distribution and size-structure of the intertidal littorinid, *Littorina unifasciata* (Gastropoda: Littorinidae) in New South Wales. Aust. J. Mar. Freshwater Res. 45: 635–642.

Chapman, M. G., 1995. Aggregation of the littorinid snail *Littorina unifasciata* in New South Wales, Australia. Mar. Ecol. Progr. Ser. 126: 191–202.

Chapman, M. G., 1998. Variability in trail-following and aggregation in *Nodilittorina unifasciata* Gray. J. exp. mar. Biol. Ecol. 224: 48–71.

Chapman, M. G. & A. J. Underwood, 1992a. Foraging behaviour of marine benthic grazers. In John, D. M., S. J. Hawkins & J. H. Price (eds), Plant–Animal Interactions in the Marine Benthos. The Systematics Association Special Volume No. 46. Clarendon Press, Oxford: 289–317.

Chapman, M. G. & A. J. Underwood, 1992b. Experimental designs for analyses of movements by molluscs. In Grahame, J., P. J. Mill & D. G. Reid (eds), Proceedings of the Third International Symposium on Littorinid Biology. The Malacological Society of London, London: 169–180.

Chapman, M. G. & A. J. Underwood, 1994. Dispersal of the intertidal snail, *Nodilittorina pyramidalis*. in response to the topographic complexity of the substratum. J. exp. mar. Biol. Ecol. 179: 145–169.

Chapman, M. G. & A. J. Underwood, 1996. Influences of tidal conditions, temperature and desiccation on patterns of aggregation of the high-shore periwinkle, *Littorina unifasciata*, in New South Wales, Australia. J. exp. mar. Biol. Ecol. 196: 213–237.

Chelazzi, G., R. Innocenti & P. Della Santini, 1983. Zonal migration and trail-following of an intertidal gastropod analysed by LED tracking in the field. Mar. Behav. Physiol. 10: 121–136.

Chelazzi, G., J. L. Deneubourg & S. Focardi, 1984. Cooperative interactions and environmental control in the intertidal clustering of *Nerita textilis* (Gastropoda: Prosobranchia). Behaviour 90: 151–166.

Chelazzi, G., S. Focardi & J. L. Deneubourg, 1988, Analysis of movement patterns and orientation mechanisms in intertidal chitons and gastropods. In Chelazzi, G. & M. Vannini (eds), Behavioural Adaptation to Intertidal Life. NATO ASI Series A, Life Sciences, Volume 151. Plenum Press, New York and London: 173–184.

Chelazzi, G., G. Santini, D. Parpagnoli & P. Della Santini, 1994. Coupling motographic and sonographic recordings to assess foraging behaviour of *Patella vulgata*. J. moll. Stud. 60: 123–128.

Connell, J. H., 1972. Community interactions on marine rocky intertidal shores. Ann. Rev. Ecol. Syst. 3: 169–192.

Connell, J. H., 1985. The consequences of variation in initial settlement versus post-settlement mortality in rocky intertidal communities. J. exp. mar. Biol. Ecol. 93: 11–46.

Connell, J. H. & W. P. Sousa, 1983. On the evidence needed to judge ecological stability or persistence. Am. Nat. 121: 789–824.

Cook, S. B. & C. B. Cook, 1981. Activity patterns in *Siphonaria* populations: heading choice and effects of size and grazing interval. J. exp. mar. Biol. Ecol. 49: 69–80.

Creese, R. G., 1980. An analysis of distributions and abundance of populations of the high-shore limpet, *Notoacmea petterdi* (Tennison-Woods). Oecologia (Berl.) 335: 252–260.

Erlandsson, J. & V. Kostylev, 1995. Trail following, speed and fractal dimension of movement in a marine prosobranch, *Littorina littorea*, during a mating and a non-mating season. Mar. Biol. 122: 87–94.

Fairweather, P. G. & A. J. Underwood, 1991. Experimental removals of a rocky intertidal predator: variations within two habitats in the effects on prey. J. exp. mar. Biol. Ecol. 154: 29–75.

Hawkins, S. J. & R. G. Hartnoll, 1983. Grazing of intertidal algae by invertebrates. Ann. Rev. Oceanogr. mar. Biol. 21: 195–282.

Kitting, C. L., 1979. The use of feeding noises to determine the algal food being consumed by individual intertidal molluscs. Oecologia (Berl.) 40: 1–17.

Levings, S. C. & S. D. Garrity, 1983. Diel and tidal movement of two co-occurring neritid snails: differences in grazing patterns on a tropical rocky shore. J. exp. mar. Biol. Ecol. 67: 261–278.

Little, C., J. C. Partrindge & L. Teagle, 1991. Foraging activity of limpets in normal and abnormal tidal regimes. J. mar. biol. Ass. U. K. 71: 537–554.

Mardia, K. V., 1972. Statistics of Directional Data. Academic Press, London, 357 pp.

MacLulich, J. H., 1987. Variations in the density and variety of intertidal epilithic microflora. Mar. Ecol. Progr. Ser. 40: 285–293.

McQuaid, C. D., 1981. The establishment and vertical size gradients in populations of *Littorina africana knysnaensis* (Philippi) on an exposed rocky shore. J. exp. mar. Biol. Ecol. 54: 77–90.

Menge, B. A., 1978. Predation intensity in a rocky intertidal community: relation between predator foraging activity and environmental harshness. Oecologia (Berl.) 34: 1–16.

Menge, B. A., 1997. Detection of direct versus indirect effects: were experiments long enough? Am. Nat. 149: 801–823.

Moran, M.J., 1985. The timing and significance of sheltering and foraging behaviour of the predatory intertidal gastropod *Morula marginalba* Blainville (Muricidae). J. exp. mar. Biol. Ecol. 93: 103–114.

Ohgaki, S., 1988. Rain and the distribution of *Nodilittorina exigua* (Dunker) (Gastropoda: Littorinidae). J. exp. mar. Biol. Ecol. 122: 213–223.

Paine, R. T., 1974. Intertidal community structure: experimental studies on the relationship between a dominant competitor and its principal predator. Oecologia (Berl.) 15: 93–120.

Petraitis, P. S., 1982. Occurrence of random and directional movements in the periwinkle, *Littorina littorea* (L.). J. exp. mar. Biol. Ecol. 59: 207–217.

Petraitis, P. R. & L. Sayigh, 1987. In situ measurements of radular movements of three species of *Littorina* (Gastropoda: Littorinidae). Veliger 29: 384–387.

Poore, G. C. B., 1994, Marine biogeography of Australia. In Hammond, L. S. & R. N. Synnot (eds), Marine Biology. Longman Cheshire, Melbourne, Australia: 189–213.

Raffaelli, D. G. & R. N. Hughes, 1978. The effects of crevice size and availability on populations of *Littorina rudis* and *Littorina neritoides*. J. anim. Ecol. 47: 71–83.

Underwood, A. J., 1977. Movements of intertidal gastropods. J. exp. mar. Biol. Ecol. 26: 191–201.

Underwood, A. J., 1979. The ecology of intertidal gastropods. Adv. mar. Biol. 16: 111–210.

Underwood, A. J., 1981. Structure of a rocky intertidal community in New South Wales: patterns of vertical distribution and seasonal change. J. exp. mar. Biol. Ecol. 51: 57–85.

Underwood, A. J., 1984. Microalgal food and the growth of intertidal gastropods *Nerita atramentosa* Reeve and *Bembicium nanum* (Lamarck) at four heights on the shore. J. exp. mar. Biol. Ecol. 79: 277–91.

Underwood, A. J., 1986, The analysis of competition by field experiments. In Kikkawa, J. & D. J. Anderson (eds), Community Ecology. Pattern and Process. Blackwell Scientific Publications, Melbourne: 240–268.

Underwood, A. J., 1988. Design and analysis of field experiments on competitive interactions affecting behaviour of intertidal animals. In Chelazzi, G. & M. Vannini (eds), Behavioural Adaptation to Intertidal Life. NATO ASI Series A, Life Sciences, Volume 151. Plenum Press, New York: 333–357.

Underwood, A. J., 1997. Experiments in Ecology: Their Logical Design and Interpretation using Analysis of Variance. Cambridge University Press, Cambridge, U.K., 504 pp.

Underwood, A. J. & M. G. Chapman, 1985. Multifactorial analyses of directions of movement of animals. J. exp. mar. Biol. Ecol. 91: 17–43.

Underwood, A. J. & M. G. Chapman, 1989. Experimental analyses of the influences of topography of the substratum on movements and density of an intertidal snail, *Littorina unifasciata*. J. exp. mar. Biol. Ecol. 134: 175–196.

Underwood, A. J. & M. G. Chapman, 1992. Experiments on topographic influences on density and dispersion of *Littorina unifasciata* in New South Wales. In Grahame, J., P. J. Mill & D. G. Reid (eds), Proceedings of the Third International Symposium on Littorinid Biology. The Malacological Society of London, London: 181–195.

Underwood, A. J. & M. G. Chapman, 1996. Scales of spatial patterns of distribution of intertidal invertebrates. Oecologia (Berl.) 107: 212–224.

Underwood, A. J. & P. S. Petraitis, 1993, Structure of intertidal assemblages in different localities: how can local processes be compared? In Ricklefs, R. E. & D. Schluter (eds), Species Diversity in Ecological Communities. Historical and Geographical Perspectives. University of Chicago Press, Chicago: 39–51.

West, L., 1986. Interindividual variation in prey selection by the snail *Nucella* (= *Thais*) *emarginata*. Ecology 67: 798–809.

Yamaguchi, M., 1975. Sea level fluctuations and mass mortalities of reef animals in Guma, Mariana Islands. Micronesia 11: 227–243.

Zann, L. P., 1973. Relationships between intertidal zonation and circatidal rhythmicity in littoral gastropods. Mar. Biol. 18: 243–250.

Hydrobiologia **426:** 89–96, 2000.
G. Liebezeit, S. Dittmann & I. Kröncke (eds), Life at Interfaces and Under Extreme Conditions.

Faecal pellet production by Arctic under-ice amphipods – transfer of organic matter through the ice/water interface

Iris Werner
Institute for Polar Ecology, Wischhofstr. 1–3, D-24148 Kiel, Germany
E-mail: iwerner@ipoe.uni-kiel.de

Key words: faecal pellets, Amphipoda, sea ice, feeding, carbon flux, Greenland Sea

Abstract

The underside of Arctic sea ice is inhabited by several autochthonous amphipod species (*Apherusa glacialis, Onisimus* spp., *Gammarus wilkitzkii*). The amphipods graze on ice-bound organic matter, such as ice algae, detritus and ice fauna, and release faecal pellets into the underlying water column, thus forming a direct link between the sea ice and the pelagic ecosystems. Experiments on faecal pellet production rates showed species-specific differences, which were related to size of the animals. The smallest species, *A. glacialis*, produced the highest mean number of pellets (15.4 pellets ind.$^{-1}$ d^{-1}), followed by *Onisimus* spp. (2.7 pellets ind.$^{-1}$ d^{-1}) and the largest species, *G. wilkitzkii* (1.1 pellets ind.$^{-1}$ d^{-1}). Relative carbon content of the pellets was very similar in all species (21.2–22.6% dry mass). Juvenile amphipods (*Onisimus* spp., *G. wilkitzkii*) produced more pellets with less POC than adults. Based on field determinations of the POC concentration in the lowermost 2 cm of the sea ice (mean: 36.4 mg C m^{-2}) and mean amphipod abundances (*A. glacialis*: 33.8 ind. m^{-2}, *Onisimus* spp.: 0.5 ind. m^{-2}, *G. wilkitzkii*: 9.4 ind. m^{-2}) in the Greenland Sea in summer 1994, the amount of POC transferred from the ice to the water by faecal pellet production was estimated (0.7 mg C m^{-2} d^{-1} or almost 2% of ice-bound carbon). Since this process probably takes place in all ice-covered Arctic regions as well as during all seasons, grazing and pellet production by under-ice amphipods contributes significantly to matter flux across the ice/water interface.

Introduction

In the Arctic Ocean, sea ice covers an area of 7 * 10^6 km^2 in summer and 14 * 10^6 km^2 in winter, with the central parts being ice-covered all year (Maykut, 1985). The sea ice is in constant motion, following large- and small-scale circulation patterns throughout the Arctic Ocean, with a major outflow through the Fram Strait into the Greenland Sea (Colony & Thorndike, 1985; Aagaard & Carmack, 1989).

The brine-channel system in the ice interior provides a unique habitat for a diverse and specialised flora and fauna, ranging from bacteria to metazoans (e.g. Horner, 1976; Carey & Montagna, 1982; Cross, 1982; Gradinger & Zhang, 1997). Major primary producers are pennate diatoms, but various flagellated taxa are also abundant (Horner, 1985; Ikävalko & Gradinger, 1997). Many sea-ice organisms have their abundance and biomass maxima in the lowermost centi- to decimetres of the ice (Booth, 1984; Grainger & Mohammed, 1986; Gradinger et al., 1991). The ice underside is inhabited by several gammaridean amphipods, of which *Apherusa glacialis, Onisimus glacialis, O. nanseni* and *Gammarus wilkitzkii* are the most common species, both under first- and multi-year ice. They have been reported from almost all areas of the Arctic Ocean, being distributed by the drifting pack ice, and abundances range from single specimens to several hundreds m^{-2} ice underside (Lønne & Gulliksen, 1991a,b; Carey, 1992; Werner, 1997a; Poltermann, 1998). These amphipods are autochthonous at the ice, since they complete their entire life-cycle in this habitat (Poltermann, 1997).

Under-ice amphipods are an important food source for the cryopelagic Arctic cod *Boreogadus saida* (Bradstreet & Cross, 1982) and for diving sea birds such as black guillemot *Cepphus grylle* and little auk *Alle alle* (Lønne & Gabrielsen, 1992). The amphipods graze on ice algae, detritus and sea-ice fauna at the ice underside (Bradstreet & Cross, 1982; Grainger et al., 1985; Poltermann, 1997). Their grazing impact on ice-algal standing stock has been estimated to be about 1–3% d^{-1} in different Arctic regions (Werner, 1997b). Upon grazing, the amphipods release faecal pellets into the water column, thus transferring ice-produced organic matter to the underlying water column, which may sink to deeper water layers or the sea floor. Faecal pellets produced by zooplankton are important vehicles for a fast and effective vertical transport of organic matter and have thus been investigated intensively (e.g. Lampitt et al., 1990; Noji, 1991). The present study aims to (1) examine the faecal pellet production by different species of Arctic under-ice amphipods, (2) estimate the amount of pellet carbon released in the field, and (3) describe external and internal characteristics of the pellets in order to provide an identification guide for sediment trap material.

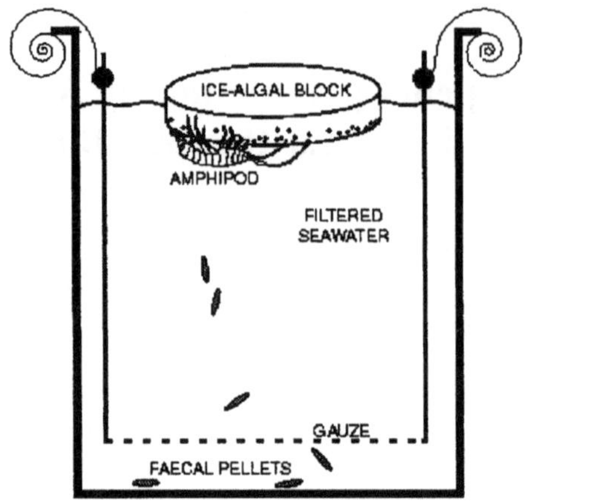

Figure 1. Experimental set up designed for experiments on faecal pellet production with Arctic under-ice amphipods. Volume of seawater in the vial is 1.5 l. The frozen ice-algal block serves as substrate and contains the food source.

Materials and methods

Experimental animals

Experiments on the faecal pellet production of Arctic under-ice amphipods were carried out onboard RV '*Polarstern*' during the expeditions ARK X/1 (1994), XI/1+2 (1995) and ARK XIII/1 (1997) to the Laptev Sea, the Arctic Ocean and the Greenland Sea (Fahrbach, 1995; Krause, 1996; Rachor, 1997; Spindler et al., 1998, respectively). Under-ice amphipods of the species *Apherusa glacialis* (Hansen, 1867), *Onisimus glacialis* and *O. nanseni* (Sars, 1900) and *Gammarus wilkitzkii* (Birula, 1897) were sampled at numerous ice stations by means of an under-ice pumping system (Werner, 1997a) and from ice cores drilled with a SIPRE corer, as well as with Bongo and RMT nets (200, 310 and 4500 μm mesh size, respectively) at partially ice-covered plankton stations (0–500 m sampling depths). The animals were reared in filtered (0.8 μm pore size) seawater, at $T = 0 \pm 1$ °C and S = 30–34. White styrofoam pieces floating on the water surface provided a substrate for the amphipods to attach to. For the experiments, only animals with a healthy and active appearance were chosen from the stock, which comprised amphipods from different stations. Species of *Onisimus* were not separated in the experiments. Prior to the experiments, the amphipods were not fed for 2 weeks.

Faecal pellet production

In order to determine the faecal pellet production rate of under-ice amphipods, experiments were conducted with mixed ice-algal cultures frozen into sea-ice blocks as food. *Gammarus wilkitzkii* were sometimes fed dead shrimp as well. Experimental animals were placed in 1.5 l plastic vials filled with filtered seawater ($T = 0 \pm 1$ °C, S = 30–34) which contained a plexiglas tube covered by gauze at the lower end to prevent coprophagy (Figure 1). A total of 118 experiments were set up with the different species and some juvenile stages as treatments (Table 1). Experiments were terminated when the ice had melted (after 36–72 h). After removal of the amphipods from the vials, faecal pellets were pipetted from the experimental media under a stereo microscope, counted and further processed for characterization.

Characterization of pellets

All faecal pellets collected from the experimental vials were visually inspected for shape and colour, as well as videotaped for later measurements of length and width by means of an image analysis programme on a NeXT workstation. Afterwards, varying amounts of pellets were either filtered on Whatman GF/F filters for determination of chlorophyll *a* equivalents, or on pre-weighed and pre-combusted (500 °C, 12 h) GF/F filters for determination of dry mass (DM), particulate organic carbon (POC) and particulate organic nitrogen (PON). The filters were stored frozen (−18 °C). Determinations of algal pigment concentrations were

Table 1. Arctic under-ice amphipods. Length and dry mass of individuals for experiments on faecal pellet production (mean ± standard deviation), number of experimental replicates for each species and number of specimens per replicate, ad.: adults, juv.: juveniles

Species	Length [mm]	Dry mass [mg]	Replicates	Specimens per replicate
Apherusa glacialis, ad.	9.0 ± 1.7	2.5 ± 2.0	20	7–8
Onisimus spp., ad.	13.7 ± 2.3	12.1 ± 5.1	20	7–8
Onisimus spp., juv.	5.7 ± 1.0	1.1 ± 0.5	4	11–13
Gammarus wilkitzkii, ad.	35.2 ± 5.2	63.7 ± 42.8	72	1
G. wilkitzkii, juv.	5.5 ± 0.6	0.3 ± 0.2	2	4

carried out onboard (after 1–2 weeks) fluorometrically (Turner Designs) according to Arar & Collins (1992) after extraction into 90% acetone. DM was measured in the home laboratory (after 2–3 months) after drying at 60 °C for 12 h, POC and PON were analysed by high-temperature combustion with a CHN-O Rapid Heraeus analyser after acidification (HCl vapour for 24 h).

Carbon transfer via pellets

Concentrations of POC in the lowermost 2 cm of the pack ice from the Greenland Sea were measured in melted ice-core samples (processing see faecal pellet samples) at six stations during ARK XI/2, and converted from chlorophyll measurements at nine stations during ARK X/1 (Gradinger, 1999) using a carbon: chlorophyll ratio of 100, as determined for mixed ice-algal cultures from the ice underside (Werner, 1997a). Amphipod abundances were determined from under-ice video recordings in the Greenland Sea at eight stations during ARK X/1 and at four stations during ARK XI/2 (Werner, 1997a). These field data were combined with experimentally derived faecal pellet production rates and pellet carbon content of adult amphipods in order to estimate the amount of organic carbon transferred from the ice to the water column by grazing and pellet production.

Results

Faecal pellet production

During experiments with frozen ice-algal cultures as food source, faecal pellets were released by all species of under-ice amphipods, but remarkable differences regarding the production rate emerged between species (Figure 2). The smallest species, *Apherusa glacialis*, produced the largest number of pellets with a mean of 15.4 pellets ind.$^{-1}$ d^{-1}. *Onisimus* spp. released on average 2.7 pellets ind.$^{-1}$ d^{-1}, followed by the largest species, *Gammarus wilkitzkii*, with only 1.1 pellets ind.$^{-1}$ d^{-1}. The relative POC content [% DM] was very similar in the faecal pellets of all species, but large species-specific differences in size and dry mass of the pellets were found (Table 2). Thus, the amount of pellet carbon released increased from *A. glacialis* (0.010 mg C ind.$^{-1}$ d^{-1}) over *Onisimus* spp. (0.014 mg C ind.$^{-1}$ d^{-1}) to *G. wilkitzkii* (0.032 mg C ind.$^{-1}$ d^{-1}) (Figure 2).

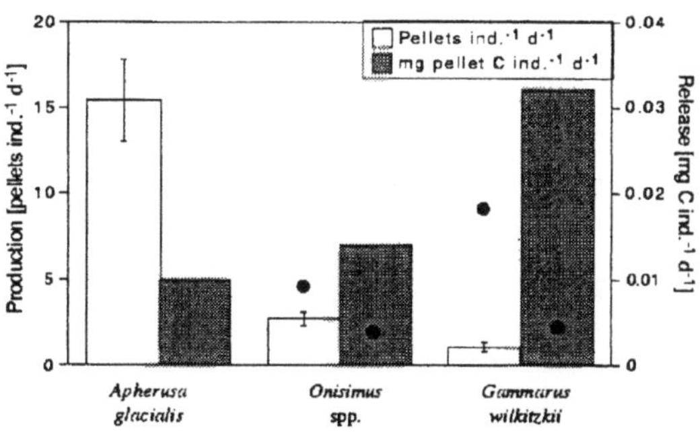

Figure 2. Arctic under-ice amphipods. Mean rates of individual faecal pellet production (experimentally derived) and release of pellet carbon into the water column (calculated). Black dots show mean values for juveniles. Error bars indicate standard error.

Comparison of the faecal pellet production rates of juvenile and adult amphipods revealed pronounced differences, both for *Onisimus* spp. and *Gammarus wilkitzkii* (Figure 2). With 4.8 pellets ind.$^{-1}$ d^{-1}, the pellet production rate of juvenile *Onisimus* spp. was almost twice the amount produced by adults. Juvenile *G. wilkitzkii* with 8.7 pellets ind.$^{-1}$ d^{-1} released one

Table 2. Characterization of faecal pellets from Arctic under-ice amphipods. Given are means ± standard deviation. *N*: number of determinations, POC: particulate organic carbon, PON: particulate organic nitrogen

	Apherusa glacialis adults	*Onisimus* spp. adults	*Onisimus* spp. juveniles	*Gammarus wilkitzkii* adults	*Gammarus wilkitzkii* juveniles
Colour	Green	Yellow-Orange-Red	Yellow Orange	Orange-Red-Brown	Yellow-Green
Length [mm]	1.1 ± 0.4	1.5 ± 0.7	0.7 ± 0.1	3.1 ± 1.9	0.6 ± 0.4
Width [mm]	0.11 ± 0.02	0.24 ± 0.06	0.14 ± 0.03	0.41 ± 0.10	0.11 ± 0.03
N	613	347	108	118	184
Dry mass [μg pellet^{-1}]	3 ± 1	21 ± 19	4 ± 2	144 ± 114	6
POC [%DM]	22.1 ± 25.9	22.6 ± 11.1	15.6 ± 5.3	21.2 ± 12.7	7.5
PON [%DM]	3.0 ± 3.9	2.2 ± 1.1	1.3 ± 0.7	2.1 ± 1.8	0.6
N	10	11	3	6	2
Chl *a* equiv. [ng pellet^{-1}]	4.9 ± 2.0	9.3 ± 5.4	7.6	4.6 ± 2.0	3.0 ± 3.0
N	5	4	1	14	4

order of magnitude more faecal pellets than adults. However, due to the small size, the small dry mass and the small carbon content of their pellets (Table 2), juvenile *Onisimus* spp. and *G. wilkitzkii* released only small amounts of pellet POC (0.003 and 0.004 mg C ind.$^{-1}$ d^{-1}, respectively).

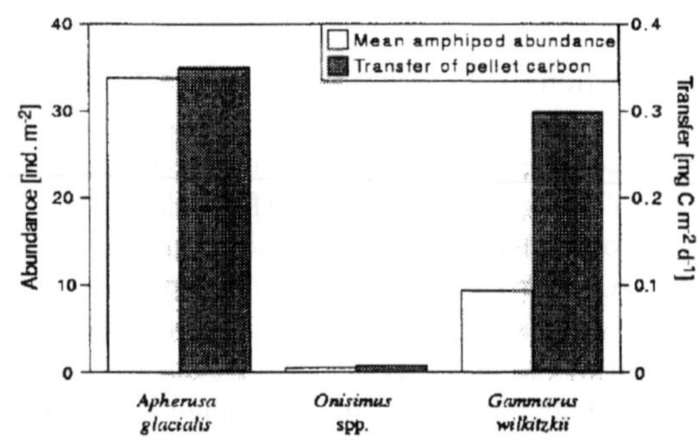

Figure 3. Arctic under-ice amphipods (adults). Mean abundances (summer, 1994) in the Greenland Sea and integrated transfer rates of pellet carbon to the water column (calculated).

Characterization of faecal pellets

Faecal pellets produced by all species of under-ice amphipods were collected during the experiments to describe external and internal characteristics of these particles (Table 2). All pellets had the same basic shape, they were cylindrical and slightly bent. None of them appeared to have a stable peritrophic membrane because they desintegrated easily. Colours were different for the individual species and seemed to depend on the type of food ingested: pellets from *Gammarus wilkitzkii* feeding on ice-algae were yellow-green-brown, whereas pellets from the same species feeding on shrimp were orange-red. Size of pellets varied according to size of species and stages (Tables 1 and 2). Dry mass (DM) of faecal pellets varied according to size of pellets and thus with species and stages, too. POC and PON (% DM) in faecal pellets did not differ much between species (Table 2). Mean values varied between 21.2 and 22.6% DM for POC in adult amphipod pellets, and between 2.1 and 3.0% DM for PON, but pellets of juvenile amphipods contained less carbon and nitrogen. The amount of chlorophyll *a* equivalents in faecal pellets varied considerably between species, independently of size of pellets (Table 2).

Carbon transfer across the ice/water interface

Concentrations of POC in the lowermost 2 cm of the pack ice in the Greenland Sea, although patchily distributed, were on average very similar during ARK X/1 in summer (1.82 ± 2.17 mg C l^{-1}, n = 9) and

ARK XI/2 in autumn (2.03 ± 2.02 mg C l^{-1}, n = 6). These values amount to 36.4 and 40.6 mg C m^{-2} ice underside, respectively. Mean abundances of under-ice amphipods during ARK X/1 varied considerably between species (Figure 3): *Apherusa glacialis* occurred with 33.8 ind. m^{-2} (SD = 25.0 ind. m^{-2}, n = 8), *Onisimus* spp. with 0.5 ind. m^{-2} (SD = 1.2 ind. m^{-2}, n = 8), and *Gammarus wilkitzkii* with 9.4 ind. m^{-2} (± 15.3 ind. m^{-2}, n = 8). During ARK XI/2 (autumn), only *G. wilkitzkii* was observed at the ice underside with an average of 5.6 ind. m^{-2} (SD = 9.8 ind. m^{-2}, n = 4). Depending on the mean abundance of a species and the individual pellet POC, the species-specific carbon transfer across the ice/water interface via pellet production was estimated for the summer situation during ARK X/1 (only adult amphipods). The under-ice population of the small but abundant species *A. glacialis* contributed 0.35 mg C m^{-2} d^{-1} to this process, followed by the large but less abundant *G. wilkitzkii* with 0.30 mg C m^{-2} d^{-1}, the carbon transfer by the very rare amphipods of the genus *Onisimus* was with <0.01 mg C m^{-2} d^{-1} negligible (Figure 3). The total amount of POC released by all under-ice amphipods of about 0.65 mg C m^{-2} d^{-1} would indicate that 1.8% of carbon from the lowermost 2 cm of the ice were transferred to the water column via faecal pellets per day during the summer. For autumn conditions with only *G. wilkitzkii* releasing 0.18 mg C m^{-2} d^{-1}, about 0.5% of the ice-bound organic carbon would enter the water column in form of their faecal pellets.

Discussion

Methods

The experimental work with the under-ice amphipods *Apherusa glacialis*, *Onisimus* spp. and *Gammarus wilkitzkii* has provided new data on their faecal pellet production and an estimation of their contribution to the process of cryopelagic carbon flux. It should, however, be kept in mind that the applicability of experimental data to nature always remains critical to a certain degree. In the present study, a thorough attempt was made to design the experimental conditions as 'natural' as possible, with natural seawater of temperatures and salinities very similar to field conditions and food offered in the same way as in the wild: frozen into sea ice. This technique has already been used studying under-ice grazing of Antarctic krill *Euphausia superba* (Stretch et al., 1988) and Arctic under-ice amphipods (Werner, 1997b). The major shortcoming of this study is the high variability of field data, e.g. abundances, which makes calculations using mean values questionable. Therefore, the quantitative results of pellet carbon transfer across the ice/water interface should rather be understood as estimations of the order-of-magnitude dimension of this process.

Faecal pellet production rate

The production rates of faecal pellets by Arctic under-ice amphipods during the present experiments showed species- and stage-specific differences, which clearly mirror the specific ingestion rates of these amphipods (Werner, 1997b), and which support the principle that specific metabolic rates decline with rising body size within one metabolic group (Peters, 1983).

Comparative data on faecal pellet production rates of amphipods are scarce. Experiments with *O. nanseni* and *G. wilkitzkii* feeding on an abundant supply of beef and crabmeat yielded very similar results as the present study with mainly ice algae as food source: 1.8 ± 0.8 pellets ind.$^{-1}$ d^{-1} for *O. nanseni*, and 0.9 ± 0.3 pellets ind.$^{-1}$ d^{-1} for *G. wilkitzkii* (Poltermann, 1997). The freshwater epibenthic amphipod *Diporeia* sp. (5–6 mm length) from the Great Lakes produced 1–10 pellets ind.$^{-1}$ d^{-1} (at 4 °C) while feeding on the filamentous diatom *Melosira varians* and a varying sediment flora (Quigley & Vanderploeg, 1991). These production rates of under-ice and benthic amphipods are in the same range as the results of the present study.

An example with a very high production rate of faecal material is the Antarctic krill, *Euphausia superba*, which released 13–40 mg DM ind.$^{-1}$ d^{-1} while feeding on natural phytoplankton in summer (Clarke et al., 1988). This is more than two orders of magnitude higher in comparison with the largest Arctic under-ice amphipod *G. wilkitzkii* with 0.03–0.10 mg DM ind.$^{-1}$ d^{-1} (Poltermann, 1997; this study), although the mean individual DM of the krill studied (597 mg) was only one order of magnitude higher. This difference probably reflects the higher metabolic expenditure of a pelagic animal like krill (Clarke & Peck, 1991).

Transfer of organic carbon through the ice/water interface

Grazing at the ice underside and release of faecal pellets by Arctic under-ice amphipods form a direct link between the sea-ice realm and the pelagic ecosystem (cryopelagic coupling) and provide a pathway for organic matter flux. Besides production rates, the composition of faecal pellets is an important factor for the significance of this transfer, as well as for the potential consumers of this material. The average content of particulate organic carbon (POC) in faecal pellets of all adult under-ice amphipods studied was approximately 22% of dry mass (DM). This is a typical value for pellets produced under sufficient feeding conditions (Lampitt et al., 1990), and has also been measured in pellets of the copepod *Calanus finmarchicus* (Noji, 1991). Faecal pellets of Antarctic krill, *Euphausia superba*, contained less POC (10–14% DM) while feeding on natural phytoplankton (Clarke et al., 1988). Thus, amphipods feeding at the ice underside produced pellets comparable to pelagic grazers.

The potential contribution of the under-ice amphipods to organic carbon flux from the ice to the underlying water column was about 0.7 mg C m^{-2} d^{-1} or 2% of standing stock d^{-1} (summer) in the ice-covered Greenland Sea, which does not seem much on a short-term basis. However, since the amphipods are present at the ice underside all year round (Melnikov, 1997; Poltermann, 1997), grazing and pellet release provide an important process of organic particle formation and flux also functioning during seasons without primary production and ice melt, which are periods of low sedimentation rates in ice-covered areas (Ramseier et al. in press). Based on sea-ice cover and flux dates in the Greenland Sea in 1994 (Martin & Wadhams, 1996), a carbon export from the ice of $1.0 * 10^5$ t C y^{-1} via amphipod faecal pellets can be estimated for this area.

The significance of ice-algal primary production for the Arctic Ocean is not fully understood as yet. It can be higher than water column production, locally in the central deep basins (Wheeler et al., 1996), and seasonally in coastal areas before ice break-up in spring (Horner & Schrader, 1982). Estimated Arctic ice-algal production ranges from 0.9 to 164.0 g C m^{-2} yr^{-1} in different regions, accounting for <3–25% of total production (Subba Rao & Platt, 1984; Legendre et al., 1992). The Arctic under-ice amphipods are important mediators between this ice-based primary production and higher trophic levels of adjacent ecosystems, and in contrast to the Antarctic krill, they play this key role during all seasons.

Fate of faecal pellets

Zooplankton faecal pellets are widely regarded as an important vehicle for an effective vertical transport of organic matter (Lampitt et al., 1990; Noji, 1991; Perissinotto & Pakhomov, 1998) and are often found as a major component in sediment-trap samples from the Arctic Ocean (Bauerfeind et al., 1994; Andreassen et al., 1996), and also in ice-covered areas (Legendre et al., 1992). Pellet sedimentation from the Arctic copepods *Calanus glacialis* and *Pseudocalanus* spp. feeding at the ice/water interface under fast ice in spring amounted to 2.3% of total standing stock of chlorophyll *a* d^{-1} (Tremblay et al., 1989). Faecal pellets from under-ice amphipods (*Onisimus litoralis*) grazing on ice algae reached the benthos in very shallow (10 m depth) coastal areas (Carey, 1987). It has been assumed that the faecal pellets released by *O. nanseni* and *Gammarus wilkitzkii* in the fjords off Franz-Josef-Land represented an important food pulse to the benthos (Poltermann, 1997). Determined mean sinking velocity of the large *G. wilkitzkii* pellets (Table 2) was 0.6 cm s^{-1} $\pm$ 0.1 cm s^{-1} (n = 18, unpubl. data), amounting to about 500 m d^{-1}. Long-term particle flux studies in the ice-covered Greenland Sea with sediment traps deployed in 500 m depth showed POC fluxes of <5–35 mg m^{-2} d^{-1}, depending on ice concentration, duration of ice cover and distance from the ice edge (Ramseier et al. in press). Pellet POC released by amphipods at the ice/water interface potentially contributed to the flux, however, the particular types of faecal pellets produced by the under-ice amphipods as described in the present study have not been recorded from the sediment-trap samples. This lack and the observation that the pellets disintegrate easily cast some doubt on the hypothesis that pellets produced by under-ice amphipods sink rapidly through the water column.

Sedimentation to deeper waters or the sea floor is not the only fate of faecal pellets in the marine environment. They can be retained in the upper water layer due to hydrographical conditions, e.g. turbulent mixing (Alldredge et al., 1987; Carey, 1987). A particle flux study in the Norwegian Sea revealed that less than 5% of the copepod faecal pellet standing stock actually sank out of the euphotic zone (Bathmann et al., 1987). Several biologically governed processes like coprophagy (ingestion of faecal material), coprorhexy (fragmentation of faecal material) and coprochaly (loosening of faecal material) have been proposed to explain this phenomenon (Bathmann et al., 1987; Lampitt et al., 1990; Noji et al., 1991). It has been clearly shown in experiments that the cyclopoid copepod *Oithona similis* feeds extensively on large copepod pellets, removing 24–70% of the pellets offered. These experimental findings and field observations on high abundances of *O. similis* (200–2000 ind. m^{-3}) in the ice-covered Barents Sea led to the conclusion that a large portion of the faecal material was retained and reworked in the upper water layers by this species (González & Smetacek, 1994; Andreassen et al., 1996). In the Greenland Sea, abundances of the ubiquitous *O. similis* directly below the ice were sometimes even higher (100–4700 ind. m^{-3}; Werner, 1997a). It is, therefore, well conceivable that under-ice amphipod pellets, which are comparatively rich in organic carbon and still contain algal pigments partially not degraded (I. Peeken, pers. comm.), are a preferred food source for *O. similis* under the ice. Their raptorial feeding mode would result in further destruction of the fragile pellets, and hence in reduced sinking velocity. A significant amount of carbon and nutrients would thus be maintained in the upper water layers in ice-covered areas. In order to verify this assumption, explicit experiments with *O. similis* are required for future research.

Conclusions

Arctic under-ice amphipods graze at the ice underside and release faecal pellets into the underlying water column, thus transferring ice-produced organic matter to the pelagic ecosystem (cryopelagic coupling). Pellet production rates vary with species and stages of amphipods and reflect their different metabolic rates.

Faecal pellet carbon released by the under-ice amphipods potentially contributes considerably to total carbon flux in Arctic ice-covered areas.

The fate of under-ice amphipod pellets in the water column is not known as yet. Future research should focus on the hypothesis that coprophagous cyclopoid copepods retain and rework the faecal material in the upper water layers.

Acknowledgements

I am very grateful to the captains, crews and colleagues onboard RV '*Polarstern*' for their constant support during ARK X/1, ARK XI/1+2 and ARK XIII/1. Particular thanks are due to G. Donner, H. Eicken, R. Gradinger, J. Ikävalko, F. Lindemann, K. Meiners, T. Mock, Q. Zhang and several other ice-group fellows for their help. H. Auel and W. Hagen provided many under-ice amphipods from their Bongonet and RMT samples, M. Krumbholz kindly conducted the analyses of DM, POC and PON of all the pellets. This manuscript benefited from the constructive comments made by R. Gradinger. This is contribution 353 of the Sonderfoschungsbereich 313, University of Kiel.

References

Aagaard, K. & E. C. Carmack, 1989. The role of sea ice and other freshwater in the Arctic circulation. J. Geophys. Res. 94: 14485–14498.

Alldredge, A. L., C. C. Gottschalk & S. MacIntyre, 1987. Evidence for sustained residence of macrozooplankton faecal pellets in surface waters off Southern California. Deep-Sea Res. 34: 1641–1659.

Andreassen, I., E.-M. Nöthig & P. Wassmann, 1996. Vertical particle flux on the shelf off northern Spitsbergen, Norway. Mar. Ecol. Prog. Ser. 137: 215–228.

Arar, E. J. & G. B. Collins, 1992. *In vitro* determination of Chlorophyll *a* and Phaeophytin *a* in marine and freshwater phytoplankton by fluorescence. U.S. EPA, Cincinnati: 14 pp.

Bathmann, U. V., T. T. Noji, M. Voss & R. Peinert, 1987. Copepod fecal pellets: abundance, sedimentation and content at a permanent station in the Norwegian Sea in May/June 1986. Mar. Ecol. Prog. Ser. 38: 45–51.

Bauerfeind, E., B. v. Bodungen, K. Arndt & W. Koeve, 1994. Particle flux and composition of sedimenting matter in the Greenland Sea. J. Mar. Syst. 5: 411–423.

Booth, J. A., 1984. The epontic algal community of the ice edge zone and its significance to the Davis Strait ecosystem. Arctic 37: 234–243.

Bradstreet, M. S. W. & W. E. Cross, 1982. Trophic relationships at high Arctic ice edges. Arctic 35: 1–12.

Carey, A. G., 1987. Particle flux beneath fast ice in the shallow southwestern Beaufort Sea, Arctic Ocean. Mar. Ecol. Prog. Ser. 40: 247–257.

Carey, A. G., 1992. The ice fauna in the shallow southwestern Beaufort Sea, Arctic Ocean. J. Mar. Syst. 3: 225–236.

Carey, A. G. & P. A. Montagna, 1982. Arctic sea ice faunal assemblage: first approach to description and source of the underice meiofauna. Mar. Ecol. Prog. Ser. 8: 1–8.

Clarke, A. & L. S. Peck, 1991. The physiology of polar marine zooplankton. Polar Res. 10: 355–369.

Clarke, A., L. B. Quetin & R. M. Ross, 1988. Laboratory and field estimates of the rate of faecal pellet production by antarctic krill, *Euphausia superba*. Mar. Biol. 98: 557–563.

Colony, R. & A. S. Thorndike, 1985. Sea ice motion as a drunkard's walk. J. Geophys. Res. 90: 965–974.

Cross, W. E., 1982. Under-ice biota at the Pond Inlet ice edge and in adjacent fast ice areas during spring. Arctic 35: 13–27.

Fahrbach, E., 1995. Die Expedition ARKTIS X/1 des Forschungsschiffes 'Polarstern' 1994. Ber. Polarforsch. 181: 1–79.

González, H. E. & V. Smetacek, 1994. The possible role of the cyclopoid copepod *Oithona* in retarding vertical flux of zooplankton faecal material. Mar. Ecol. Prog. Ser. 113: 233–246.

Gradinger, R., 1999. Vertical fine structure of the biomass and composition of algal communities in Arctic pack ice. Mar. Biol. 133: 745–754.

Gradinger, R. & Q. Zhang, 1997. Vertical distribution of bacteria in Arctic sea ice from the Barents and Laptev Seas. Polar Biol. 17: 448–454.

Gradinger, R., M. Spindler & D. Henschel, 1991. Development of Arctic sea-ice organisms under graded snow cover. Polar Res. 10: 295–307.

Grainger, E. H. & A. A. Mohammed, 1986. Copepods in Arctic sea ice. Syllogeus 58: 303–310.

Grainger, E. H., S. I. C. Hsiao, N. Pinkewycz, A. A. Mohammed & V. Neuhof, 1985. The food of ice fauna and zooplankton in Frobisher Bay. Can. Data Rep. Fish. Aquat. Sci. 558: 1–67.

Horner, R. A., 1976. Sea ice organisms. Oceanogr. Mar. Biol. ann. Rev. 14: 167–182.

Horner, R. A., 1985. Ecology of sea ice microalgae. In Horner, R. A. (ed.), Sea Ice Biota, CRC Press, Boca Raton: 83–103.

Horner, R. A. & G. C. Schrader, 1982. Relative contributions of ice algae, phytoplankton and benthic microalgae to primary production in nearshore regions of the Beaufort Sea. Arctic 35: 485–503.

Ikävalko, J. & R. Gradinger, 1997. Flagellates and heliozoans in the Greenland Sea ice studied alive using light microscopy. Polar Biol. 17: 473–481.

Krause, G., 1996. The expedition ARKTIS-XI/2 of RV 'Polarstern' in 1995. Ber. Polarforsch. 197: 1–65.

Lampitt, R. S., T. T. Noji & B. V. Bodungen, 1990. What happens to zooplankton faecal pellets? Implications for material flux. Mar. Biol. 104: 15–23.

Legendre, L., S. F. Ackley, G. S. Dieckmann, B. Gulliksen, R. A. Horner, T. Hoshiai, I. A. Melnikov, W. S. Reeburgh, M. Spindler & C. W. Sullivan, 1992. Ecology of sea ice biota. 2. Global significance. Polar Biol. 12: 429–444.

Lønne, O. J. & G. W. Gabrielsen, 1992. Summer diet of seabirds feeding in sea-ice covered waters near Svalbard. Polar Biol. 12: 685–692.

Lønne, O. J. & B. Gulliksen, 1991a. On the distribution of sympagic macro-fauna in the seasonally ice covered Barents Sea. Polar Biol. 11: 457–469.

Lønne, O. J. & B. Gulliksen, 1991b. Source, density and composition of sympagic fauna in the Barents Sea. Polar Res. 10: 289–294.

Martin, T. & P. Wadhams, 1996. Ice fluxes within the Greenland Sea and their variability. In Wadhams, P., J. B. Wilkinson & S. C. S. Wells (eds), ESOP Sci. Rep. 1: 72–79.

Maykut, G. A., 1985. The ice environment. In Horner, R. A. (ed.), Sea Ice Biota. CRC Press, Boca Raton: 21–82.

Melnikov, I. A., 1997. The Arctic sea ice ecosystem. Gordon and Breach Science Publ., Amsterdam: 204 pp.

Noji, T. T., 1991. The influence of macrozooplankton on vertical particulate flux. Sarsia 76: 1–9.

Noji, T. T., K. W. Estep, F. MacIntyre & F. Norrbin, 1991. Image analysis of faecal material grazed upon by three species of copepods: evidence for coprorhexy, coprophagy and coprochaly. J. mar. biol. Ass. U.K. 71: 465–480.

Perissinotto, R. & E. A. Pakhomov, 1998. Contribution of salps to carbon flux of marginal ice zone of the Lazarev Sea, southern Ocean. Mar. Biol. 131: 25–32.

Peters, R. H., 1983. The Ecologocal Implications of Body Size. Cambridge Univ. Press, Cambridge: 329 pp.

Poltermann, M., 1997. Biologische und ökologische Untersuchungen zur kryopelagischen Amphipodenfauna des arktischen Meereises. Ber. Polarforsch. 225: 1–170.

Poltermann, M., 1998. Abundance, biomass and small-scale distribution of cryopelagic amphipods in the Franz Josef Land area (Arctic). Polar Biol. 20: 134–138.

Quigley, M. A. & H. A. Vanderploeg, 1991. Ingestion of live filamentous diatoms by the Great Lakes amphipod, *Diporeia* sp.: a case study of the limited value of gut content analysis. Hydrobiologia 223: 141–148.

Rachor, E., 1997. Scientific cruise report of the Arctic expedition ARK-XI/1 of RV 'Polarstern' in 1995. Ber. Polarforsch. 226: 1–157.

Ramseier, R. O., E. Bauerfeind & R. Peinert, in press. Sea-ice impact on long term particle flux in the Greenland Sea's Is Odden-Nordbukta region, 1985–1996. J. Geophys. Res.

Spindler, M., W. Hagen & D. Stübing, 1998. Scientific cruise report of the Arctic expedition ARK-XIII/1 of RV 'Polarstern' in 1997. Ber. Polarforsch. 296: 1–65.

Stretch, J. J., P. P. Hamner, W. M. Hamner, W. C. Michel, J. Cook & C. W. Sullivan, 1988. Foraging behavior of Antarctic krill *Euphausia superba* on sea-ice microalgae. Mar. Ecol. Prog. Ser. 44: 131–139.

Subba Rao, D. V. & T. Platt, 1984. Primary production of Arctic waters. Polar Biol. 3: 191–201.

Tremblay, C., J. A. Runge & L. Legendre, 1989. Grazing and sedimentation of ice algae during and immediately after a bloom at the ice-water interface. Mar. Ecol. Prog. Ser. 56: 291–300.

Werner, I., 1997a. Ecological studies on the Arctic under-ice habitat – colonization and processes at the ice–water interface. Ber. Sonderforsch. 313 Univ. Kiel 70: 1–167.

Werner, I., 1997b. Grazing of Arctic under-ice amphipods on sea-ice algae. Mar. Ecol. Prog. Ser. 160: 93–99.

Wheeler, P. A., M. Gosselin, E. Sherr, D. Thibault, D. L. Kirchman, R. Benner & T. E. Whitledge, 1996. Active cycling of organic carbon in the central Arctic Ocean. Nature 380: 697–699.

Hydrobiologia **426**: 97–103, 2000.
G. Liebezeit, S. Dittmann & I. Kröncke (eds), Life at Interfaces and Under Extreme Conditions.

High abundance of *Myriochele fragilis* Nilsen & Holthe, 1985 (Polychaeta: Oweniidae) in the deep sea of the Eastern Mediterranean

Dieter Fiege[1], Ingrid Kröncke[2] & Ruth Barnich[1]
[1]*Forschungsinstitut Senckenberg, Senckenberganlage 25, D-60325 Frankfurt, Germany*
[2]*Forschungsinstitut Senckenberg, Abteilung für Meeresforschung, Schleusenstr. 39a, D-26382 Wilhelmshaven, Germany*
E-mail: dfiege@sng.uni-frankfurt.de

Key words: Eastern Mediterranean Sea, deep sea, Oweniidae, *Myriochele fragilis*, high abundance

Abstract

An exceptionally high abundance of *Myriochele fragilis* Nilsen & Holthe, 1985 (Polychaeta: Oweniidae) is reported for the Ierapetra Basin in the Eastern Mediterranean Sea from depths of 4159 and 4260 m, representing the first record outside the Norwegian and Arctic Sea and extending the depth range by more than 1600 m. Specimens are described and compared to closely related species.

Introduction

During a cruise to the Eastern Mediterranean with R/V Meteor (cruise 40, leg 3) from December 1997 to January 1998 benthos was sampled in the northern Sporades Basin (1200 m), Cretan Sea (1800 m) and the Ierapetra Basin (4200 m), south of Crete, using an Usnell giant box corer, a beam trawl and an epibenthic sledge (Fiege et al., 1999). In line with the present knowledge of extreme oligotrophy of the eastern Mediterranean Sea (Fredj & Laubier, 1985; Janssen, 1989; Tselepides & Eleftheriou, 1992) only small numbers of macrobenthic animals were collected in the various areas.

In the Ierapetra Basin, however, a high abundance of sessile, tube-building polychaetes was found from depths between 4159 and 4260 m compared to data of a previous cruise (Meteor cruise 25, leg 1, 1993). These flexible tubes were inhabited by *Myriochele fragilis* Nilsen & Holthe, 1985 (Polychaeta: Oweniidae), a species hitherto only reported from the Norwegian and Arctic Sea from depths between 800 and 2600 m (Nilsen & Holthe, 1985).

Material and methods

Area of investigation

The area of investigation was located in the Ierapetra Basin south of Crete. The water depth ranged between 4159 m and 4263 m (Table 1). Sampling took place in January 1998. Sediments were very silty, light grey to beige, containing a high number of empty pteropod shells, otoliths and pumice stones.

Sampling

Samples were taken with a 0.25-m^2 Usnell giant box corer (for station data see Table 1). One core per station was taken for taxonomy, a second one for quantitative investigation of abundance. Therefore, data presented in the Results section were determined from a single box core. Samples were sieved over 0.5-mm mesh size, the residue was fixed in 4% buffered formalin. Biomass is given as wet weight.

In addition, sediment samples were taken from the box core for total organic carbon (TOC) analyses. The samples were dried on board ship at 55 °C. An aliquot of 10 mg (total sediment) was analysed using an 'elementar' CHNS analyser. Prior to analysis the samples were kept in a desiccator with concentrated HCl underneath to eliminate the carbonate.

Preparation for SEM

For SEM-investigations, specimens were dehydrated via graded ethanol series and critical point drying

Table 1. Station data

Station	Coordinates		Water depth	Worm density[a]	TOC[b]
	N	E	(m)	(m^{-2})	[mg g^{-1} dw]
4	34° 26.02′	026° 04.03′	4159	116	5.4
5	34° 25.00′	026° 04.00′	4260	72	5.7
6	34° 24.45′	026° 04.52′	4261	Empty tubes only	5.8
7	34° 24.85′	026° 05.79′	4263	Absent	No sample

[a] Based on a single sample.
[b] Surface sediment.

using CO_2, coated with Au-Pd and examined in a CamScan CS 24 SEM.

Deposition of specimens

Specimens are deposited in the collections of the Senckenberg Museum Frankfurt (SMF).

Results

Systematics

Myriochele fragilis Nilsen & Holthe, 1985 (Figures 1A–F and 2A,B).

Myriochele fragilis Nilsen & Holthe, 1985; 25–27; Figures 9A–D and 12D.

Material

Eastern Mediterranean Sea, Ierapetra Basin, south of Crete. Meteor cruise 40, leg 3: stn. 4, 4159 m, >20 spms., stn. 5, 4260 m, >20 spms, and stn. 6, 4260 m, only empty tubes. Length of largest complete specimen 25 mm, width 0.4 mm, for 35 chaetigers. As already mentioned by Nilsen & Holthe (1985) and suggested by the specific name, specimens are extremely fragile and it is almost impossible to remove them from their tube without breaking them.

Description

Prostomium truncate, ca. 1.5–2 times longer than wide. Mouth opening densely ciliated, situated anteroventrally with medioventral groove reaching about two thirds to proximal end of prostomium (Figures 1A,B). Proboscis eversible (Figure 1C). No eyespots visible. Prostomium separated from thorax by groove and low collar at first chaetiger (Figure 1A). Thorax with pronounced constriction between first and second chaetiger. Prostomium and first chaetiger smooth, second to sixth chaetiger secondarily ringed. Thoracic chaetigers 1–3 short, relative length 1:2:1; with only capillary notochaetae present. Ventral tori with numerous uncini present from fourth chaetiger to posterior end. Sixth to eighth chaetiger longest (up to 3 mm long). Chaetigers near posterior end shortened, condensed, with notochaetae about twice as long as in anterior chaetigers. Pygidium with three to five lobes, one to three small dorsal and two larger lateral (Figure 1D). All capillary chaetae minutely spinous with spines adhering close to chaetae (Figure 1E); no acicular chaetae present. Uncini bidentate, one tooth on top of the other (Figure 1F). Tube much longer than animal, up to 65 mm length and ca. 0.5–0.8 mm in diameter, ends slightly tapering, sometimes flattened; tube flexible, leathery, but can be easily torn, densely encrusted with small Foraminifera, mostly globigerinids, and occasional small sponge spicules; no specific pattern recognizable.

Remarks

Myriochele fragilis appears to be close to *M. heeri* Malmgren, 1867, *M. danielsseni* Hansen, 1879, *M. picta* Southern, 1921, and *M. oculata* Zachs, 1923 (Table 2). *M. oculata* and *M. heeri* have repeatedly been recorded from the Mediterranean, e.g., by Laubier & Paris (1962: *M. heeri*); Cognetti-Variale (1979: *M. heeri* and *M. oculata*); Martìn (1989: *M. oculata*, as *Galathowenia oculata*), and Ben-Eliahu & Fiege (1995: *M. oculata*). However, *M. oculata* clearly differs from *M. fragilis* in the presence of eye spots, the arrangement of uncinal teeth and the lower number of pygidial lobes, while *M. heeri* has the same arrangement of uncinal teeth as *M. fragilis* but additional acicular chaetae in notopodia, and only two pygidial lobes. *M. danielsseni* also has only one pair of pygidial lobes, and additional acicular chaetae in notopodia, arrangement of uncinal teeth is side by side. Judging from the description given by Southern (1921) *Myri-*

Table 2. Characters of *Myriochele* species (based on Nilsen & Holthe (1985), Martín (1989) and respective original descriptions; n.d., no data)

Character/species	*M. heeri* Malmgren, 1867	*M. danielsseni* Hansen, 1882	*M. picta* Southern, 1921	*M. oculata* Zachs, 1923	*M. fragilis* Nilsen & Holthe, 1985, Norwegian Sea	*M. fragilis* Nilsen & Holthe, 1985 Ierapetra Basin
Prostomium	Cylindrical, rounded	Short with dorsolateral sulcus, collarlow and thick	Set off from thorax by constriction; patch of reticulate, purplish brown pigment dorsally and transverse dorsal band at posterior end of prostomium	Truncate, without constriction	Long, usually with high collar	Truncate, ca. 1.5–2 times longer than wide; separated from thorax by narrow groove; 1st chaetiger forming low collar
Mouth	Ventral, forming longitudinal slit	n.d.	Ventrally, surrounded by thick folded lip	Terminal with lobes, midventral cleft	n.d.	Anteroventrally, densely ciliated; medioventral groove reaching almost posterior end of prostomium; proboscis eversible
Eyes	No eyespots	Eyespots may be present	n.d.	2 ventrolateral eyespots	No eyespots	No eyespots
Thorax	First 2 to 3 chaetigers; relative length 1:1:3; no mediodorsal groove	First 3 chaetigers with relative length 2:4:1	Chaetigers 1–3	Chaetigers 1–3 short, relative length 1:1:1; mediodorsal groove present	Chaetigers 1–3; relative length 1:2:1; pronounced constriction between 1st and 2nd chaetiger	Chaetigers 1–3 short, constriction between 1st and 2nd chaetiger; relative length 1:2:1
Abdomen	6th or 7th chaetiger longest, segments decreasing gradually in length towards posterior end; body short, nearly cylindrical	7th chaetiger longest	5th chaetiger longest, segments gradually decreasing in length posteriorly	5th chaetiger longest, segments decreasing gradually in length towards posterior end	5th and 6th chaetiger longest	6th to 8th chaetiger longest; posteriormost chaetigers short, condensed with capillary notochaetae about twice as long as in anterior chaetigers. 2nd to 6th chaetiger secondarily ringed
Notochaetae	Distally spinous capillaries and smooth acicular chaetae from 5th or 6th chaetiger	1st and 2nd chaetiger with long, narrow, weakly spinous capillaries; from 3rd with additional short acicular chaetae	Spinous capillaries; thoracic chaetae slightly thicker than abdominal	Only spinous capillaries; few but very long in posterior chaetigers	Only spinous capillaries	Only spinous capillaries, basal part smooth, spines adhering closely to chaetae
Uncini	From 3rd or 4th chaetiger, bidentate, one tooth on top of the other	From 4th chaetiger, teeth side by side, nearly at same level	From 4th chaetiger, bidentate, one tooth above the other, distinctly constricted neck	From 4th chaetiger; bidentate, teeth obliquely side by side	From 4th chaetiger, bidentate, one tooth on top of the other, tori shorter and with fewer uncini than *M. oculata*. Uncini thicker than *M. heeri* and *M. danielsseni*	From 4th chaetiger, bidentate, one tooth on top of the other

Continued on p. 100

Table 2. Continued

Character/species	*M. heeri* Malmgren, 1867	*M. danielsseni* Hansen, 1882	*M. picta* Southern, 1921	*M. oculata* Zachs, 1923	*M. fragilis* Nilsen & Holthe, 1985, Norwegian Sea	*M. fragilis* Nilsen & Holthe, 1985 Ierapetra Basin
Pygidium	2 weakly developed lobes	1 pair of lateral lobes	Conical	2–3 pronounced lobes	5–6 lobes; 1–2 dorsal, 2 lateral, and 2 ventral	3–5 lobes; 1–3 small, dorsal and 2 larger, lateral
Length (mm)/chaetigers	Up to 35/ca. 30	up to 15. 2/15	3–4/16	To 35/up to ca. 30	To 30/25	24. 5/35
Width (mm)	1.5–2.0	0.65	n.d.	0.25–0.6	0.8	0.65
Tube	Length/width 65–75/2 mm, ends tapering; rigid, encrusted with sand-grains or sponge spicules, often in helical pattern	Length at least up to 20 mm, width 0.9 mm; rigid, easily fragmented, encrusted with foraminiferans, sand grains or sponge spicules in regular pattern	Length/width 7–11/2 mm; tough inner membrane covered with small quartz grains	Length/width 55–90/0.5–1.0 mm; encrusted with firmly cemented oblong sand-grains in characteristic transverse pattern	Length/width 50/1.2 mm; slightly tapering towards hind end, front end conical; covered with particles, no specific pattern	Length/width up to 65/0.5–0.8 mm; ends slightly tapering, flattened; tube not very tough, encrusted with dense layer of foraminifera and some small sponge spicules, no specific pattern
Distribution	N Atlantic to N of Spitzbergen, N Pacific, ?W Atlantic, ?California, ?Subantarctic seas; 120–2600 m. Mediterranean Sea; 115 m	North Sea, Norwegian Sea, Polar Sea; 104–1700 m	Chilka Lake, India; 1.2–4.6 m	Atlantic: SW of British Isles to N of Spitzbergen, Bering Sea, N Pacific; 12–2500 m. Mediterranean Sea; 100–1400 m	N Atlantic to N of Spitzbergen; 800–2600 m	Eastern Mediterranean Sea; 4159–4260 m

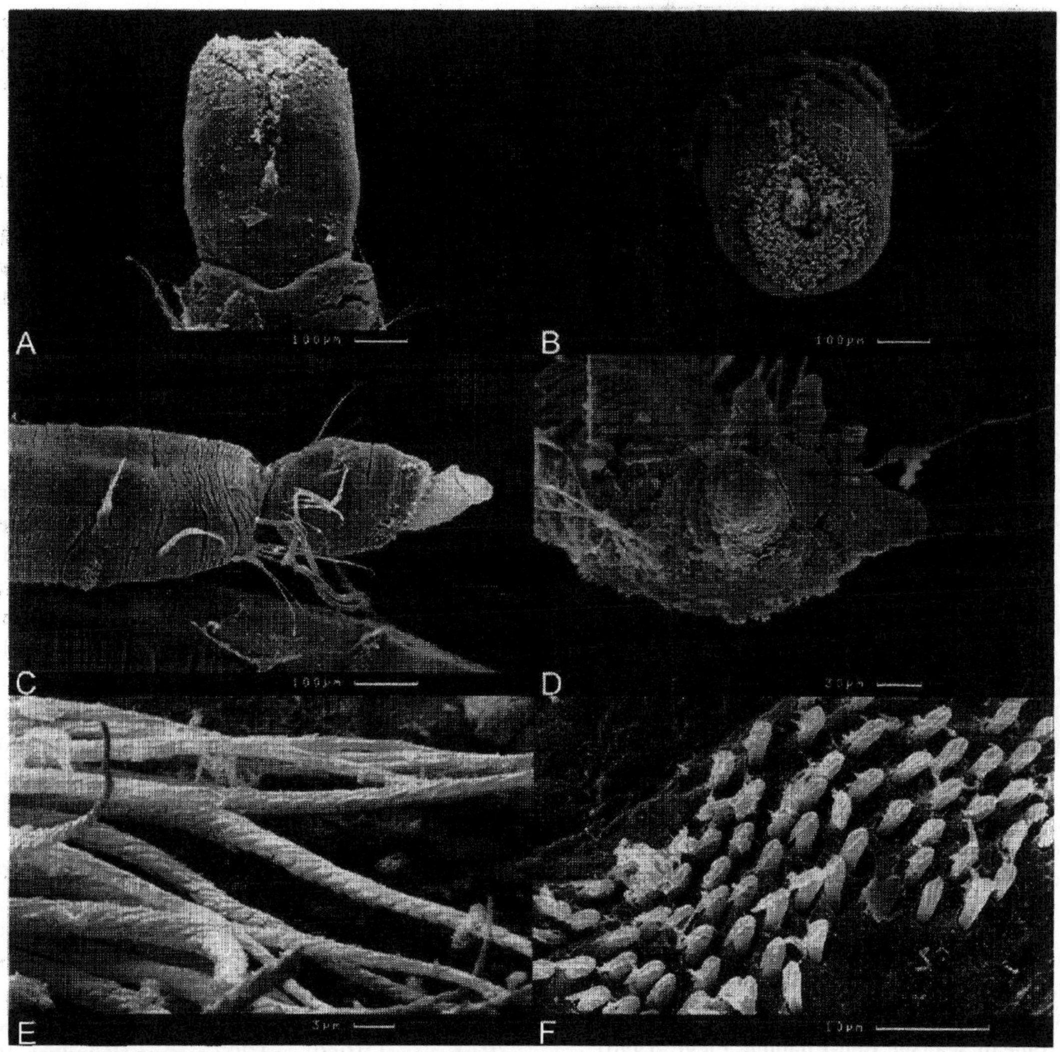

Figure 1. Scanning electron micrographs of *Myriochele fragilis* Nilsen & Holthe, 1985, collected in the Ierapetra Basin. (A) Prostomium, ventral view; (B) prostomium, frontal view, ventral side up; (C) anterior region, proboscis extended, lateral view; (D) pygidium; (E) notopodial spinous capillaries; (F) uncini of abdominal chaetiger.

ochele picta appears to be very close to *M. fragilis*. But *M. picta* is only known from Chilka Lake, India, a lake that is connected to the Bay of Bengal only by a narrow opening, thus, representing a special habitat with reduced salinity (Southern 1921). The peculiarity of the hydrographical conditions might be reflected in the small size of the specimens from Chilka Lake (3–4 mm; 16 chaetigers).

The present Mediterranean specimens of *M. fragilis* differ from those of the North Atlantic, as described by Nilsen & Holthe (1985), in the following two characters: longest chaetigers are sixth to eighth instead of fifth to sixth, and pygidial lobes number 3–5 versus 5–6 in Mediterranean and North Atlantic specimens, respectively (Table 2). Considering the present knowledge of the taxon and the unrevised status of the genus *Myriochele*, these differences may rather represent variations between populations than the basis for the description of a new taxon. A comparison of the 20 species currently described for the genus *Myriochele* revealed that many descriptions are not very detailed and a revision of the group is badly needed.

Distribution

North Atlantic to North of Spitzbergen, 800–2600 m;

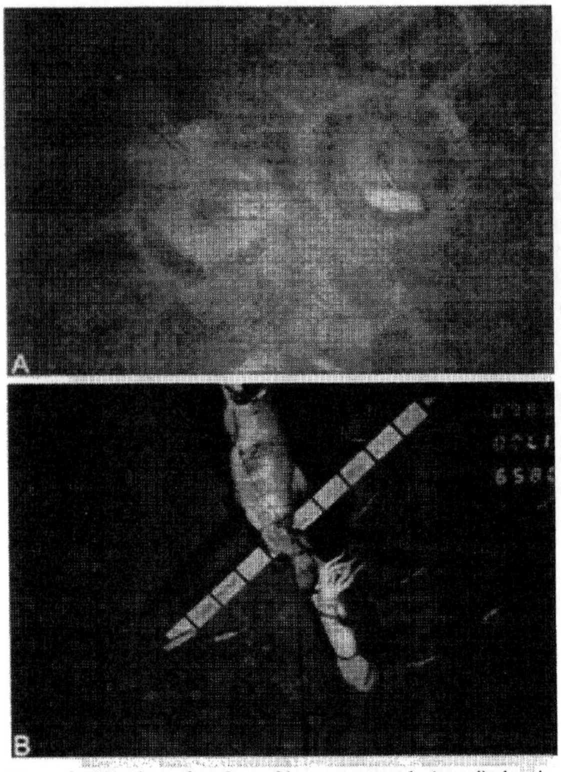

Figure 2. (A) view of surface of box core sample (stn. 4) showing tubes of *Myriochele fragilis* and grazed surroundings; diameter of circles ca. 4–5 cm; (B) picture of the sea bottom (34° 25.89′ N, 26° 08.79′ E; 4172 m) showing areas grazed by *Myriochele fragilis* in the background (photo courtesy of Emma Jones, Aberdeen).

eastern Mediterranean Sea, 4159–4260 m. Like *Protis arctica*, *M. fragilis* is a North Atlantic deep water species with its distribution extending into the eastern Mediterranean Sea (see Ben-Eliahu & Fiege, 1996).

Biological notes

The tubes rise up to 3 cm above the sediment surface. Each tube is surrounded by a conspicuous, slightly depressed circle showing a yellowish-coloured surface in contrast to the adjacent greyish deep sea sediment (Figure 2A). These differences in color probably result from the worms feeding activities on the sediment surface within these circles, as can be seen from the intestine being filled with fine sediment and the radius of the circles equaling the length of the free part of the tubes.

Abundance of Myriochele fragilis

The abundance of *M. fragilis* ranged between 116 individuals per m^2 at station 4 and 72 individuals per m^2 at station 5. At station 6 only empty tubes were found and at station 7 the species did not occur. Biomass varied between 344 and 212 mg wet weight per m^2 at stations 4 and 5.

Total organic carbon content in surface sediments

The total organic carbon content was measured in the surface sediments. 5.4 mg g^{-1} sediment (dry weight) were determined at station 4, 5.7 mg g^{-1} at station 5 and 5.8 mg g^{-1} at station 6.

Discussion

The occurrence of up to 116 individuals per m^2 of *Myriochele fragilis* calculated from one sample is notable due to the general impression of low faunal density in the deep basins of the eastern Mediterranean Sea (Fredj & Laubier, 1985; Janssen, 1989, Tselepides & Eleftheriou, 1992). Moreover, these polychaetes were not found in the Ierapetra Basin during a previous cruise (Meteor cruise 25, leg 1) in summer 1993 using the same gear. Resampling of station 7 during our recent cruise – a location visited also during the earlier Meteor cruise – similarly did not find such polychaetes.

Similar or even higher densities for *Myriochele* species have so far only been reported for *Myriochele oculata* in shallow waters (18–22 m) of the Øresund (Hornbaek Bay) with 173 specimens per m^2 (Kirkegaard, 1978) and in the vicinity of the Doggerbank in the North Sea with up to 348 per m^2 (Kirkegaard, 1969) and 707 per m^2 (Künitzer, 1990).

Since circles of grazed sediment around worm tubes are also visible in pictures of the sea bottom taken by a camera during two large foodfall experiments at 34° 24.85′ N, 26° 8.00′ E (4264 m) and 34° 25.89′ N, 26° 08.79′ E (4172 m) (Figure 2B) (E. Jones, pers. com.), conducted in parallel to our benthos sampling at the eastern slope of the basin, the possibility of circles being artificially formed by swaying movements of the box core during heaving can be neglected.

The distance of about 4 nautical miles between our benthos stations and the location of the two foodfall experiments suggests that these polychaete assemblages may be widely distributed in the Ierapetra Basin. But because no individuals were found at stations 6 and 7, which are located in the more plain part of the basin, the worms seem to occur mainly in a patchy distribution at the bottom of the slopes and not in the plain basin. Therefore, this distribution pattern might be caused by the food supply via lateral transport of organic matter or by turbidity currents. The TOC contents of the surface sediments are similar to the ones measured in the same area in summer 1993 (Boetius et al., 1996). Nevertheless, a supply of fresh organic material cannot be detected by analysing TOC content alone, but Boetius et al. (1996) found also higher values of chloroplastic pigments. Therefore in accordance with these authors, who found that the microbial activity in the deep trenches, e.g., the Ierapetra Basin, were higher than at the abyssal plains, we assume that the increased abundance of *Myriochele fragilis* may be dependent on the lateral influx of fresh organic material.

Beside *M. fragilis* only few individuals of other macrofaunal groups were found thus expressing low diversity. This result confirms the findings of Tselepides & Eleftheriou (1992) that macrofauna decreases with increasing depth along the slope north of Crete. Biomass values including *M. fragilis* specimens are similar to those in the Arctic deep sea under permanent ice cover (Kröncke, 1994, 1998). Excluding the polychaete, biomass values would be much lower than in the central Arctic Ocean and other deep sea areas (Rowe, 1983). This confirms the extreme oligotrophy of the Mediterranean deep sea and the importance of the lateral flux of fresh organic material into these abyssal regions.

Acknowledgements

We thank Captain Papenhagen and crew of RV Meteor, all our colleagues on board for their help in sampling and Iannis Karakassis from the Institute for Marine Biology in Crete, and an anonymous reviewer for their helpful comments. Emma Jones, University of Aberdeen, kindly provided Figure 2B.

References

Ben-Eliahu, M. N. & D. Fiege, 1995. Polychaeta from the continental shelf and slope of Israel collected by the 'Meteor' 5 expedition (1987). Senckenberg. marit. 25: 85–105.

Ben-Eliahu, M. N. & D. Fiege, 1996. Serpulid tube-worms (Annelida: Polychaeta) of the central and eastern Mediteranean with particular attention to the Levant Basin. Senckenberg. marit. 28: 1–51.

Boetius, A., S. Scheibe, A. Tselepides & H. Thiel, 1996. Microbial biomass and activities in deep-sea sediments of the eastern Mediterranean: trenches are benthic hotspots. Deep Sea Res. I. 43: 1439–1460.

Cognetti-Variale, A. M., 1979. Su due policheti owenidi di sabbie infralitorali del Golfo di Follonica. Atti Soc. Tosc. Sci. Nat. Mem., ser. B 86, suppl.: 263–267.

Fiege, D., A. Allspach, M. Apel, R. Barnich, C. O. Coleman, S. Doulgeraki, T. Jellinek, M. Kitsos, A. Koukouras, I. Kröncke & W. Rosenboom, 1999. Macro- and megabenthos. In: Hieke, W., Ch. Hemleben, P. Linke, M. Türkay & H. Weikert (eds), Mittelmeer 1997/98, Cruise No. 40, 28 October 1997–10 February 1998. METEOR-Berichte, Universität Hamburg, 99-2: 114–116.

Fredj, G. & L. Laubier, 1985. The deep Mediterranean benthos. In: Moraitu-Apostolopoulou, M. & V. Kiortsis (eds), Mediterranean Marine Ecosystems. Plenum Press, New York: 109–145.

Janssen, R., 1989. Benthos-Mollusken aus dem Tiefenwasser des östlichen Mittelmeeres, gesammelt während der 'Meteor'-Fahrt 5 (1987). Senckenberg. marit. 20: 265–276.

Kirkegaard, J. B., 1969. A quantitative investigation of the central North Sea. Polychaeta. Spolia 29: 1–285.

Kirkegaard, J. B., 1978. Settling, growth and life span of some common polychaetes from Danish waters. Medd. Danmarks Fisk. Havundersog. N. S. 7: 447–496.

Kröncke, I., 1994. Macrobenthos composition, abundance and biomass in the Arctic Ocean along a transect between Svalbard and the Makarov Basin. Polar Biol. 14: 519–529.

Kröncke, I., 1998. Macrofauna communities in the Amundsen Basin, at the Morris Jesup Rise and at the Yermak Plateau (Eurasian Arctic Ocean). Polar Biol. 19: 383–392.

Künitzer, A., 1990. Zur Verbreitung und Populationsdynamik der Bodenfauna der zentralen Nordsee. Dissertation, Universität Bremen, 218 pp.

Laubier, L. & J. Paris, 1962. Faune marine des Pyrénées-Orientales. IV. Annélides polychètes. Vie Milieu 13: 11–80.

Martìn, D., 1989. Revisión de las especies de Owenidae (Annelida, Polychaeta) de la Península Ibérica. Sci. mar. 53: 47–52.

Nilsen, R. & T. Holthe, 1985. Arctic and Scandinavian Oweniidae (Polychaeta) with a description of *Myriochele fragilis* sp. n., and comments on the phylogeny of the family. Sarsia 70: 17–32.

Rowe, G. T., 1983. Biomass and production of the deep sea macrobenthos. In: Rowe, G. (ed.) Deep-Sea Biology. Wiley Interscience, New York: 97–121.

Southern, R., 1921. Fauna of the Chilka Lake. Polychaeta of the Chilka Lake and also of fresh and brackish waters in other parts of India. Mem. Indian Museum Calcutta 5: 563–693.

Tselepides, A. & A. Eleftheriou, 1992. South Aegean (Eastern Mediterranean) continental slope benthos: macroinfaunal-environmental relationships. In: Rowe, G. T. & V. Pariente (eds), Deep-Sea Food Chains and the Global Carbon Cycle. Kluwer Academic Publishers, Dordrecht, Boston, London: 139–156.

Hydrobiologia **426:** 105–112, 2000.
G. Liebezeit, S. Dittmann & I. Kröncke (eds), Life at Interfaces and Under Extreme Conditions.

The vertical distribution of *Chthamalus montagui* and *Chthamalus stellatus* (Crustacea, Cirripedia) in two areas of the NW Mediterranean Sea

F. G. Pannacciulli* & G. Relini
Laboratori di Biologia Marina ed Ecologia Animale, Dipartimento per lo Studio del Territorio e delle sue Risorse, Università di Genova, Via Balbi 5, Genova, 16126, Italy
*E-mail: fgp@unige.it (*author for correspondence)*

Key words: barnacle, *Chthamalus montagui*, *Chthamalus stellatus*, distribution, rocky shore, Mediterranean Sea

Abstract

The Mediterranean Sea is characterised by a small tidal range (0.3–1 m). Despite this, intertidal communities are well established and their upper limits often extend above mean high water level. Organisms living in the intertidal region and in the supralittoral zone rely on both tides and wave action to perform their biological functions. Lack of food, desiccation and predation are common stresses in such a harsh environment. The present study deals with the vertical distribution of two species of intertidal barnacles, *Chthamalus montagui* Southward and *Chthamalus stellatus* (Poli), which are the main constituents of the barnacle belt along Mediterranean rocky shores. Previous work, carried out in the Atlantic, showed that where the distribution ranges of the two *Chthamalus* species overlap, *C. montagui* is more common in the upper barnacle zone while *C. stellatus* is dominant lower down. The main aims of our study are: (1) to establish if there is a relationship between position and extension of the barnacle belt on the shore and tidal range and/or wave exposure, (2) to test the hypothesis that in the study areas *C. montagui* is more abundant than *C. stellatus* high on the shore, and that the pattern is reversed lower down. Barnacle populations were monitored in summer 1998 in the Gulf of Genoa (Ligurian Sea) and in the Gulf of Trieste (North-Adriatic Sea). The two areas differ in tidal range and hydrodynamism, the former presenting quite strong wave action and a tidal range of 30 cm, the latter having limited wave action and 1 m tidal range. Three shores were randomly selected in each gulf and two transects on each shore. Counts of barnacles in 10 * 10 cm quadrats were done at different shore heights along each transect. The data was subjected to analysis of variance. Results showed that a more pronounced hydrodynamic regime corresponded to a shift of the barnacle belt towards the higher shore (Gulf of Genoa), while in more sheltered areas (Gulf of Trieste), the barnacle distribution was confined to the intertidal region. The relative spatial distribution of *C. montagui* and *C. stellatus* within the barnacle belt varied locally, even between transects on the same shore, and this obscured the distribution pattern along the vertical gradient. Nevertheless, it was still possible to conclude that at mid and high shore in Genoa, *C. stellatus* was more abundant than *C. montagui*, while in Trieste the pattern was reversed.

Introduction

Barnacles are one of the dominant components of littoral communities. Because of their abundance and accessibility on the shore, their ecology has been the subject of studies for over 100 years. Darwin (1851, 1854) fully realised that many species of cirripedes had a strong intertidal preference. According to Lewis (1964), barnacles are almost ubiquitous and are the dominant and most characteristic organisms of the eulittoral zone throughout the world.

Among different species of intertidal barnacles, those belonging to the genus *Chthamalus* are very common along the littoral zone of the NE Atlantic and the Mediterranean shores. Until 1976, *Chthamalus stellatus* (Poli) was considered to be the only representative of this genus in Europe (with the exception of *Euraphia depressa* (Poli) which in the past was erroneously considered to belong to it). Several authors (Moore & Kitching, 1939; Southward, 1950, 1963; Connell, 1961) described distribution of the *C. stellatus*, along latitudinal and wave exposure gradients, and related its vertical distribution on the shore to several factors including competitive interactions with other species (e.g. *Semibalanus balanoides* (L.) in the Atlantic). Other authors (Barnes & Barnes, 1964; Kensler et al., 1965; Klepal, 1971; Relini, 1983) concentrated their attention on the horizontal and vertical distribution of *C. stellatus* in the Mediterranean and its competition with *Euraphia depressa*.

With regard to *Chthamalus* distribution, the conclusion emerging from these studies is that physical factors limit the distribution at the upper part of the shore, while the combination of physical and biological factors operates at the lower level. In fact, brief exposure to seawater and consequent problems with desiccation and poor food supply would be the main constraints for individuals living higher on the shore, while competition for space, predation and strong wave action would be limiting factors at mid and low shore. Furthermore, according to Hawkins & Hartnoll (1982), the presence of algal turf would set the lower limit of the barnacle distribution.

Southward (1976) separated the species *Chthamalus montagui* Southward from *C. stellatus*. Following this separation, Crisp et al. (1981) and Burrows (1988) investigated the relative distribution of the two *Chthamalus* species along geographical and vertical gradients. From their work, it appears that the two species can occasionally be separated by habitat, with *C. stellatus* being dominant on exposed shores and *C. montagui* on moderately sheltered ones. Where the two species overlap, *C. montagui* is more common in the upper barnacle zone (MHWS and MHWN), while *C. stellatus* is dominant lower down (MTL and below). Occasionally, in wave-washed, wet or shaded places, *C. stellatus* can occur higher up on the shore. According to Burrows (1988), processes of larval dispersal, larval development and settlement appear to be the most important factors producing the different adult distributions in the two species. Juvenile mortality could also account for this. In fact, due to morphological differences in the shape of the opercular plates, *C. stellatus* is believed to be more susceptible to desiccation during early life stages than *C. montagui* (Foster, 1971). In addition, *C. montagui* juveniles seem to require exposure to air in order to be able to consolidate their shell plates (Burrows, 1988).

The Mediterranean Sea, in common with most enclosed seas, is characterised by a small tidal range (0.3–1 m). Despite this, shore communities are well established in this area and their upper limits are often extended above high water level. Organisms living in the intertidal region and in the supralittoral zone rely on tides and wave action to perform their biological functions.

The present study was carried out on six shores along the northern Italian coast (Mediterranean Sea) and deals with the vertical distribution of the main constituents of the barnacle belt: *C. montagui* and *C. stellatus*. *Euraphia depressa*, a species endemic to this basin, is the third barnacle species found along Mediterranean shores. In the areas of study, *Euraphia* lives in crevices and shadowed areas of the supralittoral zone and seldom enters in the formation of the barnacle belt on open rocks. For this reason, *Euraphia* has not been included in the investigations.

This study had two main aims:

1. to establish if there is a relationship between position and extension of the barnacle belt on the shore and tidal range and/or wave exposure;
2. to test the hypothesis that in the study areas, *C. montagui* is more abundant than *C. stellatus* high on the shore, and that the pattern is reversed lower down.

Materials and methods

Study sites

The study was carried out in two areas of the NW Mediterranean Sea: the Gulf of Genoa (44° 24′ N, 8° 54′ E) in the Ligurian Sea and the Gulf of Trieste (45° 39′ N, 13° 46′ E) in the north part of the Adriatic Sea, Italy (Figure 1). The two areas differ in tidal range and hydrodynamism, the former presenting quite strong wave action and a tidal range of 30 cm, the latter having more limited hydrodynamism and 1 m tidal range. Wave action is determined by several factors including wind direction, local topography and hydrography. In Genoa, dominant winds blow offshore (from north) and inshore (from south-west); the inshore wind, in conjunction with the open coast and a steep continental shelf, generates high hydrodynamic forces in the area. In contrast, the more embayed situation of the Gulf of Trieste, together with dominant winds blowing offshore (from north-east) and very shallow waters (50 m on average) create a more sheltered condition.

Table 1. ANOVA on the effect of shore height on *Chthamalus* species distribution

Source of variation	DF	MS	F
Species = Sp	1	66.67	
Height = He	2	86 443.56	
Locality = Lo	1	154 989.80	
Shore (Lo)	4	7875.12	
Transect (Lo × Shore)	6	5910.06	
Sp × He	2	2011.10	
Sp × Lo	1	173 853.63	
Sp × Shore (Lo)	4	34 751.61	
Sp × Transect (Lo × Shore)	6	3138.32	
He × Lo	2	18 052.59	
He × Shore (Lo)	8	11 330.16	
He × Transect (Lo × Shore)	12	3195.66	
Sp × He × Lo	2	42 473.14[b]	51.33**
Sp × He × Shore (Lo)[b]	8	827.47	
Sp × He × Transect (Lo × Shore)	12	4164.42[a]	5.16**
Residual[a]	144	807.76	
Cochran's test	$C = 0.1091$, not significant		
Transformation	none		

Superscripts are used to identify the appropriate denominator for each test. The term in the model sharing the superscript of any mean square is the term whose mean square is used in the denominator of the F-ratio for the test.

Methods

The study was carried out in summer 1998. To test for local spatial variation, three shores (Figure 1) were randomly selected in each gulf (S. Chiara, Pontetto and P. Chiappa in the Gulf of Genoa and Barcola, S. Croce and Aurisina in the Gulf of Trieste) and two transects (A and B) on each shore. Shores were selected from those presenting barnacles but not showing peculiarities; they were a few kilometres apart and covered the range of exposure typical of the area. With regard to this last feature, shores in Trieste were always quite sheltered, while those in Genoa had a wide range of exposure (in our study S. Chiara was the most sheltered one, while P. Chiappa was the most exposed). Transects were selected on open rock, were 10–100 m apart and stretched from the lower to the upper limit of the barnacle belt. Barnacles were considered to have reached their distribution limit where densities of less than 1 ind. m^{-2} were recorded.

The heights of the upper and lower *Chthamalus* spp. zone were set with reference to Chart Datum (CD), values were adjusted in relation to the air pressure present at the time of monitoring. In microtidal environments, variation of the mean sea level due to air pressure can sometimes be greater than that due to tides.

Sampling, to determine the density of the two *Chthamalus* spp., was carried out at five levels set at equal vertical intervals along the transect and within the barnacle zone. Due to the very narrow barnacle distribution present at S. Chiara, it was only possible to sample three levels out of five on this shore. In each transect, three randomly placed 10 * 10 cm replicate quadrats were used to estimate densities for each species separately at each of the five levels. In fact, for data to be independent (one of ANOVA's assumptions) values for one species had to come from replicates different from the ones used for the other species.

To avoid bias related to time of sampling, only adult individuals, belonging to class 1+ or above, were included in the counts.

The distribution pattern of *C. montagui* and *C. stellatus* along the vertical gradient was investigated using a 5-way Analysis of Variance (ANOVA). The factors involved in the analysis were: species (2 levels), shore height (3 levels, from now onwards referred to as height), locality (2 levels), shore (3 levels) and transect (2 levels). The number of replicates for each treatment was three. Species, height and locality were fixed orthogonal factors, while shore was random and nested

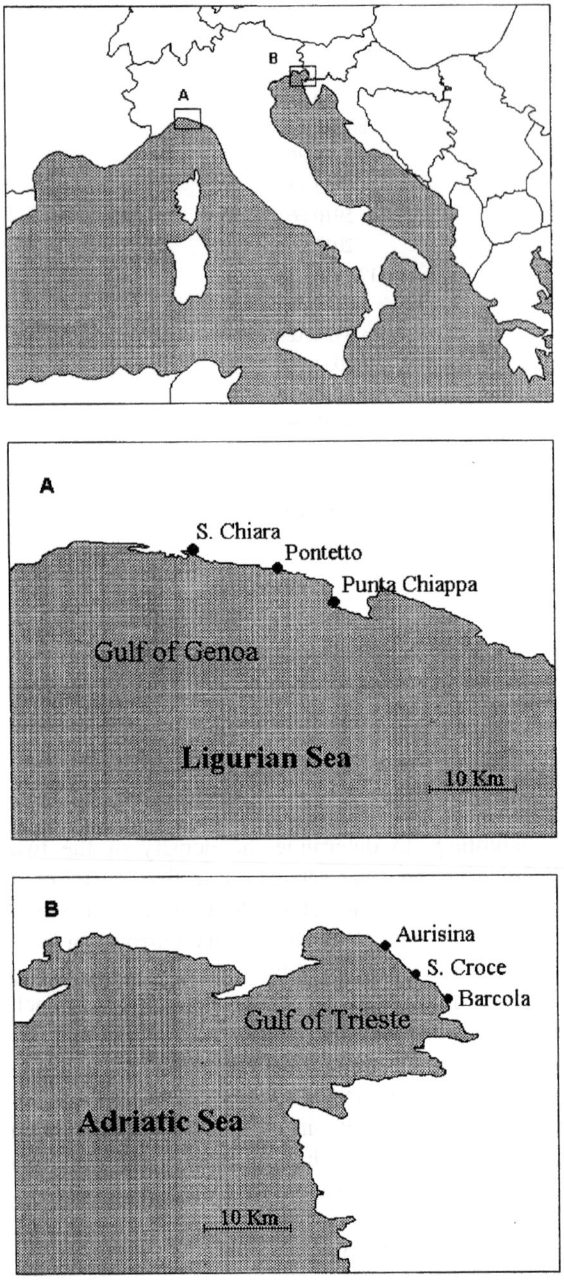

Figure 1. Maps showing the two study sites (Gulf of Genoa (A) and Gulf of Trieste (B)) and the six shores (S. Chiara, Pontetto, etc.) where the study was carried out.

in locality and transect random and nested in shore and locality. The number of levels of factor height was reduced from 5 to 3 due to S. Chiara where only three levels were monitored.

Homogeneity of variance was assessed by Cochran's *C* test (Underwood, 1997). Student-Newman-Keuls (SNK) test ($\alpha = 0.05$) was employed for multiple comparisons of the means. Data were analysed with the programme GMAV5 (Underwood & Chapman, 1998).

Results

The first aim of our study was to establish whether there is a relationship between the position of the barnacle belt on the shore and tidal range and/or wave exposure.

The extension and position of the barnacle belt along a given transect varied between the shores of the two study areas (Figure 2). In the Gulf of Genoa, the vertical distribution of barnacles stretched from 10 cm (S. Chiara transect A) to 205 cm above CD (P. Chiappa transect A). In this area, nearly all barnacles were found above extreme mean high water level (EMHW) which, due to the very limited tidal range, is situated at only 30 cm above CD. In addition, as the exposure gradient increased, the barnacle distribution shifted further up-shore. Barnacle distribution in the Gulf of Trieste was different, the belt stretching from −3 to 105 cm above CD, hence confining *Chthamalus* spp. to the intertidal region. No trend relative to exposure emerged from the data.

The second aim was to test the hypothesis that in the study areas *C. montagui* is more abundant than *C. stellatus* high on the shore, and that the pattern is reversed lower down.

Figures 3 and 4 show the relative distribution of the two species in Genoa and in Trieste. To allow comparisons across shores and localities, vertical levels were related to their position within the barnacle belt (high, mid high, mid, mid low and low) and not to Chart Datum. In the Gulf of Genoa (Figure 3), *C. stellatus* was more abundant than *C. montagui* at most levels. Here, barnacles reached their peak density around mid and mid low shore. In the Gulf of Trieste (Figure 4), individuals of *C. montagui* were far more abundant than those of *C. stellatus* (several hundreds against a few tens). The highest abundance of *C. montagui* was found around mid high level on all shores, while the greatest densities of *C. stellatus* were recorded at low/mid low levels.

The levels high, mid and low were employed in the Analysis of Variance (Table 1). The interaction Sp * He * Transect (Lo * Shore) was shown to be highly

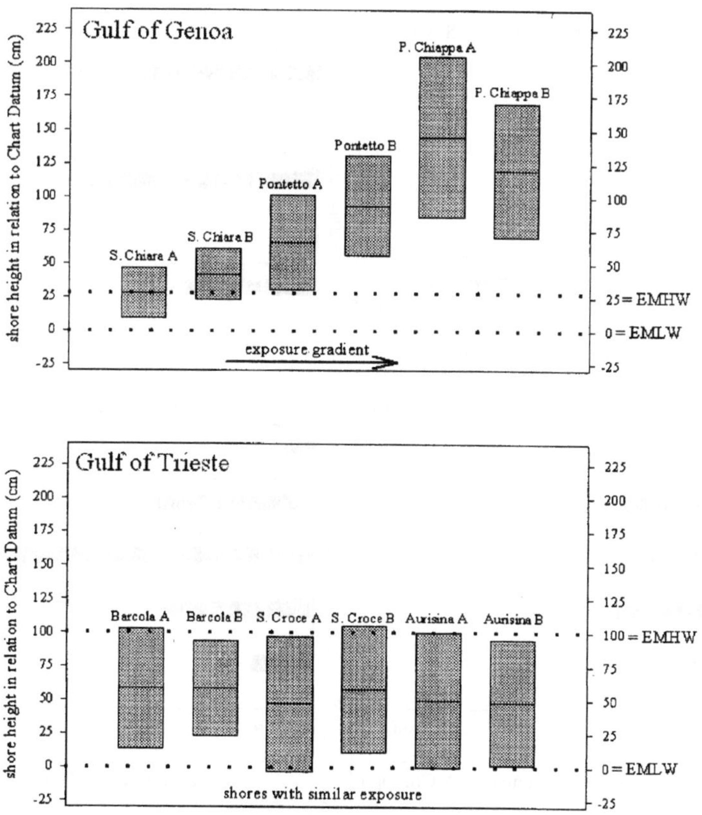

Figure 2. Box plots represent position and extension of the barnacle belt along a given transect (A or B) at one shore (e.g. S. Chiara, Pontetto, etc.). The area between dotted lines represents the intertidal region where: EMHW = Extreme Mean High Water, EMLW = Extreme Mean Low Water.

significant, therefore indicating that even between transects of the same shore the interaction Sp * He changed. Despite this, the interaction Sp * He * Lo was still significant, suggesting that variation between localities was greater than variation between transects. Multiple comparisons carried out on the interaction Sp * He * Lo showed that at mid and high shore in Genoa, *C. stellatus* was significantly more abundant than *C. montagui*, while in Trieste the pattern was reversed. Thus, *C. montagui* was not the dominant barnacle species higher on the shore in both study areas nor was *C. stellatus* the most common one on the lower shore.

Discussion

The position and extension of barnacle belts on the shore are generally related to tidal range and grade of exposure of the coast. In a microtidal environment, such as the Mediterranean Sea with a mean tidal range of 30 cm, wave exposure can be considered to be the main factor responsible for the vertical distribution of barnacles.

Two areas with different tidal range and exposure were selected for this study: the Gulf of Genoa, which can be considered as representative of the NW Mediterranean, and the Gulf of Trieste in the north Adriatic Sea. Genoa is characterised by strong wave action and tidal range of 30 cm, Trieste presents shallow waters, moderate wave action and 1 m tidal range (an exceptional value for the Mediterranean Sea). Results showed that barnacles in Trieste were confined to the intertidal region, while in Genoa, their distribution stretched further up-shore as the rate of exposure increased.

The upper limit of the *Chthamalus* distribution in the Mediterranean is considered to be set by physical factors, as Barnes & Barnes (1964) pointed out. Mediterranean barnacles living on the upper shore have to deal with an extreme environment where a scarce water supply, and consequently a lack of food, together

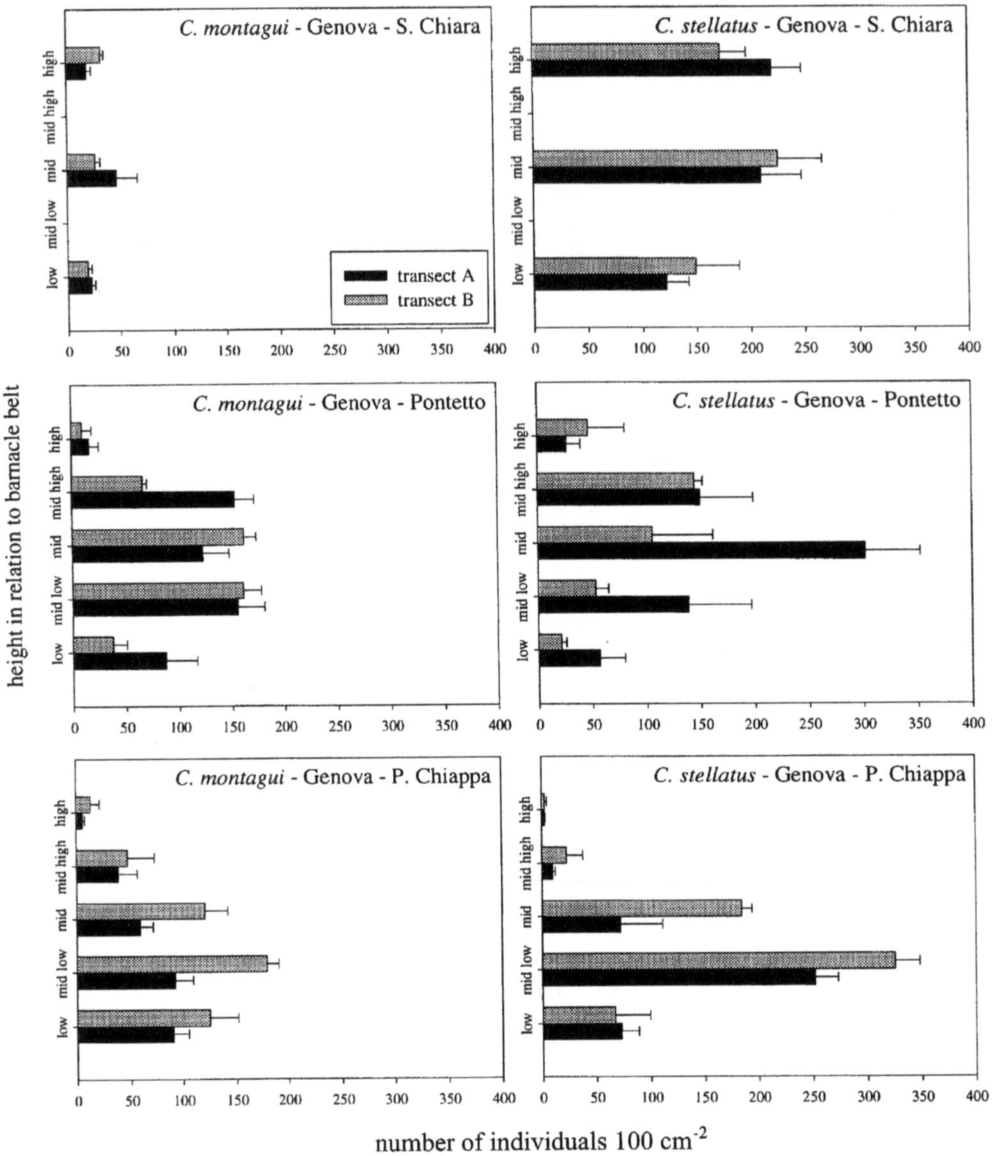

Figure 3. Vertical distribution of *C. montagui* (left) and *C. stellatus* (right) on three shores in the Gulf of Genoa. In each transect, three 10 * 10 cm replicate quadrats were used to estimate densities for each species separately at each of the five levels within the barnacle belt. Mean values and standard errors are reported.

with desiccation and high air temperature limit the distribution. This could explain why *Chthamalus* was absent from the area above the upper tidal level in Trieste, which is seldom reached by water, while it was present in Genoa, where wave action is more intense.

The factors setting the lower limit of *Chthamalus* distribution have been suggested to be either physical or biological. Moore & Kitching (1939) could not decide whether excessive immersion or competition with sublittoral organisms, such as coralline algae or other rock-encrusting forms, determined this limit. Barnes & Barnes (1964) supported the theory of competition, but suggested that local topography and wave action could also be involved. Thus, there can be several reasons why the distribution of *Chthamalus* does not extend lower on the shore in our study areas and various models can be proposed to explain the observed pattern. Among these, competition with algae,

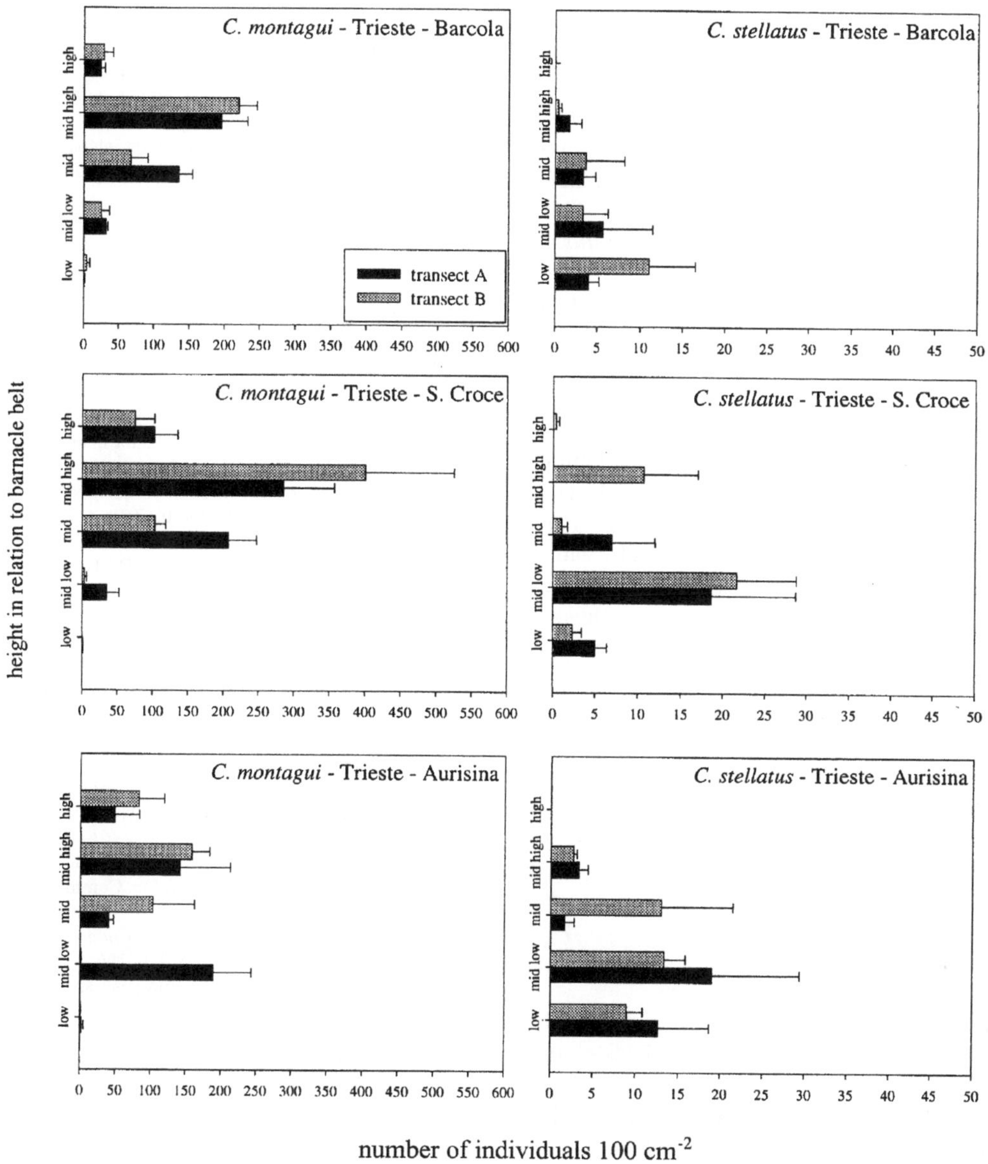

Figure 4. Vertical distribution of *C. montagui* (left) and *C. stellatus* (right) on three shores in the Gulf of Trieste. In each transect, three 10 * 10 cm replicate quadrats were used to estimate densities for each species separately at each of the five levels within the barnacle belt. Mean values and standard errors are reported.

supply-side processes, settlement behaviour and predation should be considered and investigated in future studies.

Little is known as to the relative distribution of *C. montagui* and *C. stellatus* along the vertical gradient on Mediterranean rocky shores. Since the separation of *C. montagui* from *C. stellatus* by Southward in 1976, most of the studies related to the distribution of the two species were carried out in the Atlantic (Southward, 1976; Crisp et al., 1981, Burrows, 1988; Pannacciulli, 1995). A brief account on the distribution in the Mediterranean was published by Crisp et al. (1981) and Relini (1983). Our work, therefore, aimed to expand their investigations and verify whether the pattern of *Chthamalus* distribution described for the Atlantic was also valid in the Mediterranean.

Our investigations showed that the spatial variation in the two study areas was high with interactions between species and shore height changing even from one transect to the other. Nevertheless, some conclusions could still be drawn.

In Genoa, where wave action prevails over tides, the two *Chthamalus* spp. appeared to be equally distributed along the vertical gradient, although *C. stellatus* was in general more abundant at all levels. This result contrasts with the assumption that *C. stellatus* is dominant at MTL and below (Southward, 1976). On the other hand, the same author also quotes that in wave-beaten places, such as the west coast of Ireland, *C. stellatus* can form 30–70% of the total *Chthamalus* population. This last consideration agrees well with our observations on the wave exposed coast of Genoa.

In contrast, in Trieste, *C. montagui* was the dominant species at all levels and showed peaks in density in the upper part of the shore. Conversely, *C. stellatus* appeared to be very scarce above MTL, while its abundance increased lower down. These results are quite consistent with the suggestion Southward (1976), i.e. *C. montagui* is the dominant species in sheltered places and where the two *Chthamalus* spp. co-occur, *C. montagui* is usually more abundant at MHW.

A more extensive study, on a broader geographical scale, is needed to clarify the pattern of vertical distribution of *C. montagui* and *C. stellatus* on the shore.

Acknowledgements

The authors wish to thank Prof. Mario Specchi for the logistic support provided in Trieste and Maura Maselli, Maria Chiara Casagrande and Paula Moschella for help with the fieldwork. We are also grateful to Prof. A. J. Southward and Prof. A. J. Underwood for useful comments on the manuscript. This study was funded by the E.U. (EUROROCK project - MAS3-CT95-0012).

References

Barnes, H. & M. Barnes, 1964. Répartition et écologie générale des Cirripèdes *Chthamalus depressus* (Poli) et *C. stellatus* (Poli) le long des côtes méditerranéennes d'Espagne et de France. Bull. Inst. océanogr. Monaco 62: 1–19.

Burrows, M. T., 1988. The comparative biology of *Chthamalus stellatus* (Poli) and *Chthamalus montagui* Southward. Ph. D. Thesis, University of Manchester: 318 pp.

Connell, J. H., 1961. The influence of interspecific competition and other factors on the distribution of the barnacle *Chthamalus stellatus*. Ecology 42: 710–723.

Crisp, D. J., A. J. Southward & E. C. Southward, 1981. On the distribution of the intertidal barnacles *Chthamalus stellatus*, *Chthamalus montagui* and *Euraphia depressa*. J. mar. biol. Ass. U.K. 61: 359–380.

Darwin, C., 1851. A Monograph on the sub-class Cirripedia. The Lepadidae; or, pedunculated cirripedes. Ray Society, London: 400 pp.

Darwin, C., 1854. A Monograph on the sub-class Cirripedia. The Balanidae; the Verrucidae, etc. Ray Society, London: 684 pp.

Foster, B. A., 1971. Desiccation as a factor in the intertidal zonation of barnacles. Mar. Biol. 8: 12–29.

Hawkins, S. J. & R. G. Hartnoll, 1982. Settlement patterns of *Semibalanus balanoides* (L.) in the Isle of Man (1977–1981). J. exp. mar. Biol. Ecol. 62: 271–283.

Kensler, C. B., K. M. Bhatnagar & D. J. Crisp, 1965. Distribution and ecological variation of *Chthamalus* species in the Mediterranean area. Vie Milieu (sér B) 16: 271–294.

Klepal, W., 1971. *Chthamalus stellatus* (Poli) und *C. depressus* (Poli) in der Adria. J. exp. mar. Biol. Ecol. 7: 271–294.

Lewis, J. R., 1964. The Ecology of Rocky Shores. The English Universities Press Ltd., London: 323 pp.

Moore, H. B. & J. A. Kitching, 1939. The biology of *Chthamalus stellatus* (Poli). J. mar. biol. Ass. U. K. 23: 521–541.

Pannacciulli, F. G., 1995. Population ecology and genetics of European species of intertidal barnacles. Ph. D. Thesis, University of Liverpool: 143 pp.

Relini, G., 1983. Remarks on the ecology of chthamalids in the Ligurian Sea. Rapp. Proc.Verb. Réun. Comm. Int. Explor. Scient. Mer Méditerranée 28: 273–275.

Southward, A. J., 1950. Distribution of *Chthamalus stellatus* on the shores of North-East Ireland. Nature 166: 311.

Southward, A. J., 1963. On the European species of *Chthamalus* (Cirripedia). Crustaceana 6: 241–254.

Southward, A. J., 1976. On the taxonomic status and distribution of *Chthamalus stellatus* (Cirripedia) in the North-East Atlantic region: with a key to the common intertidal barnacles of Britain. J. mar. biol. Ass. U. K. 56: 1007–1028.

Underwood, A. J., 1997. Experiments in Ecology – Their Logical Design and Interpretation using Analysis of Variance. Cambridge University Press, Cambridge: 504 pp.

Underwood, A. J. & M. G. Chapman, 1998. GMAV5 for Windows – An analysis of variance programme. Institute of Marine Ecology, University of Sydney, Australia.

Hydrobiologia **426:** 113–121, 2000.
G. Liebezeit, S. Dittmann & I. Kröncke (eds), Life at Interfaces and Under Extreme Conditions.

First survey of sessile communities on subtidal rocks in an area with hydrothermal vents: Milos Island, Aegean Sea

Silvia Cocito[1,*], C. Nike Bianchi[1], Carla Morri[2] & Andrea Peirano[1]
[1]*Marine Environment Research Centre, ENEA Santa Teresa, P.O. Box 316, I-19100 La Spezia, Italy*
*(*author for correspondence)*
[2]*DIPTERIS, Università di Genova, Corso Europa 26, I-16132 Genova, Italy*

Key words: hard substrate epibenthos, community zonation, shallow hydrothermal vents, bioconstruction, Aegean Sea

Abstract

The major epibenthic communities on subtidal rocks of Palaeochori Bay and the marine tract on the southern coast of Milos Island (Greece) were described down to 44 m depth. Six sites were investigated by snorkelling and SCUBA diving. Samples, photographs and video images were also taken to integrate information. Three out of the six sites were close to hydrothermal vents, a common feature in the area. In total, nine major epibenthic communities were found, most of which were characterised by a diverse algal growth down to the maximum depth explored. Macrobenthic cover was severely reduced only in the close proximity of vents, a white flocculent bacterial mat covering the rock at the point from which fluid escaped. Large-scale effects of vents on the epibenthic communities were not detected. However, the abundance of species with warm-water affinity was recognisable in both algal and animal dominated communities, which may be related to higher winter temperature in the vent area. Epifaunal communities under overhangs were composed of distinct groups of suspension feeders at vent as compared to non-vent sites: this might indicate differences in trophic conditions. Mounds of the bioconstructional coralline alga *Mesophyllum lichenoides* were conspicuous only at vent sites, thus suggesting enhanced biodeposition of carbonates due to vent activity.

Introduction

The discovery of deep-water hydrothermal vents in the last decades revealed unusual benthic communities dependent on chemosynthesis (Tunnicliffe, 1991). Communities were more recently found also around vents in shallow-water and continental shelf areas, which showed analogies with their deep-sea counterparts but also contrasting patterns (Fricke et al., 1989; Tarasov et al., 1990; Dando et al., 1991; Giménez & Marín, 1991; Hashimoto et al. 1993; Kamenev et al., 1993; Acunto et al., 1996).

In the Aegean Sea, hydrothermal vents occur along the so-called Hellenic Volcanic Arc, an area with strong volcanic activity (Fytikas et al., 1984). Most published studies on benthic communities around vents in this area deal with soft-bottoms (Diapoulis et al., 1994a; Dando et al., 1995a; Thiermann et al., 1997), whereas epibenthic communities on hard substrates have received little attention.

In this paper, we describe the sessile epibenthic communities on subtidal rocks at Milos, an island in the Hellenic Volcanic Arc where hydrothermal activities have been more intensively studied (Dando et al., 1995b). Previous studies on hard-bottom benthos at Milos took into account algal communities only, and the effect of vents was not investigated (Coppejans, 1974; Diapoulis et al., 1994b).

Materials and methods

Sessile epibenthic communities were investigated in June 1996 in Palaeochori Bay and in the near off-shore tract on the southern coast of Milos (Bianchi et al., 1997). Palaeochori Bay is a sandy bay approximately 1.4 km wide. The bottom of the bay is covered by sea-

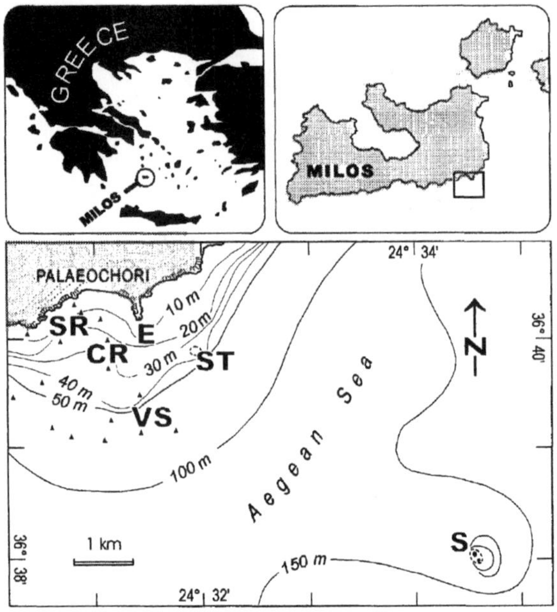

Figure 1. Geographical setting of the study area and location of study sites.

grass beds, *Cymodocea nodosa* in the inner part (4 m – 15 m depth) and *Posidonia oceanica* in the outer part and extending further outside (Aliani et al., 1998).

Communities were identified physiognomically by snorkelling and SCUBA diving at depths ranging from the surface down to about 45 m. Divers annotated on a diving slate biological observations, together with slope, bearing and depth (Bianchi et al., 1991). Samples were collected to identify cryptic species and epiphytes and species not easily recognisable underwater (Bianchi et al., 1999). UW photographs and video records, both by hand camera and ROV, were also taken to integrate information on the distribution of conspicuous species, easily recognisable on images. Species were grouped according to their role as space occupiers: primary space occupiers, cryptic fauna and epiphytes. Epibenthic communities identification was mainly based on primary space occupiers, generally corresponding to conspicuous species.

Six sites were studied (Figure 1): SR (36° 40.20′ N, 24° 30.64′ E); CR (36° 39.73′ N, 24° 31.24′ E); VS (36° 39.55′ N, 24° 31.37′ E); E (36° 40.22′ N, 24° 31.61′ E); ST (36° 40.01′ N, 24° 32.14′ E); and S (36° 38.14′ N, 24° 34.50′ E). The position of main vent systems is mapped in Dando et al. (1995b). Although all sites were located in a hydrothermally active area, the first three, i.e. SR, CR and VS, were closest to actual vents, and emission of gas bubbles was observed there during the dives.

Results

Description of communities:

Nine major epibenthic communities were recognised by divers and/or on photos and video records and through integration with sample identification. They were mainly distinguished by algal or animal dominance and were differently distributed according to depth and substrate slope. These communities were identified following Pérès & Picard (1964), Boudouresque (1984), Bellan-Santini (1985) and Giaccone et al. (1994a, b) and named according to Augier (1982).

The community of Exposed or Semi-Exposed Photophilic Algae (ESEPA) was characterised essentially by the two algal species *Cystoseira compressa* and *Corallina elongata* and was found in the first meter of depth.

The community of Infralittoral Photophilic and Thermophilic Algae (IPTA) occurred in sheltered and well-lit water, especially on boulders and low rocks partly covered with sand. In some cases, this community was dominated by *Cystoseira crinita*, a species widely distributed in the Aegean (Huvé, 1972). Other important algal species were *Amphiroa rigida*, *Cladostephus spongiosus*, *Padina pavonica*, *Acetabularia acetabulum*, *Anadyomene stellata*, and *Dasycladus vermicularis.*

The community of relatively Calm-water Infralittoral Photophilic Algae (CIPA) preferred horizontal or gently sloping rocks in depths within 25 – 30 m. Mostly brown algae were represented, especially with different species of *Cystoseira* (*C. barbata*, *C. brachycarpa*, *C. corniculata*, *C. schiffneri*, *C. spinosa*) according to their specific depth-ranges (Giaccone & Bruni, 1973). Other important algal species were *Amphiroa rigida*, *Jania rubens*, *Dictyopteris polypodioides*, *Dictyota dichotoma*, *Halopteris scoparia*, *Padina pavonica*, and *Sargassum vulgare*, while the only conspicuous invertebrate was the sponge *Axinella verrucosa.*

The community of Circalittoral Calm-water Sciaphilic Algae (CCSA) occurred on deeper substrates, 30 to more than 40 m depth. The most conspicuous species was *Sargassum acinarium*, which sheltered a lower-storey with red (*Beckerella mediterranea*, *Jania adhaerens*, *Mesophyllum lichenoides*, *Peyssonnelia*

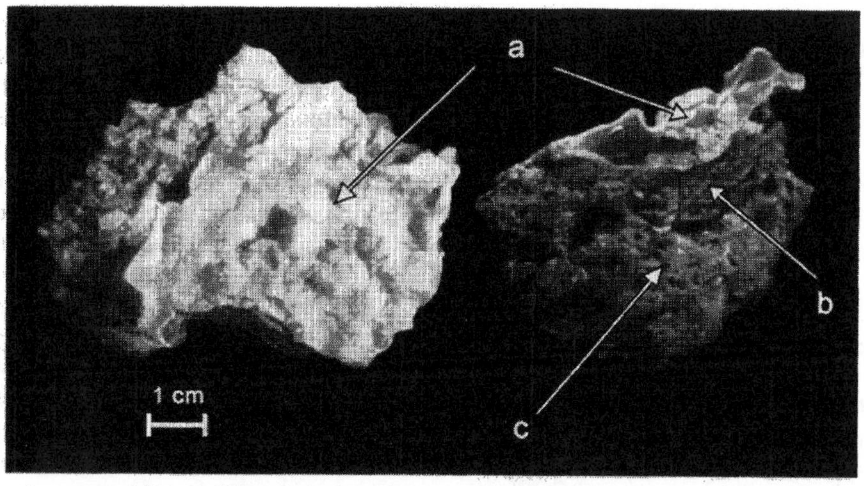

Figure 2. Fragments of a mound of *Mesophyllum lichenoides* collected at the vent site VS: **a:** living algal laminae on the outer portion of the fragments; **b:** dead laminae undergoing early diagenesis; **c:** inner lithified portions.

polymorpha, *Rhodophyllis divaricata*, *Sphaerococcus coronopifolius*), brown (*Asperococcus bullosus*, *Halopteris filicina*, *Spatoglossum solieri*, *Stypopodium schimperi*, *Zanardinia prototypus*, *Zonaria tournefortii*) and green (*Codium effusum*) algae. Animals were poorly represented among the primary space occupiers and included sponges (*Axinella cannabina*, *Cliona nigricans*), sea-urchins (*Sphaerechinus granularis*) and ascidians (*Molgula manhattensis*). The encrusting coralline alga *Mesophyllum lichenoides*, an important builder in the Mediterranean (Sartoretto, 1994), was responsible for the construction of multilayered mounds, several cm thick (Figure 2), sheltering a rich cryptic fauna. Heavily calcified colonies of the encrusting bryozoans *Monoporella nodulifera* and *Calpensia nobilis* throve below the laminae of the alga, whereas *Mollia patellaria* settled on the outer surface. Intertwined network of calcareous tubes of serpuloidean polychaetes (*Spirobranchus polytrema*, *Janita fimbriata*, *Semivermilia crenata*, *Vermiliopsis labiata*) alternated with wrapping encrusting bryozoans (*Puellina arrecta*, *Schizomavella auriculata auriculata*), thus participating in the bioconstruction. Brachiopods (*Megathiris detruncata*, *Neocrania anomala*) and bivalves (*Hiatella arctica*) inhabitated the numerous small cavities inside the mounds, adding their calcareous shells to the bioconstruction. Lithified portions, with some mineral inclusions, formed the inner parts of the mounds.

The community of Infralittoral Calm-water Sciaphilic Algae (ICSA) typically throve on steep to vertical substrates. The most important algal species were *Flabellia petiolata* and *Halimeda tuna*, together with *Acrosymphyton purpuriferum*, *Amphiroa cryptarthrodia*, *Peyssonnelia* sp., *Codium bursa*, and *Pseudochlorodesmis furcellata*. The only important animals were the sponge *Spongia officinalis* and the hydroid *Eudendrium armatum*.

The community of Encrusting Calcareous Rhodophyceae and Urchins (ECRU) was found on rocks grazed by the sea urchins *Arbacia lixula* and *Paracentrotus lividus*. The biological cover was scarce and algae were poorly represented, with the exceptions of the encrusting coralline *Lithophyllum incrustans*, the green alga *Codium bursa* and the brown alga *Stypopodium schimperi*. The latter is said to be resistant to sea urchin grazing (Verlaque & Boudouresque, 1991). Sponges (*Cliona copiosa*, *Crambe crambe*, *Ircinia oros*, *I. variabilis*, *Petrosia ficiformis*), corals (*Balanophyllia europaea*), vermetids (*Serpulorbis arenaria*) and bryozoans (*Reptadeonella violacea*, *Schizoporella longirostris*) were the major components of the sessile fauna.

Two different aspects of the community of Semi-Dark Grottoes (SDG) were observed under overhangs. The first (SDG-1) was characterised mostly by sponges (*Agelas oroides*, *Axinella damicornis*, *Chondrosia reniformis*, *Phorbas tenacior*), corals (*Caryophyllia inornata*, *Hoplangia durotrix*, *Leptopsammia pruvoti*, *Madracis pharensis*, *Phyllangia mouchezii*, *Polycyathus muellerae*) and bryozoans (*Myriapora truncata*, *Sertella septentrionalis*, *Smittina cervicornis*), the only alga being *Palmophyllum crassum*, a species adapted to low light conditions (Sartoni et

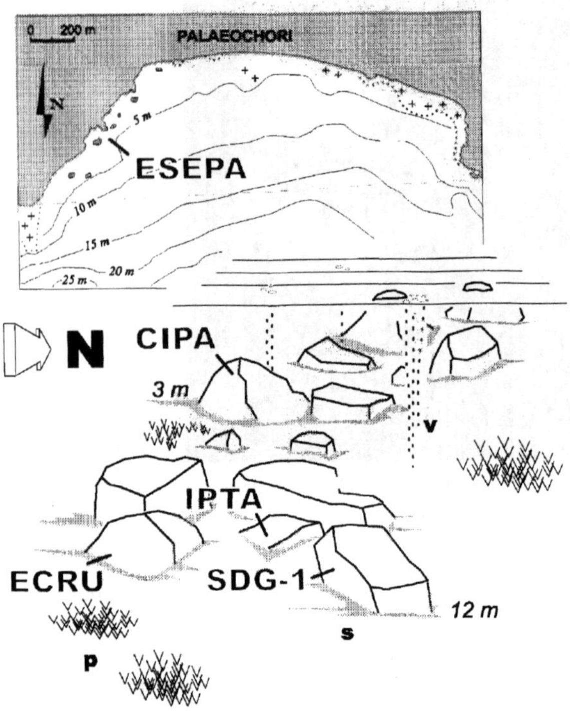

Figure 3. A pictorial view of the vent site SR, exemplifying the distribution of the main epibenthic communities: **ESEPA:** Exposed or Semi-Exposed water Photophilic Algae; **IPTA**: Infralittoral Photophilic and Thermophilic Algae; **CIPA:** Calm-water Infralittoral Photophilic Algae; **ECRU:** Encrusting Calcareous Rhodophyceae and Urchins; **SDG-1:** Semi-Dark Grottoes dominated by sponges, corals and bryozoans; **v:** vent; **s:** sandy floor; **p:** *Posidonia* bed.

al., 1993). The second (SDG-2) was characterised by serpulids (*Protula tubularia, Salmacina dysteri*) and especially by several species of ascidians (*Halocynthia papillosa, Microcosmus sabatieri, Microcosmus savignyi, Polycarpa gracilis, Polyclinum aurantium, Pyura microcosmus, Pyura squamulosa, Pyura tessellata*), whereas sponges were represented only by *Agelas oroides* while corals were absent.

The community of Coastal Detritus bottoms (CD) was recognised on biodetritic bottoms at 44 m. The two most conspicuous species were the red alga *Osmundaria volubilis* and the sea urchin *Spatangus purpureus*: both species are known to indicate bottom currents (Pérès & Picard, 1964).

Distribution of communities in the sites

Site SR (Figure 3) was located in shallow water inside Palaeochori Bay. It included rocks and large boulders on a gravely floor 2 – 3 m deep near the beach, together with rocks below the western point, ending at 12 m on the sandy floor. Hydrothermal activity was observed throughout the bay and appeared to be most intense in the northwest and the centre. Hydrothermal vents were seen springing from a sandy bottom and extensive white patches of minero-bacterial mats occurred in their proximity. ESEPA community was present near the water surface on rocks, whereas blocks near the sandy bottom harboured IPTA community, with the green algae *Acetabularia acetabulum* and *Dasycladus vermicularis* emerging from an algal mat. CIPA community, mainly composed by Dictyotales and *Cystoseira*, was alternating with patches of ECRU community down to about 12 m depth, whereas components of ICSA community were present under blocks or in the lower-storey of CIPA community. Overhangs and crevices were inhabited by sponges belonging to SDG-1 community. On the whole, this site was characterised by algal-dominated communities, which explains the great numbers of algal species and of their epiphytic sessile fauna sampled there (Figure 4).

Site CR was situated in front of Palaeochori, on rocks surrounded by a sandy bottom at about 32 m depth (Figure 5). Basaltic rocks and boulders lay along a north to south axis, partially encircled by the *Posidonia oceanica* meadow. Vents sprung continuously from the outcrops, and a flocculant bacterial mat covered the rock around the fluid springs. Where fluid emission was intermittent, the white bacterial patches were not present. Horizontal, well-lit substrates harboured CIPA community, with *Cystoseira brachycarpa* and *C. schiffneri*. It was replaced by IPTA community near the sandy bottom and by ECRU community on more inclined substrates. CCSA community, together with few elements of ICSA community, was present on vertical substrates, and either community SDG-1 or SDG-2 under overhangs.

Site VS was further off shore, on a rock outcrop emerging from the sandy floor at 44 m depth (Figure 6). The outcrop, about 500 m^2 in area, had a north-south extension and was mainly composed of two different formations. The base was made by overlying layered slabs, up to several tens of cm thick and extending to 1–2 m above the sediment surface. Calcareous concretions, apparently of biological origin, lay upon the slabs, containing protrusions, crevices and occasionally small cavities which gave shelter to a rich cryptic fauna (Figure 4). The central portion of the outcrop was occupied by a sandy depression. Gas bubbles flowed intermittently both from this depression and the sandy floor around. CIPA community, with *Cystoseira corniculata, C. brachicarpa* and *C.*

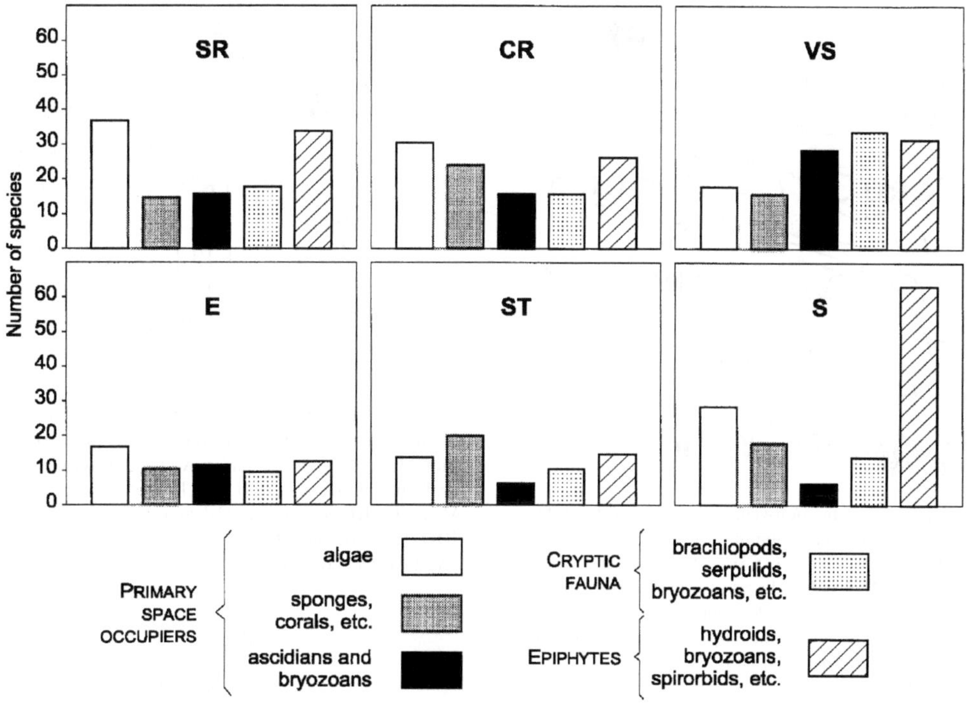

Figure 4. Number of species of different sessile epibenthic groups sampled in the six sites.

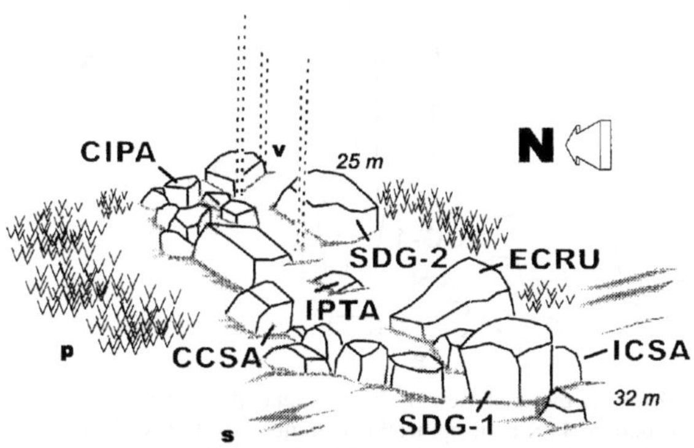

Figure 5. A pictorial view of the vent site CR, exemplifying the distribution of the main epibenthic communities: **IPTA:** Infralittoral Photophilic and Thermophilic Algae; **CIPA:** Calm-water Infralittoral Photophilic Algae; **CCSA:** Circalittoral Calm-water Sciaphilic Algae; **ICSA:** Infralittoral Calm-water Sciaphilic Algae; **ECRU:** Encrusting Calcareous Rhodophyceae and Urchins; **SDG-1:** Semi-Dark Grottoes dominated by sponges, corals and bryozoans; **SDG-2:** Semi-Dark Grottoes dominated by serpulids and ascidians; **v:** vent; **s:** sandy floor; **p:** *Posidonia* bed.

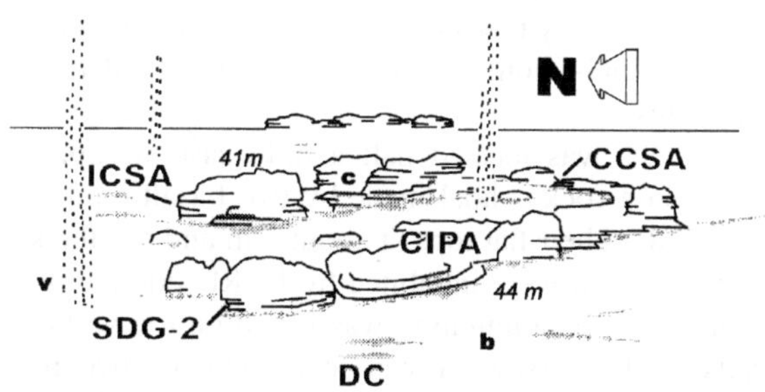

Figure 6. A pictorial view of the vent site VS, exemplifying the distribution of the main epibenthic communities: **CIPA:** Calm-water Infralittoral Photophilic Algae; **CCSA:** Circalittoral Calm-water Sciaphilic Algae; **ICSA:** Infralittoral Calm-water Sciaphilic Algae; **SDG-2:** Semi-Dark Grottoes dominated by serpulids and ascidians; **CD:** Coastal Detritus bottoms; **v:** vent; **b:** bioclastic sandy floor; **c:** layeredcarbonate crusts.

schiffneri, covered most of the substrate, but was replaced by CCSA community, with elements of ICSA community, on sub-vertical portions. SDG-2 community occurred inside small cavities within layered slabs, CD community on the biodetritic sandy floor around the outcrop.

Site E corresponded to the base of Spathi Point, closing Paleochori Bay to the East (Figure 7). Rocky blocks and boulders were present at around 7–13 m depth. Horizontal or gently sloping substrates were colonised by IPTA community, especially near the sandy bottom, and by elements of CIPA community. No species of *Cystoseira* was conspicuous. ECRU

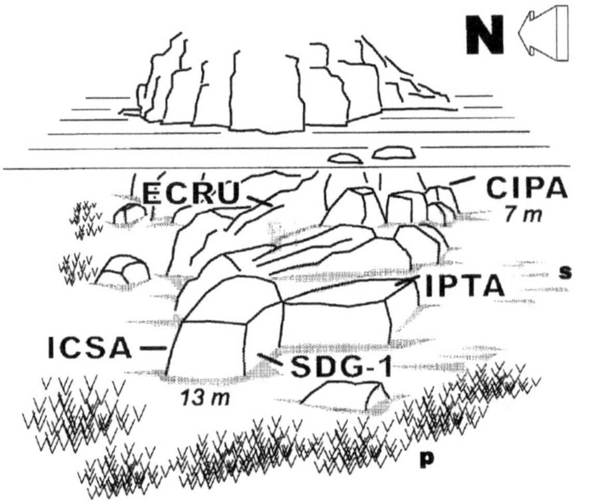

Figure 7. A pictorial view of the non-vent site E, exemplifying the distribution of the main epibenthic communities: **IPTA:** Infralittoral Photophilic and Thermophilic Algae; **CIPA:** Calm-water Infralittoral Photophilic Algae; **ICSA:** Infralittoral Calm-water Sciaphilic Algae; **ECRU:** Encrusting Calcareous Rhodophyceae and Urchins; **SDG-1:** Semi-Dark Grottoes dominated by sponges, corals and bryozoans; **s:** sandy floor; **p:** *Posidonia* bed.

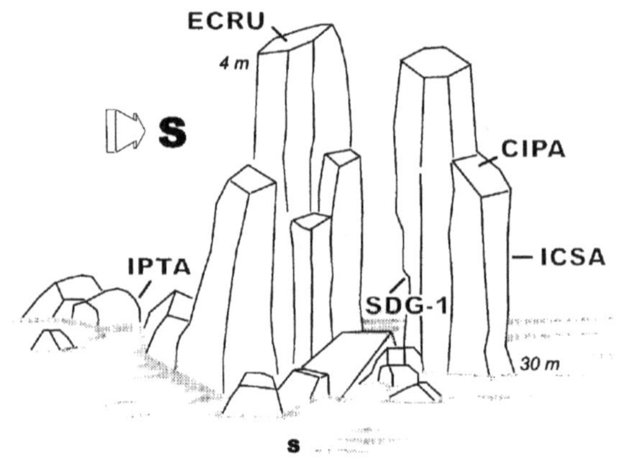

Figure 8. A pictorial view of the non-vent site ST, exemplifying the distribution of the main epibenthic communities: **IPTA:** Infralittoral Photophilic and Thermophilic Algae; **CIPA:** Calm-water Infralittoral Photophilic Algae; **ICSA:** Infralittoral Calm-water Sciaphilic Algae; **ECRU:** Encrusting Calcareous Rhodophyceae and Urchins; **SDG-1:** Semi-Dark Grottoes dominated by sponges, corals and bryozoans; **s:** sandy floor.

community was found on inclined substrates. Vertical portions of rocky blocks harboured ICSA community, whereas SDG-1 community was well developed under overhangs.

Site ST was located half a mile east off Spathi Point, on a rocky shoal made by two basaltic pillars (Figure 8). Their flat tops at about 4 m and the flanks down to 15 m were inhabited by ECRU community, whereas ICSA community was found on the vertical walls of the narrow canyon separating the two pillars. CIPA community covered (sub-)horizontal rocks in depth less than 25 m, with *Cystoseira schiffneri* forming a low canopy over *Halopteris scoparia* and other algae. SDG-1 community under overhangs was particularly rich in coral species. Boulders and cobbles at the base of the pillars, ending on the sandy floor at about 30 m depth, were inhabited by IPTA community.

Site S was positioned at two stately rocky pinnacles named Vrakhos Ktenia, situated about three miles offshore to the south-east (Figure 9). The upper portion of the pillars, which stood about 10 m above the sea surface, was made of granite, whereas the rest was composed of black basaltic rock. Horizontal or gently sloping substrates at 10 – 15 m depth were covered by CIPA community, with dense canopies of *Sargassum vulgare*, *Cystoseira barbata* and *C. brachycarpa*. Within the same depth range, ICSA

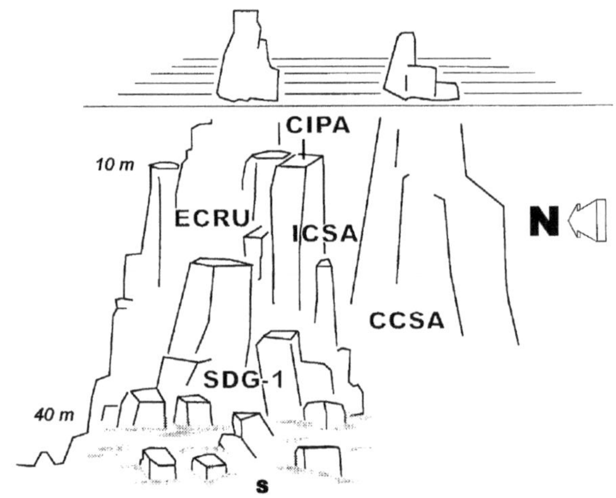

Figure 9. A pictorial view of the non-vent site S, exemplifying the distribution of the main epibenthic communities: **CIPA,** Calm-water Infralittoral Photophilic Algae; **CCSA,** Circalittoral Calm-water Sciaphilic Algae; **ICSA:** Infralittoral Calm-water Sciaphilic Algae; **ECRU:** Encrusting Calcareous Rhodophyceae and Urchins; **SDG-1:** Semi-Dark Grottoes dominated by sponges, corals and bryozoans; **s:** sandy floor.

and ECRU communities inhabited shadowed crevices. Deeper walls of the pillars and boulders at their base, at about 40 m depth, exhibited a rich algal cover dominated by CCSA community: *Sargassum acinarium* and other brown algae were the most important species and provided the substratum for many species of epiphytes (Figure 4). SDG-1 community was found under overhangs, at depths between 20 and 40 m.

Discussion

The epibenthic communities identified at Milos correspond rather well to those already described for the western Mediterranean (Ros et al., 1985) and the Adriatic (Gamulin-Brida, 1974).

Most communities were characterised by a luxuriant and diverse algal growth, with a canopy dominated by species of *Cystoseira* in shallow water, by *Sargassum acinarium* and other brown algae in deeper water, down to 44 m depth, and by sciaphilic green algae on (sub-) vertical rocks. Sea-urchin barrens, common in the western Mediterranean (Sala, 1997), were occasionally present. There, algal cover was almost completely restricted to encrusting corallines and *Stypopodium schimperi*. The latter species, recently introduced in the eastern Mediterranean Sea (Verlaque & Boudouresque, 1991), was widespread at Milos (Sartoni & De Biasi, 1999). It also participated in CCSA community in relatively deep waters, thus suggesting adaptability to a wide ecological range. Being a conspicuous species, *Stypopodium schimperi* is able to shape the physiognomy of epibenthic assemblages, and its further expansion in the rest of the Mediterranean might alter patterns of species dominance in different algal assemblages.

Apart from the present abundance of *S. schimperi*, two other aspects differentiated the epibenthic communities we observed at Milos from those typically described for the Mediterranean: 1. the penetration of photophilic algal communities into rather deep water; 2. and the abundance of sponges and corals in shadowed situations. The first aspect had been already underlined by Giaccone (1968) and is probably due to the great water transparency. The second was described by Pérès & Picard (1958) as an "eastern *facies* of coralligenous biocenosis", and was assumed to be determined by the comparatively high water temperature in the Aegean Sea. Hydrological measurements showed that winter temperature in the study area is even higher than expected, due supposedly to the hydrothermal activity (S. Aliani, pers. comm.). This agrees with the fact that IPTA community, containing thermophilic algae, mostly occurred in inshore stations. Sartoni & De Biasi (1999) also noticed the occurrence of warm-water algal flora in the area.

Besides the putative influence of warmer water, we did not find evidence that vents had any large-scale effects on the epibenthic communities. In other areas, Bright (1977) and Tarasov et al. (1990) found only local effects. At Milos, we easily observed small scale effects: a white flocculent bacterial mat often covered the substrate at the point from which fluid escaped, and for a short distance (30 cm) the rock was deprived of macrobiota. A similar observation was made by Bianchi & Morri (1983) around hydrothermal vents at Kos, also part of the Hellenic Arc.

However, there were a few differences between the communities at vent and non-vent sites, which deserve further comment. Under overhangs, for example, the sessile epifaunal community was dominated by ascidians in vent sites, by sponges and corals in non-vent sites. All these invertebrate groups are suspension feeders, but exploit particles of different size using different mechanisms (Riedl, 1966; Gili & Coma, 1998). Cocito et al. (1997) related change in guild dominance within suspension-feeding communities to particulate matter quality in the water column. Should a different trophic composition of suspended matter close to vent sites be hypothesised? A similar effect was found by Southward et al. (1996) in a submarine cave influenced by hydrothermal vents.

Another difference was that mounds of *Mesophyllum lichenoides* were by far more conspicuous at vent sites. This apparently contrasts with early studies in the Hellenic Arc, which reported reduced bioconcretion near hydrothermal vents (Laborel, 1960). However, the peculiar geochemistry of the water near vents, high Ca^{2+} and CO_2 contents (Fitzsimons et al., 1997), is considered to cause the precipitation of carbonate cements and the local lithification of marine sediments (Hovland, 1990). In many areas of the world, carbonate slabs and crusts have been described from near cold seeps, where water geochemistry is similarly altered. The layered slabs we observed in VS (Figure 6) were similar to those described by Jensen et al. (1992) and, partly, to the smaller authigenic carbonate structures found by Kulm & Suess (1990). Similarly, well developed at VS were the bioconstructed mounds of *Mesophyllum lichenoides*. Thus, it might be hypothesised that seawater enrichment in Ca^{2+} and CO_2 by vent activity enhances the biological deposition of carbonates.

Acknowledgements

This study is part of the project AG-HY-FL (Hydrothermal fluxes and biological production in the Aegean) and was supported by the EC under contract MAS3-CT95-0021. S. Varnavas (Patras) took charge of logistics at Milos. P.R. Dando (Bangor), coordinator of the project, kindly revised the manuscript. S. Aliani, F. Degl'Innocenti, R. Meloni, late R. Palmer, S. Schwabe, C. Smith and S. Sgorbini helped during field work. U. Ugolini took the photo of Figure 2. Thanks are also due to F. Ferdeghini (La Spezia), M. Pansini (Genoa), M. Pestarino (Genoa) and G. Sartoni (Florence), for identifying bryozoans, sponges, ascidians and algae, respectively.

References

Acunto, S., F. Maltagliati & F. Cinelli, 1996. Osservazioni sui popolamenti bentonici di un'area interessata da attività idrotermale nei pressi dell'Isola di Panarea (Isole Eolie). Biol. mar. Medit. 3: 434–436.

Aliani, S., C. N. Bianchi, S. Cocito, P. R. Dando, R. Meloni, C. Morri, A. NieMeyer, A. Peirano & W. Ziebis, 1998. A map of seagrass meadows in Palaeochori Bay (Milos Island, Greece), a marine area with hydrothermal activity. Rapp. Comm. int. Mer Médit. 35: 512–513.

Augier, H., 1982. Inventory and classification of the marine benthic biocoenoses of the Mediterranean. Council of Europe, Strasbourg, Nat. Envir. Ser. 25: 1–57.

Bellan-Santini, D., 1985. The Mediterranean benthos: reflections and problems raised by a classification of the benthic assemblages. In Moraitou-Apostolopoulou, M. & V. Kiortsis (eds), Mediterranean Marine Ecosystems. Plenum Press, New York: 19–48.

Bianchi, C. N. & C. Morri, 1983. Note sul benthos marino costiero dell'isola di Kos (Egeo sud-Orientale). Natura 74 (1–2): 96–114.

Bianchi, C. N., S. Cocito, C. Morri & S. Sgorbini, 1991. Rilevamento bionomico subacqueo. In Abbiati, M. (ed.), Lezioni del Corso Formativo per Ricercatore Scientifico Subacqueo. Intern. School for scient. Diving, Pisa: 67–93.

Bianchi, C. N., S. Cocito, A. Peirano, C. Morri, S. Aliani, A. M. De Biasi & P. Dando, 1997. The epibenthic communities of an area with hydrothermal vents at Milos island (Aegean Sea). International Conference "Progress in Oceanography of the Mediterranean Sea", Rome, Abstracts Volume: 391–392.

Bianchi, C. N., S. Aliani, A. Balduzzi, M. Boyer, S. Cocito, A. M. De Biasi, F. Ferdeghini, C. Morri, M. Pansini, A. Peirano & M. Pestarino, 1999. Biodiversità dell'epifauna marina sessile in un'isola del Mar Egeo: Milos, Cicladi. Biol. mar. medit. 6 (1): 179–183.

Boudouresque, C. F., 1984. Groupes écologiques d'algues marines et phytocénoses benthiques en Méditerranée nord-occidentale: une revue. Giorn. bot. ital. 118 suppl. 2: 7–42.

Bright, T. J., 1977. Coral reefs, nepheloid layers, gas seeps and brine flows on hard-banks in the Northwestern Gulf of Mexico. In Taylor, D. L. (ed.), Proc. 3rd Intern. Coral Reefs Symp., Rosenstiel School Mar. Atmospher. Sci., Miami: 39–46.

Cocito, S., S. Sgorbini & C. N. Bianchi, 1997. Zonation of a suspension-feeder assemblage on a temperate rocky shoal: the influence of water current and bottom topography. In Hawkins, L. E. & S. Hutchinson (eds), The Responses of Marine Organisms to Their Environments, University of Southampton, U. K.: 183–192.

Coppejans, E., 1974. A preliminary study of the marine algal communities on the Island of Milos and Sikinos (Cyclades-Greece). Bull. Soc. r. Bot. Belg. 107: 387–406.

Dando, P. R., J. A. Hughes & F. Thiermann, 1995a. Preliminary observations on biological communities at shallow hydrothermal vents in the Aegean Sea. In Parson, L. M., C. L. Walker & D. R. Dixon (eds), Hydrothermal Vents and Processes. Spec. Publ. 87, Geol. Soc., London: 303–317.

Dando, P. R., J. A. Hughes, Y. Leahy, S. J. Niven, L. J. Taylor & C. Smith, 1995b. Gas venting rates from submarine hydrothermal areas around the island of Milos, Hellenic Volcanic Arc. Cont. Shelf. Res. 15 (8): 913–929.

Dando, P. R., M. C. Austen, R. J. Burke, M. A. Kendall, M. C. Kennicutt, A. G. Judd, D. C. Moore, S. C. M. O'Hara, R. Schmaljohann & A. J. Southward, 1991. Ecology of a North Sea pockmark with an active metane seep. Mar. Ecol. Prog. Ser. 70: 49–63.

Diapoulis, A., T. Koussouris, C. Bogdanos, G. Fotis & I. Bertahas, 1994a. Distribution of benthic organisms in a volcanic sea area, Aegean Sea, Greece. Fresenius' envir. Bull. 3: 18–23.

Diapoulis, A., A. Conides & T. Koussouris, 1994b. Interaction between macrophytobenthic groups on the sublittoral belt of a Greek island. PSZN I: Mar. Ecol. 15: 41–50.

Fitzsimons, M. F., P. R. Dando, J. A. Hugues, F. Thiermann, I. Akoumianaki, S. M. Pratt, 1997. Submarine hydrothermal brine seeps off Milos, Greece: observations and geochemistry. Mar. Chem. 57: 325–340.

Frike, H., O. Giere, K. Stetter, G. A. Alfredsson, J. K. Kristjansson, P. Stoffers & J. Svarsson, 1989. Hydrothermal vent communities at the shallow subpolar Mid-Atlantic ridge. Mar. Biol. 102: 425–429.

Fytikas, M., F. Innocenti, P. Manetti, R. Mazzuoli, A. Peccerillo & L. Villari, 1984. Tertiary to Quaternary evolution of volcanism in the Aegean region. In Dixon, J. E. & A. H. F. Robertson (eds), The Geological Evolution of the Eastern Mediterranean. Geol. Soc. London Spec. Publ. 17: 687–699.

Gamulin-Brida, H., 1974. Biocoenoses benthiques de la mer Adriatique. Acta Adriat.15: 1–102.

Giaccone, G., 1968. Contributo allo studio fitosociologico dei popolamenti algali del Mediterraneo Orientale. Giorn. bot. ital. 102: 485–506.

Giaccone, G., G. Alongi, F. Pizzuto & A. Cossu, 1994a. La vegetazione marina bentonica fotofila del Mediterraneo: II. Infralitorale e Circalitorale. Proposte di aggiornamento. Boll. Acc. Gioenia Sci. nat. 27: 111–157.

Giaccone, G., G. Alongi, F. Pizzuto & A. Cossu, 1994b. La vegetazione marina bentonica sciafila del Mediterraneo: III. Infralitorale e Circalitorale. Proposte di aggiornamento. Boll. Acc. Gioenia Sci. nat. 27: 201–227.

Giaccone, G. & A. Bruni, 1973. Le Cistoseire e la vegetazione sommersa del Mediterraneo. Atti Istituto Veneto Scienze, Lettere e Arti 81: 59–103.

Gili, J. -M. & R. Coma, 1998. Benthic suspension feeders: their paramount role in littoral marine food webs. Trends Ecol. Evol. 13 (8): 316–321.

Giménez, F. & A. Marín, 1991. Los Anelidos poliquetos de una solfatara submarina en el Golfo de Napoles. An. Biol. (Biol. ambiental) 17: 143–151.

Hashimoto, J., T. Miura, K. Fujikura & J. Ossaka, 1993. Discovery of vestimentiferan tube-worms in the euphotic zone. Zool. Sci. 10: 1063–1067.

Hovland, M., 1990. Do carbonate reefs form due to fluid seepage? Terra Nova 2: 8–18.

Huvé, H., 1972. Aperçu sur la distribution en mer Egée de quelques espèces du genre *Cystoseira* (Phéophycées, Fucales). Soc. Phycol. France 17: 22–37.

Jensen, P., I. Aagaard, R. A. Burke Jr., P. R. Dando, N. O. Jørgensen, A. Kuijpers, T. Laier, S. C. M. O'Hara & Schaljohann, 1992. 'Bubbling reefs' in the Kattegat: submarine landscapes of carbonate-cemented rocks support a diverse ecosystem at methane seeps. Mar. Ecol. Prog. Ser. 83: 103–112.

Kamenev, G. M., V. I. Fadeev, N. I. Selin & V. G. Tarasov, 1993. Composition and distribution of macro- and meiobenthos around sublittoral hydrothermal vents in the Bay of Plenty, New Zealand. N. Z. J. mar. Freshwat. Res. 27: 407–418.

Kulm, L. D. & E. Suess, 1990. Relationship between carbonate deposits and fluid venting: Oregon accretionary prism. J. geophys. Res. 95: 8899–8915.

Laborel, J., 1960. Contribution à l'étude directe des peuplements benthiques sciaphiles sur substrat rocheux en Méditerranée. Rec. Trav. St. mar. Endoume 20 (33): 117–174.

Pérès, J. M. & J. Picard, 1958. Recherches sur les peuplements benthiques de la Méditerranée Nord-Orientale. Ann. Inst. Océanogr. 34: 213–291.

Pérès, J. M. & J. Picard, 1964 - Nouveau manual de bionomie benthique de la Mér Mediterranée. Rec. Trav. St. mar. Endoume 31: 1–137.

Riedl, R., 1966. Biologie der Meereshöhlen. Verlag Paul Parey, Hamburg: 636 pp.

Ros, J. D., J. Romero, E. Ballestreros & J. M. Gili, 1985. Diving in the blue water. The benthos. In Margalef, R. (ed.), Western Mediterranean. Pergamon Press, Oxford: 233–295.

Sala, E., 1997. Fish predators and scavengers of the sea urchin *Paracentrotus lividus* in protected areas of the north-west Mediterranean Sea. Mar. Biol. 129: 531–539.

Sartoni, G. & A. M. De Biasi, 1999. A survey of the marine algae of Milos Island, Greece. Cryptogam-Algol. 20 (3): 271–283.

Sartoni, G., F. Cinelli, T. Hirata, N. Katayama & Y. Yokohama, 1993. On the lack of green light-harvesting pigments and the extremely high chlorophyll *b/a* ratio in the deep-water green alga, *Palmophyllum crassum* (Chlorosphaerales). Jpn. J. Phycol. 41: 327–331.

Sartoretto, S., 1994. Structure et dynamique d'un noveau type de bioconstruction à *Mesophyllum lichenoides* (Ellis) Lemoine (Corallinales, Rhodophyta). C. r. hebd. Séanc. Acad. Sci., Paris, Sciences de la vie 317: 156–160.

Southward, A. J., M. C. Kennicutt II, J. Alacalà-Herrera, M. Abbiati, L. Airoldi, F. Cinelli, C. N. Bianchi, C. Morri & E. C. Southward, 1996. On the biology of submarinecaves with sulphur springs: appraisal of $^{13}C/^{12}C$ ratios as a guide to trophic relations. J. mar. biol. Ass. U. K. 76: 265–285.

Thiermann, F., I. Akoumianaki, J. A. Hughes & O. Giere, 1997. Benthic fauna of a shallow-water gaseohydrothermal vent area in the Aegean Sea (Milos, Greece). Mar. Biol. 128: 149–159.

Tarasov, V. G., M. V. Propp, A. V. Zhirmunsky, B. B. Namsarev, V. M. Gorlenko & D. A. Starynin, 1990. Shallow-water gaseohydrothermal vents of Ushishir Volcano and the ecosystem of Kraternaya Bight (The Kurile Islands). PSZN I: Mar. Ecol. 11: 1–23.

Tunnicliffe, V., 1991. The biology of hydrothermal vents: ecology and evolution. Oceanogr. Mar. Biol. ann. Rev. 29: 319–407.

Verlaque, M. & C. F. Boudouresque, 1991. *Stypopodium schimperi* (Buchinger Ex Kützing) Verlaque et Boudouresque Comb. Nov. (Dictyotales, Fucophyceae), algue de Mer Rouge récemment apparue en Méditerranée. Cryptogam-Algol. 12 (3): 195–21.

Hydrobiologia **426:** 123–137, 2000.
G. Liebezeit, S. Dittmann & I. Kröncke (eds), Life at Interfaces and Under Extreme Conditions.

Influence of temperature, oxygen and food availability on the migrational activity of bathyal benthic foraminifera: evidence by microcosm experiments

Onno Gross
Hegestrasse 46 b, D-20251 Hamburg, Germany
E-mail: onnogross@compuserve.com

Key words: deep-sea, foraminifera, migration, laboratory experiments, oxygen

Abstract

Foraminifera are a dominant group of amoeboid protists in the deep-sea and play possibly a significant role in decomposition processes of incoming organic matter. In order to study the poorly known ecology of these protozoans, microcosm experiments with living bathyal benthic foraminifera have been conducted. Foraminifera from 2880 m and shallower water depths were successfully maintained and their movement patterns investigated. By video documentation, high mean migration speed of 20,02 ($N = 22$) and 24,48 μm min^{-1} ($N = 10$) at 4 °C were found for species such as *Hoeglundina elegans* and *Quinqueloculina seminula* from 1847 m and 1921 m water depth, respectively. The results demonstrate that some bathyal foraminifera have migration speeds comparable to those of shallow water species. Environmental factors such as temperature, food concentration and oxygen content showed a marked influence on the migration of some species. An increase in temperature from 10 °C to 15 °C resulted in an increase of 35% in the migration speed of *Allogromia* spp. However, other species reacted differently. Higher food concentration resulted in a decrease in speed of some species. While in *Quinqueloculina lamarckiana* speed was not greatly affected by a low oxygen content in the sediment, other foraminifera responded to oxygen depletion by migration to the surface layers. Observations of benthic foraminifera in the laboratory microcosms are discussed in relation to microhabitats and the fate of organic matter on the sea floor and in the sediment.

Introduction

Within the biocoenoses of the deep-sea, protozoa play a major role in the turn-over of incoming organic material from the overlying water column (Thiel, 1983; Turley et al., 1988; Alongi, 1992). Seasonal deposition of primary production (Thiel et al., 1988; Pfannkuche, 1993; Graf et al., 1995) and various disturbances (food pulses, benthic storms, slope runoffs, moulds) create small scale heterogeneity in the deep-sea sediment which is important for biological communities (Thistle, 1983; Levin, 1991) and enables manifold biotic interactions and niche separations (Bernstein & Meadow, 1979; Smith, 1994; Gage, 1996; Rex, 1997). In particular, Foraminiferida (Sarcodina, Class: Granuloreticulosea) are assumed to be ecologically relevant owing to the presence of their pseudopodial network (Lee, 1974; Gooday et al., 1992).

The seasonal and vertical distribution patterns of benthic foraminifera found in field samples have become a focus of recent research (Gooday et al., 1992; Linke & Lutze, 1993; Kaminski et al., 1994; Silva et al., 1996). Changes in biomass, size-structure and species diversity suggest that foraminifera are able to respond quickly to the incoming organic material (Gooday, 1988; Gooday & Lambshead, 1989; Gooday & Turley, 1990; Altenbach, 1992; Linke et al., 1995). Benthic foraminifera are not restricted to the uppermost layer of the sediment column, but live down to a depth of 15 cm and more in both shallow and deep water settings (e.g. Boltovskoy 1966; Moodley & Hess, 1992; Bernhard, 1993).

In order to understand the diversity of shell morphology and life habits, foraminifera have been classified by micropaleontologists into morphological groups with restricted microhabitat preferences (e.g. epifaunal, semi-infaunal, deep infaunal). This obviously helped to simplify the heterogeneity in foraminiferal life styles for paleoecological purposes (e g. Corliss, 1985, 1991; Mackensen et al., 1993; Loubere et al., 1993, 1995). However, the concept of well defined foraminiferal microhabitats has proven valid only for some epifaunal benthic foraminifera, e.g. mostly those living firmly attached to plant surfaces or to elevated hard substrates.

Microhabitat preferences are now believed to be dynamic and to reflect an adaptation to the varying environmental gradients found in the deep-sea (Linke & Lutze, 1993; Jorissen et al., 1995). It has been discussed whether the impact of abiotic factors (porewater, nutrients, oxygen) or possible ecological responses during the life cycle of the foraminifera are responsible for the irregular vertical distribution patterns of these protozoans (cf. Gooday, 1994). The key-factors responsible for the vertical distribution patterns of benthic foraminifera still remain uncertain. The microhabitats of foraminifera lie within the seawater–sediment interface which is undergoing constant changes owing to biological activity and geochemical processes. Furthermore, the protistan cells are able to respond to environmental gradients by migration in order to avoid unfavourable conditions.

In standard micropaleontological studies (e.g. horizontal splitting of cores, rose Bengal staining), this migratory ability of sarcodines has been mostly neglected. Although previous studies have investigated locomotion in benthic species (Kitazato, 1981, 1988; Wetmore, 1988), the migration activity of many foraminifera is hardly known. Studies of locomotion have dealt with the cytochemical mechanisms or general biology of the reticulopodial network engaged in the movement (see review by Travis & Bowser, 1991). Only few ecological studies exist and they concentrate either on the survival of foraminifera to oversedimentation (Severin, 1987) or on the ecology of foraminifera from shallow water depths (Severin & Erskian, 1981; Severin et al., 1982; Kitazato, 1988; Bornmalm et al., 1997; Hannah & Rogerson, 1997).

This study investigates the little known life activities of bathyal benthic foraminifera keeping them in their natural sediments in the laboratory. Observations of 'normal' and experimentally altered conditions revealed many autecological informations. Migration behaviour is considered to be a primary factor in the life cycle of foraminifera and may reflect ecological strategies among benthic protists. A distance of 2 mm for a 0.2 mm long foraminifera corresponds to approximately 200 m for a human being, and movement across such distances have to overcome steep environmental gradients in the sediments (e.g. in grain size, oxygen and sulphide concentrations, competition). However, the ability to move within the sediment confers a number of ecological advantages. For example, an individual foraminifera is able to avoid physical disturbance and predation on the surface or benefits from food sources and reduced predation pressure in deeper sediment layers. The purpose of the present study was to obtain quantitative information on the migration activity of foraminifera. It was investigated to what extent environmental factors influence the migration rates of benthic foraminifera and in turn which influence foraminiferal activity contributes to the mineralisation of organic matter in the sea bed.

Materials and methods

The majority of foraminifera recovered from greater bathymetric depths could only be maintained successfully in natural sediment, i.e. with their accompanying microfauna (Anderson et al., 1991; Chandler et al., 1996; pers. observation). Therefore, our laboratory studies were performed with intact sediment communities. The sediments were collected from various depths from continental margins and shelf seas down to abyssal plains in three areas: the North Atlantic Ocean, the Mediterranean Sea and the Arabian Sea (Table 1). Samples were taken at one transect of stations in the Atlantic Ocean along the Goban Spur Ridge in vicinity to the Porcupine Abyssal Plain during the R.V. *'METEOR'* cruise M27-1 (Pfannkuche et al., 1994). Sampling in the eastern Mediterranean Sea was conducted during cruise M25-1 (Hieke et al., 1995), mainly along a depth transect along the Gulf of Torrent and at stations near the Crete and Cyprus Trenches, in the Levantine Basin and off the coast of Israel and Egypt. Samples from the northwestern Mediterranean Sea were collected from the French vessel *'GEORGE PETIT'*. Stations followed a depth transect along the Lacaze-Duthiers Canyon, in the Gulf of Lyons. The latter samples are designated Banyuls-sur-Mer after the nearby marine research station.

Sediment was collected with a multicorer (Barnet et al., 1984), equipped with up to eight polycarbonate core tubes of 9.6 and 5.1 cm inner diameter, respectively, which recovered virtually undisturbed sediment surfaces. At some shelf break stations, box cores and Flucha-grab samples were also subsampled. The

Table 1. Sampling areas, research vessels, cruise numbers and dates, water depths of collected sediments and bottom salinities used for these investigations

Sampling locations	Ship/cruise	Year	Water depth [m]	Salinity [PSU]
Western Mediterranean Sea (Banyuls)	George Petite	1992/93	40–1477	37–38
Eastern Mediterranean Sea	Meteor M25-1	1993	29–2880	37–38
Northern Atlantic	Meteor M11-1	1989	1640–2806	34–35
Northern Atlantic (Goban Spur)	Meteor M72-1	1994	2231–2243	34–35
Northern Atlantic (Goban Spur)	Meteor M30-1	1994	224–3665	34–35
Red-/Arabian Sea	Meteor M31-3	1995	174–1730	36
Arabian Sea	Meteor M33-1	1995	1921	35

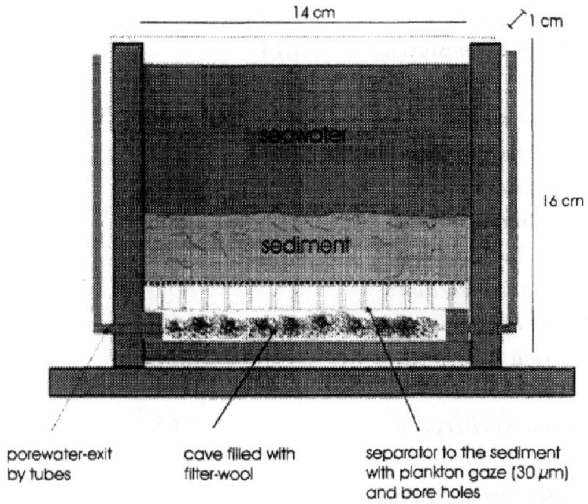

Figure 1. Schematic drawing of the special observation aquarium modified after Kiatazato (1983) with a partition separating the sediment and the porewater reservoir. The tubes on the sides are used for porewater exchange.

samples were immediately transferred to a laboratory adjusted to the ambient bottom-water temperature (Table 1). About 200 culture samples were set up by transferring the top layers of the core into 150 ml glass vessels (Schott). These cultures were stored under ambient temperature on board the ship and later transported back to the home laboratory. There, the vessels were stored in the dark and cooled to 4 ± 0.56 °C using a cooling room and to 10 ± 1.24 °C and 15 ± 0.32 °C using culture incubators.

For maintaining and studying benthic foraminifera, a special 'ant-farm' aquarium (Figure 1), modified after Kitazato (1983), was used. The aquaria (16 cm high, 14 cm wide) were filled with a sediment layer of approximately 3–4 cm thickness and the remaining 2/3 of the depth with seawater of in situ salinity. A 30 μm gauze separated the sediment from an underlying porewater reservoir (Kitazato, pers. comm.). Plastic microtubes (1.0 mm in diameter), mounted at the sides, enabled porewater exchange in order to avoid porewater stagnation within the lower sediment layers. Additionally, the overlying sea water was regularly aerated and exchanged. All cultures were studied at normal atmospheric pressure.

Locomotion of benthic foraminifera was documented by using a stereo microscope, photography or video. In order to compare migration rates of specific individuals within the sediment or along the glass walls, specimens were marked by a pen-pointer and distances over time were measured (expressed as μm min^{-1}). In the climate chamber, an automatic video system was installed. The system worked with a self-built timer in order to record undisturbed migration sequences with a video recorder (Panasonic AG5700-E). The interval between images could be set between 30 s and 10 min. Time-lapse sequences were later analysed using a video-printer (Mitsubishi P67).

Cultures or aquaria used for experiments to examine the effects of food input, temperature and oxygen were examined daily for foraminiferal movements. In the temperature experiment, eight samples from the continental shelf off Banyuls-sur-Mer were incubated at two different temperatures (10 °C and 15 °C). In order to quantify the species response to food enrichment, an experiment was set-up using culture material from 1021 m water depth from the Eastern Mediterranean Sea (M25-1 KG35). In the aquaria, the sediment was divided into halves using a polycarbonate barrier. One half of the culture was organically enriched by adding freeze-dried *Chlorella* algae on March 10, 1995, while the other half served as an untreated control. The migratory response of foraminifera to organic enrichment was recorded over a period of 2 months. The standard arrangements for seawater and porewater exchange were kept the same as described above, ensuring that the porewater did not become oxygen depleted. The experiment was run at 10 °C.

The oxygen content of the overlying seawater in the glass-cultures was monitored with a oximeter (WTW Oxi 96). Dissolved oxygen profiles in the sediments were measured using a 20 cm needle polarographic oxygen sensor (Van Gemerden et al., 1989). The micro-electrode was polarised at −800 mV against a reference electrode inserted in the overlying water. The needle tip of approximately 1.5 mm resolved oxygen concentrations on a millimetre scale. The needle was gently lowered into the sediment with a micromanipulator (Märzhauser M33). Calibration of the oxygen sensor was achieved by comparison with oversaturated seawater (1 h constant aeration of seawater) or deoxygenated seawater (using $Na_2\ SO_3$ as oxygen acceptor).

All data on foraminiferal migrations were processed using the commercially distributed statistics package SIGMA PLOT 3.0. A Mann-Whitney-*U*-test was used to test for differences between migration speeds. A χ^2 -test was used to discriminate migration speed classes. In all cases, the level of significance was 5%.

Results

Migration of benthic foraminifera

The majority of foraminifera which survived in culture originated from depth shallower than 1000 m. However, some individuals from samples obtained down to 2240 m water depth (Goban Spur, North Atlantic), 1900 m (Arabian Sea) at 4 °C and from 2880 m water depth from the eastern Mediterranean Sea at 10 °C survived and showed locomotion. All foraminifera examined moved within the sediments. Thus, they were generally not restricted to the sediment-water interface. They also migrated up above the sediment surface onto the glass walls and into deeper sediment layers (>3 cm).

The observed migration speeds of benthic foraminifera varied according to temperature, substrate (glass or sediment) and observation methods. Benthic foraminifera showed mean migration speeds of 0.06–32.4 μm min^{-1}. Table 2 summarises some of the observed migration speeds of selected species. High speeds were measured by the accurate video method and also microscopic measurements. Some species moved continuously, while others had intermittent resting phases or stopped migration while building dormant cysts. No species was permanently sessile, in a classical sense. Instead, the foraminifera (e.g. *Gavelinopsis translucens*, *Planorbulina mediterranea*, *Rosalina floridana*) rested within cysts for periods of weeks or month, after which period the cysts were abandoned and new ones constructed. Such resting cysts were build of detritus or silt material.

The foraminifera were grouped according to their mean migration rates into speed classes: slow, fast and very fast (Figure 2). This was derived from a fre-

Table 2. Migration speed classes of selected foraminiferal species: 'slow' (<2.5 μm min^{-1}, 'fast' (2.51–5.1 μm min^{-1}) and 'very fast' (>5.0 μm min^{-1})

Species	Migration speed classes		
	'slow' <2.5 μm min^{-1}	'fast' >2.5 μm min^{-1}	'very fast' >5.0 μm min^{-1}
Trochammina shannoni	0.35		
Globocassidulina subglobosa	0.46		
Adercotryma glomerata	1.36		
Amphicoryna scalaris	1.80		
Lenticulina cultrata	1.26		
Textularia porrecta	1.23		
Bathysiphon spp.	1.61		
Gavelinopsis praegeri	1.24		
Gavelinopsis translucens	1.14		
Nonion commune		3.01	
Ammoscalaria pseudospiralis		2.58	
Allogromia spp.		3.86	
Cassidulina leavigata		2.58	
Cibicidoides floridanus		2.55	
Uvigerina mediterranea		2.55	
Bulimina marginata		2.67	
Bigenerina nodosaria		2.70	
Ammonia beccarii		4.00	
Fontbotia wuellerstorfi		2.65	
Rosalina cf. *bradyi*		3.21	
Spiroplectinella wrightii		2.98	
Saccammina sp. 1			6.99
Planorbulina mediterranensis			7.23
Globobulimina affinis			9.31
Pyrgo murrhina			9.93
Quuinqueloculina lamarckiana			8.74

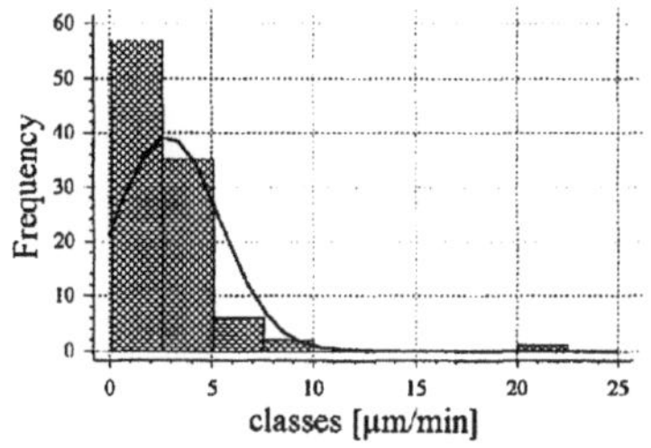

Figure 2. Frequency distribution (bars) and distribution (curve) of mean migration speeds of observed foraminiferal species (N = 103). Three frequency classes of migration speed – slow (0.1–2.5 μm/min, fast (2.51–5.0 μm min^{-1}) and very fast (>5.1 μm min^{-1}) – can be distinguished by statistical analysis ($\chi^2 = 37.42$, $p < 0.05$).

quency distribution analysis of the foraminiferal data set of migration speeds within the sediment at different temperatures (103 categories, >5 observations) and resulted in a classification into three frequency classes of migration speed: slow (0.1–2.5 μm min^{-1}), fast (2.51–5.0 μm min^{-1}) and very fast (>5.1 μm min^{-1}). The width and borders of the classes proved to be statistically significant ($\chi^2 = 37.42$, $p < 0.05$).

Slow mobile species with mean migration rates <2.5 μm min^{-1} included *Trochammina shannoni*, *Globocassidulina subglobosa* and *Gavelinopsis translucens* (North Atlantic) or species like *Textularia porrecta*, *Lenticulina cultrata* and *Gavelinopsis praegeri* (Mediterranean Sea). In the fast speed class, a variety of species occurred with average migration speeds ranging between 2.5 μm min^{-1} and 5.0 μm min^{-1}. The third class included very fast bathyal species such as *Pyrgo murrhina* and *Globobulima affinis* from the eastern Mediterranean Sea.

Comparing all foraminiferal migration rates revealed no positive correlation of foraminiferal migration speed with water depth (Figure 3). Despite the comparatively large data set (N = 3745; Gross, 1998), no differences in migration speed between shallow water species and bathyal species could be recognised. High migration speeds (>20.0 μm min^{-1}) were found in species from continental shelf samples as well as bathyal samples (down to 2200 m). This suggests no particular influence of pressure on the migration activity of benthic foraminifera. Furthermore, there does not seem to exist a relation between the geographic origin of the sampled species and migration speed. However, differences in species composition throughout the samples made it difficult to draw general conclusions.

Influence of temperature

Foraminifera showed different survival rates when cultured under different temperatures. All Mediterranean species survived at 10 °C or 15 °C, and shelf species could withstand the lower culturing temperature of 4 °C. However, bathyal North Atlantic species could not be maintained under elevated temperatures >15 °C. Presumably, the Atlantic deep-sea species are low temperature-adapted and need a cultivating temperature close to the ambient conditions during sampling at bathyal water depth.

Some foraminiferal taxa could be cultivated under different temperatures. *Allogromia* spp. showed an increase in migration speed with increasing temperature (Figure 4). The mean migration speed grew from 2.48 μm min^{-1} at 4 °C to 6.39 μm min^{-1} at 15 °C on the glass substrate. Migration rates of allogromiids within the sediment increased from 1.36 μm min^{-1} at 4 °C to 2.10 μm min^{-1} at 10 °C. Although the box-whisker plots showed high standard deviations, this temperature dependent increase was also observed for other dominant foraminiferal species (e.g. *Bolivina* spp., *Bulimina* spp.; Gross, 1998). However, *Bathysiphon* spp. showed no such trend between culture temperature and migration speed (Figure 4). While the mean migration rate at the glass wall slightly increased, it decreased in the sediment from 1.99 μm min^{-1} (10 °C) to 1.61 μm min^{-1} (15 °C).

An experimental set up using eight cultures of nearly identical species composition, from two shelf localities off Banyuls-sur-Mer, were chosen for further investigations. Four of the culture vessels were maintained at 10 °C, while the other four were stored at 15 °C. Ventilation and exchange rates of seawater were identical.

Besides two exceptions, the mean migration speed of most foraminiferal species increased at higher maintenance temperatures (Figure 5). *Bulimina marginata*, for instance, increased its mean migration speed by 59.4% between 10 °C and 15 °C. Other species, *Allogromia* spp. (6.7%), *Bathysiphon* spp. (56.7% glass), *Cassidulina laevigata* (13.7%), *Nonion commune* (31.7%) and *Trochammina inflata* (52.2%), showed less increase in migration speed.

The response to higher temperatures was statistically significant for most species (U-test, $p > 0.05$). However, some species showed differences when observed moving along the glass walls or within the sediment layers. While along the glass walls, the migration speed of *Bathysiphon* spp. was higher at 15 °C,

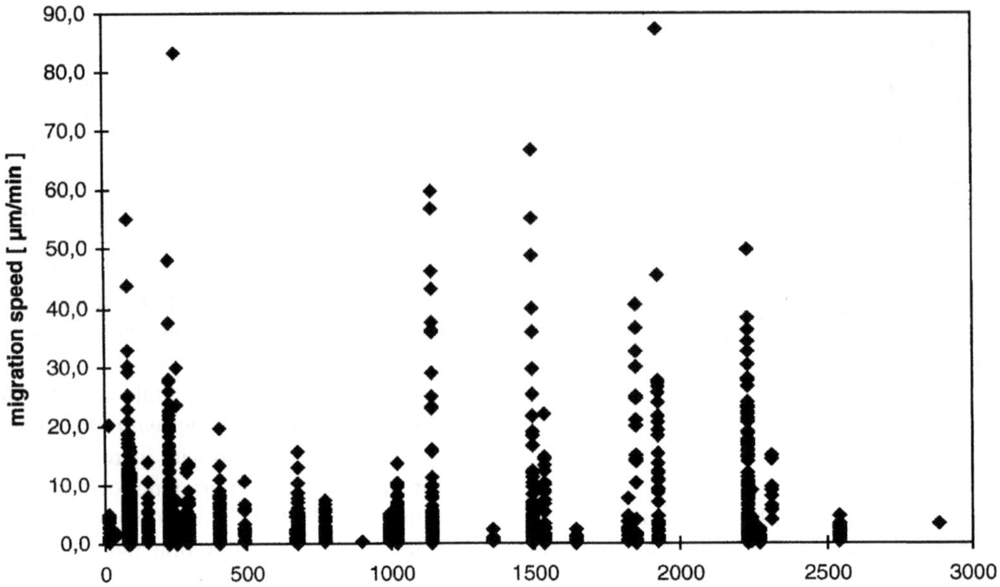

Figure 3. Migration speed [μm min^{-1}] of benthic foraminifera in relation to water depth of the sampling site (N = 3745).

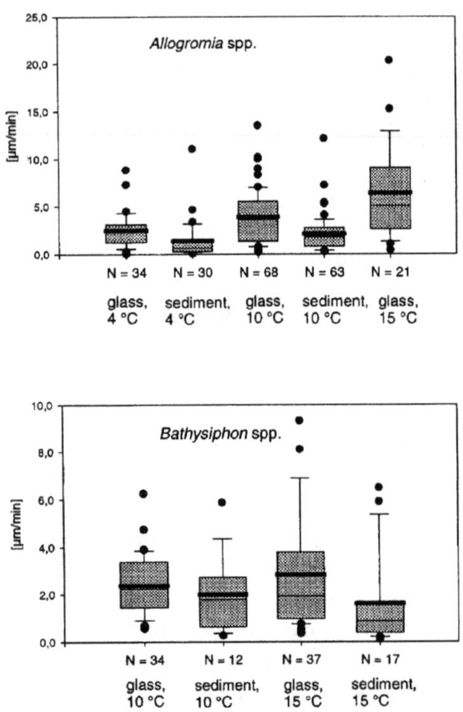

Figure 4. Diagrams showing migration speed [μm min^{-1}] of (a) *Allogromia* spp. and (b) *Bathysiphon* spp. measured on the glass side of the aquarium and within the sediment at different temperatures.

in the sediment it was higher at 10 °C (4.6%). Also, *Gyroidinoides soldanii* moved 6.9% faster along the glass at the lower culture temperature. Two species which occurred mainly on the glass walls of the culture vessels, and thus were rarely observed within the sediments, showed clearly accelerated migration speeds at the lower temperature of 10 °C. *Planorbulina mediterranea* showed a 39.0% higher mean migration speed at 10 °C and the migration speed of *Spiroplectinella wrightii* was also 19.8% higher by the lower temperature.

The observed migration speeds of benthic foraminifera at 10 °C (N = 347) and 15 °C (N = 496) were used to analyse the distribution of migration speed classes (Figure 6). The migration speeds of foraminifera were grouped into the following categories 0–0.5, 0.5–1.0, 1.0–2.5, 2.5–5.0, 5.0–15.0 and >15.0 μm min^{-1}. This classification followed the above mentioned three frequency classes of slow, fast and very fast migrating speeds of benthic foraminifera. The frequency of these speed classes provides an overview of the overall activity of the species assemblages under the different temperatures.

Generally, slow migration speeds were more common in the cultures maintained at the lower temperature (10 °C), while at the higher temperature the fast and very fast migration speed classes prevailed. At the lower temperature, 9.4% and 11.8% of all observations were classified in the slow speed classes 0.5–1.0

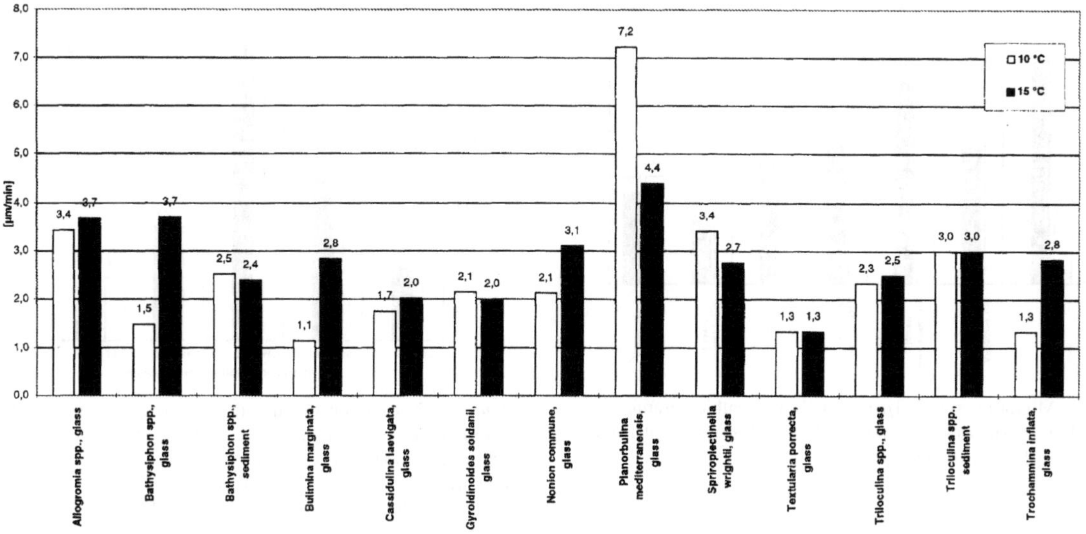

Figure 5. Migration speeds [μm min^{-1}] of benthic foraminifera measured at 10 °C and 15 °C (Banyuls-sur-Mer, 80 and 88 m water depth).

μm min^{-1} and 1.0–2.5 μm min^{-1}. At the higher temperature, the percentage of individuals moving with fast and very fast migration speeds was elevated. Maximum speed of 29.33 μm min^{-1} was observed at 15 °C, compared to 16.55 μm min^{-1} at 10 °C, indicating the increased percentage of very fast species (>15.0 μm min^{-1}) at higher temperatures. In conclusion, the effect of temperature on migrations was observed only for some foraminiferal species and at the community level higher, higher temperatures increased the overall migration activities.

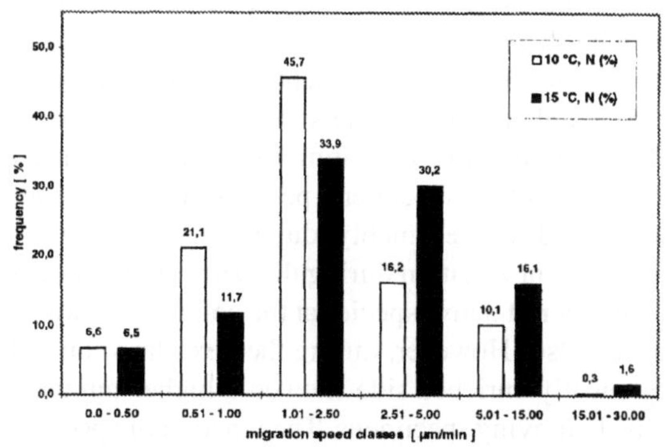

Figure 6. Distribution of migration speed classes [μm min^{-1}] of foraminifera measured in the temperature experiment (Banyuls-sur-Mer, 80 and 88 m water depth).

Influence of food enrichment

The foraminifera in the cultures ingested added algae relatively quickly within the first few days. Species such as *Bolivina lowmanni*, *Bulimina marginata* and *Uvigerina mediterranea* gathered *Chlorella* algae at the sediment surface by phagocytosis. The cytoplasm became green due to the ingestion of chloroplasts with the colour persisting for 2–3 days. Even shallow infaunal species like *Ammoscalaria pseudospiralis* or *Bigenerina nodosaria*, normally rarely seen at the surface, responded within days to a food pulse simulation. Some individuals moved into deeper sediment layers after successful ingestion of food, suggesting a specific response to the simulated surface food pulse. However, it was difficult to follow these movements since individuals were hidden within the sediment.

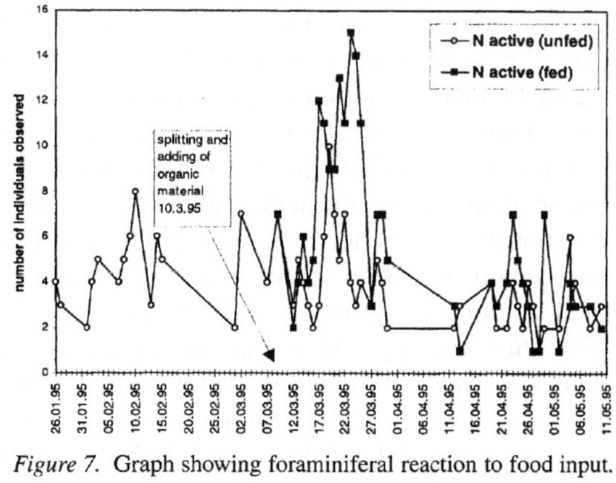

Figure 7. Graph showing foraminiferal reaction to food input.

Some foraminifera respond to organic enrichment with enhanced migration activity. This could

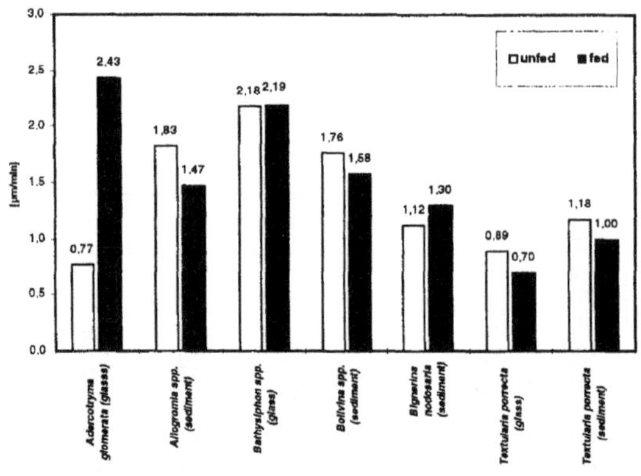

Figure 8. Migration speeds [μm min^{-1}] of benthic foraminifera observed under starving and organically enriched sediment conditions (M25-1 KG35, 1050 m, 10 °C).

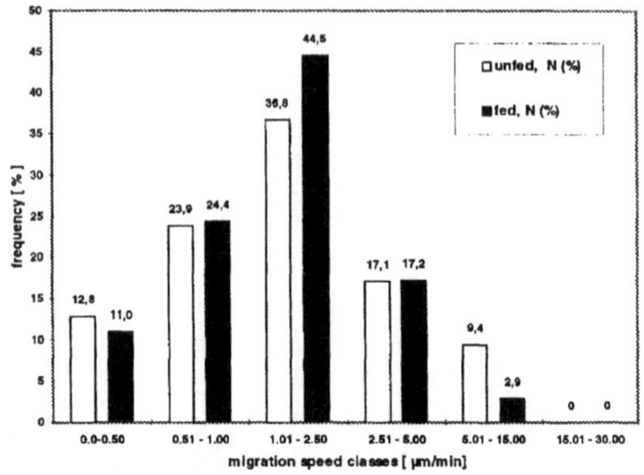

Figure 9. Distribution of migration speed classes [μm min^{-1}] of foraminifera observed in the feeding experiment (M25-1 KG35, 1050 m). Comparison between organically impoverished ($N = 234$) and enriched ($N = 209$) sediment conditions showed increased percentage of slow moving species under the organically enriched regime.

be demonstrated using an aquarium, which was divided into halves using a polycarbonate partition with only one side organically enriched. Six days after the start of the experiment, the number of observed active individuals increased on both sides (Figure 7). The response to organic enrichment could be clearly seen in the number of migrating specimen on the enriched side: the number of migrating foraminifera increased from 4–6 active specimens per day to 12–15 specimens per day. Presumably, due to the disturbance of the sediment zonation during the partitioning process, the number of active species at the control side was enhanced also. However, during the next days, only the organically enriched side showed a higher number of actively moving specimens. The reaction of specimens to the input of food was relatively fast and persisted for about 10 days (March 17–26). Thereafter, the activity of specimens became equal on both sides, and after 1 month the number of migrations was lower in the enriched sediment than in the untreated sediment.

Considering species differences between treatments (Figure 8), only two species showed an overall increase in mean migration speed under enriched sediment conditions. *Adercotryma glomerata* and *Bigenerina nodosaria* exhibited higher (68.3% and 13.9% increase, respectively) mean migration speeds following exposure to a richer food supply. Other species reduced their migration speeds under organically enriched conditions by 19.4% (*Allogromia* spp.) and 10.3% (*Bolivina* spp.) while moving within the sediment. *Textularia porrecta* reduced its speed at both substrates (glass = 21.5%; sediment = 15.2%). Only *Bathysiphon* spp. did not show much difference while moving along the glass walls. The maximum speed observed for *Allogromia* spp. decreased from 6.21 μm min^{-1} to 4.12 μm min^{-1} after food enrichment, but the maximum migration speed of *Bathysiphon* spp. was not influenced by the experimental enrichment.

Comparison of the enriched (N = 209) and untreated (N = 234) data sets showed statistically significant differences (U-test, p >0.05). The maximum migration speed observed in the enriched treatment was higher (V_{max} = 13.5 μm min^{-1}) compared with the control (V_{max} = 8.3 μm min^{-1}). These differences are expressed in the frequency distribution of migration speed classes (Figure 9). The differences between sites are clearly visible for slow moving foraminifera within the classes <0.5 and 1.0–2.5 μm min^{-1} with decreases in the numbers of observations of 1.8% and 7.7%, respectively. Migration speeds >5.0 μm min^{-1} were observed 6.5% more often under the untreated conditions. Only on the untreated side of the aquarium did foraminifera move at >10 μm min^{-1}.

Thus, a response by benthic bathyal foraminifera to organic enrichment was only apparent in the case of the two species *Adercotryma glomerata* and *Bigenerina nodosaria*, both of which showed an increase in mean migration speed. It was, however, not detected at the community level where there was an overall reduction in migration activity following experimental addition of food substances.

Influence of dysoxic conditions

The oxygen content in the sediment is believed to decisively control the benthic foraminiferal migra-

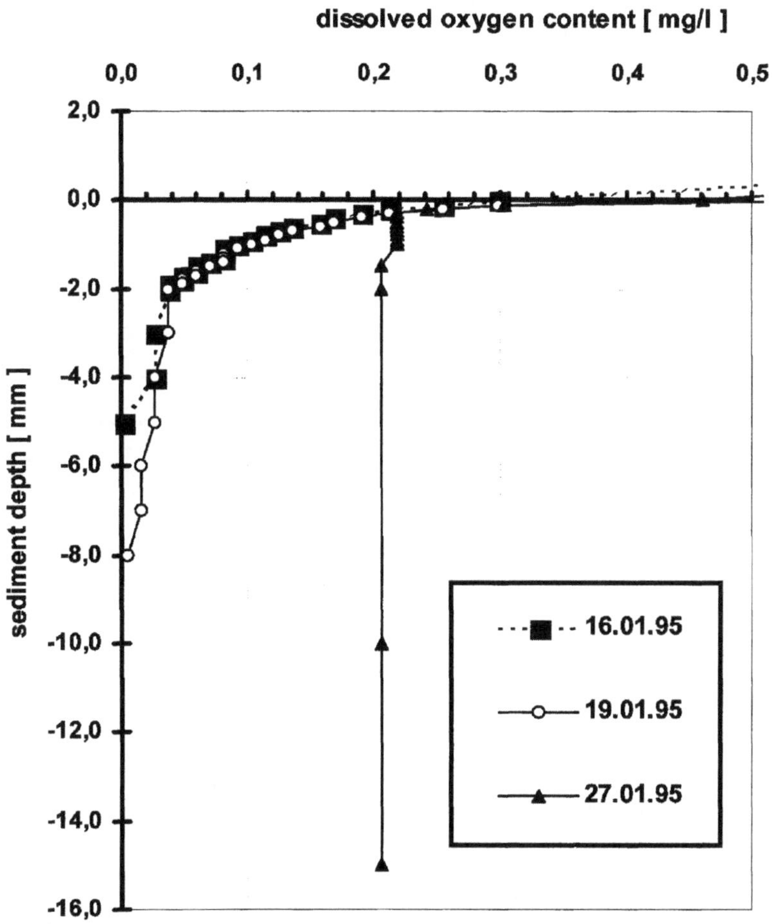

Figure 10. Dissolved oxygen content in the sediment of aquarium BAN AQ1 (80 m water depth, culture temperature 4°C) before (16 and 19 January 1995) and after porewater treatments (27 January 1995). The maximum oxygen penetration depth had increased from 4 mm to 8 mm sediment depth after continuous aeration (21–25 January 1995) and to more than 15 mm after repeated porewater exchange (25 January 1995).

tion activity (e.g. Moodley & Hess, 1992; Alve & Bernhard, 1995). Also, in the present cultures, a response in behaviour of foraminifera to changing oxygen conditions was apparent. Generally, the number of foraminifera in deeper sediment layers decreased when the dissolved oxygen concentrations dropped below <0.1 mg/l. Under these dysoxic conditions actively moving foraminifera were rarely observed, although some species could tolerate such conditions in dormant stages or cysts for periods of weeks and months. With the raising of the redoxcline owing to a reduced oxygen concentration in the sediment, most foraminifera migrated to the sediment–water interface. Some foraminifera also tried to migrate up the glass walls of the culture vessels, even when the pseudopodial network was too weak to hold the heavy test as in the case of *Ammoscalaria pseudospiralis* and *Bigenerina nodosaria*.

As foraminifera responded either with migration to the surface layers or dormancy, it was not possible to determine the influence of dissolved oxygen content on individual migration speeds with the methods used. However, one experiment with reduced and experimentally increased content of dissolved oxygen in the aquarium sediment provided the opportunity to quantify the influence of oxygen on foraminiferal migration speed. An oxygen-profile from the top sediment layer of the aquarium BAN AQ9 (80 m) at 4°C

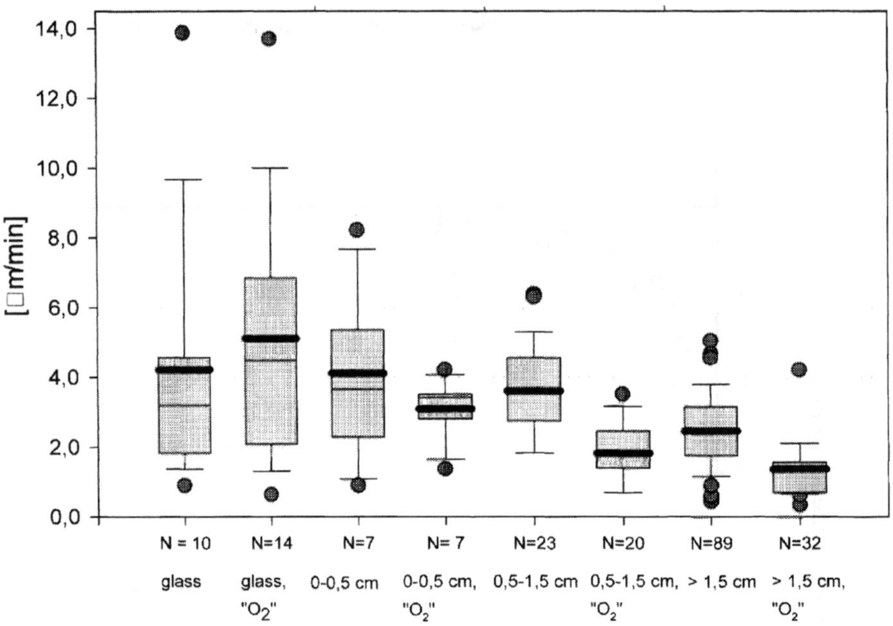

Figure 11. Mean migration speed of *Quinqueloculina lamarckiana* (BAN AQ1, 80 m, 4°C) under strong and less (O_2) hypoxic conditions. Migration speed varied on the glass, in the top (0–0.5 cm); intermediate (0.5–1.5 cm) and lower sediment horizons(>1.5 cm) and showed no difference in the top layers but increased under oxygen undersaturated sediment conditions.

on January 1995, showed a decreased content of dissolved oxygen below the sediment surface (Figure 10). In December the culture had been fed with *Chlorella* algae and, consequently, oxygen (<0.1 mg/l) penetrated to only 5 mm sediment depth. By 19 January 1995 oxygen had penetrated down to 8 mm sediment depth because of constant aeration and porewater exchange. The porewater exchange was repeated a week later on the 25 January 1995. Within two days the stagnant level of oxygenated sediment had returned to around 0.2 mg/l down to 24 mm sediment depth.

Two month before the oxygen fluctuations described above, a reproduction of one *Quinqueloculina lamarekiana* had produced 150 juveniles. At the beginning of the observation (16 January) all foraminifera which had settled to the bottom of the aquarium migrated in the oxygen depleted (<0.1 mg/l) sediment layer. A comparison of the different migration rates in the deep (>1.5 cm), intermediate (1.0–0.5 cm) and shallow (>0.5 cm) sediment layers under dysoxic (before the 25 January 1995) and under ventilated hypoxic sediment conditions (after the 25 January 1995) revealed astonishingly few differences (Figure 11). Only the migration speeds of *Quinqueloculina lamarckiana* moving along the glass and within the top 5 mm of the sediment were significantly higher after aeration (U-test, $p < 0.05$). Migration in the deeper layers, on the other hand, was faster before the aeration with 3.61 μm/mm (O_2 = 1.82 μm/mm) in the 0.5–1.0 cm horizon and 2.46 μm/mm (O_2 = 1.37 μm/mm) in the deepest layer > 1.5 cm sediment depth. Thus, migration speeds in the sediment were always higher before aeration than afterwards. In conclusion, although a vertical movement towards the surface layers could be documented for most foraminiferal species in response to very low dissolved oxygen levels in the sediment, no evidence of reduction in speed was found for the migration activity of *Quinqueloculina lamarekiana*.

Discussion

Benthic foraminifera are abundant in the deep-sea and contribute substantially to the benthic biomass (cf. Gooday, 1994). The reaction of foraminifera to food pulses has been documented by biochemical methods (Linke, 1992; Linke et al., 1995) and population studies (Gooday, 1993; Kitazato & Ohga, 1995; Stitger, 1996; Silva et al., 1996). The present study investigated the behavioural response of living bathyal benthic foraminifera under changing sediment conditions.

By time-lapse video recording, the average migration speed of foraminifera from water depths <2880 m has been found to range between 0.1 and 32.4 μm min^{-1}, with maximum speeds reaching 87.8 μm min^{-1} in the case of *Hoeglundina elegans* (M33-1, 1920 m). Hemleben & Kitazato (1995) measured a mean migration speed of 10.6 μm min^{-1} for *Gavelinopsis translucens* from 1847 m water depth. For *Cibicidoides pachyderma*, Bornmalm et al. (1997) found migration speeds of only 3.8 μm min^{-1} in specimens from 220 m water depth. Some species are able to move as fast as their shallow water relatives, reflecting no pressure effect or high pressure tolerance similar to that of deep-sea flagellates (Turley et al., 1988; Turley & Carstens, 1991). Thus, the inferred lower activity rate of deep-sea species (Weinberg, 1991; Hemleben & Kitazato, 1995) has to be rejected.

Response to temperature

One possible factor causing the observed variations in migration speed of benthic foraminifera may be the temperature conditions during culturing. To investigate the effects of temperature, eight samples from the continental shelf off Banyuls-sur-Mer were exposed to two different temperatures. In the temperature experiment, most species showed decreased migration rates when cultured at lower temperature with a range from 0.5% (*Textularia porrecta*) and 59.4% (*Bulimina marginata*), with the exception of two epifaunal hemisessile species (*Planorbulina mediterranea*, *Spiroplectinella wrightii*). Thus, most of the species investigated showed a generally decreased activity at lower temperatures. Also, the frequency distribution of migration speed classes was biased towards the slower migration speeds at the reduced temperature (10 °C).

Maximum speed for a shallow water foraminifera (*Pseudorotalia gaimardii*) cultured at 20 °C reached values of 500 μm min^{-1} as demonstrated by photographic sequences (Kitazato, 1981). Using similar methods, Wetmore (1988) observed high migration speeds in the two subtropical species *Peneroplis pertusus* (175.0 μm min^{-1}) and *Quinqueloculina* sp. 1 (138.3 μm min^{-1}) when cultured at an elevated temperature of 25 °C. This is in agreement with studies of Langer et al. (1989) on Mediterranean shallow water species which moved at speeds of 200–400 μm min^{-1} at room temperature. Kitazato (1988) showed a clear increase in migration speed in *Pseudorotalia gaimardii* and *Rectobolivina raphana* according to whether they were cultured at 17 °C or 21 °C: both foraminifera increased their migration rate at the higher temperature from 27.3 μm min^{-1} to 63.0 μm min^{-1} and from 8.0 μm min^{-1} to 16.5 μm min^{-1}, respectively.

This correlation between migration rates and temperature may reflect enhanced metabolism at higher temperatures ('van't Hoff rule'). Another indicator for elevated cell metabolism is the respiration rate. Hannah et al. (1994) showed that foraminifera increase their respiration rate drastically during migration. At lower temperatures, higher respiration rates increase the general cell metabolism (Lassere, 1976; Price & Warwick, 1980). Possibly the differences in migration speed at 10 °C and 15 °C can be attributed to the enhanced energy demand of species migrating at lower temperatures.

Response to food input

Individuals with ingested algae could be easily distinguished for a few days from others by the green colour of their cytoplasm. Such green specimens were not restricted to the enriched sediment surface but could also be found in deeper layers, suggesting vertical movement after feeding. During the feeding experiments, some foraminifera reacted with enhanced migration to food enrichment. Within the first few days, the number of migrating foraminifera increased significantly. After a week, this response declined.

Only two species (*Adercotryma glomerata*, *Bigenerina nodosaria*) showed an elevated mean migration speed. This is surprising as one might expect more activity after feeding. The reason for this lack of response is probably connected to the life strategy of many foraminifera. When food has been captured by the pseudopodial network, the individual tries to withdraw into a resting or feeding cyst for undisturbed phagocytosis. Inactivity of fed foraminifera was also observed by Langer & Gehring (1993). They proposed a bacterial farming strategy for *Rosalina* sp. which was observed to visit previous grazing sites repeatedly. They did not demonstrate different migration speeds. However, Bernhard & Bowser (1992) showed significant differences in migration rates for *Allogromia* sp. grazing on biofilms attached to glass plates. They suggested different nutritional values of the substrate responsible for reduced migration speeds of the foraminifera.

Although deep-sea foraminifera are adapted to longer periods of starvation (Linke, 1992; Graf & Linke, 1992), freshly deposited phytodetritus material can be quickly incorporated. In a copiotroph adaptation, the inactive starving population changes to an active population as also shown for bacteria (Poremba, 1994) and invertebrates (Lampitt, 1990; Smith et al., 1994). The present results suggest that the response of foraminifera to a simulated food pulse is not only an 'awakening response' in ATP-metabolism activity (Linke, 1992) but, at least, also in some species in the migratory activity.

Response to oxygen gradients

The influence of oxygen gradients on migration rates was observed in cultures where the redoxcline reached the surface. In such cases, escaping foraminifera migrated to the sediment surface and climbed up the glass walls, trying to avoid anoxia within the sediment. Migration to an elevated habitat has been previously documented for foraminifera (Moodley & Hess, 1992; Langer & Bagi, 1994; Moodley et al. 1997) and other meiobenthic animals (Meyers et al., 1988). Alve & Bernhard (1995) demonstrated that benthic foraminifera distribute themselves randomly in the different sediment layers tubes following longer periods of anoxia after having concentrated before at the sediment surface or attached to polychaete tubes. A number of species, however, could withstand prolonged dysoxic conditions. *Ammonia beccarii* and *Rosalina* sp. were observed in the microcosm to live in cysts for longer periods. Dormancy under steep subsurface oxygen gradients may be one way to reduce possible damage to the cell metabolism (Bernhard, 1996). Resting stages and inactivity as possible responses to unfavourable conditions have been suggested for *Elphidium incertum* living in severely oxygen depleted *Beggiatoa* mats (Linke & Lutze, 1993; Bernhard, 1996).

The experimental investigations of *Quinqueloculina lamarckiana* showed, however, that foraminifera are not necessarily affected in migrational activity by oxygen depletion in the sediment. Despite dysoxic conditions, the juveniles of this species showed very active locomotion within the sediment, no negative geotactic migration to the surface layer and no decrease in mean migration speed compared to those migrating under less steep oxygen gradients. The high tolerance of this species to anoxic or dysoxic conditions is consistent with the observations of Bernhard (1993, 1996) and Bernhard & Alve (1996). Since foraminifera are widely distributed in oxygen-depleted regions, such as oxygen minimum zones, fjords and polluted areas, they are considered as facultative anaerobic protists, being able to withstand severe dysoxic conditions for longer periods (Bernhard, 1996).

Nevertheless, there is an ongoing debate about the extent to which these protists are adapted to reduced oxygen levels in the sediment (Alve & Bernhard, 1995; Moodley et al, 1996; Bernhard & Alve, 1996). The oxygen content of the bottom seawater has been considered to be a main decisive factor controlling the lateral and vertical distribution of foraminifera in the sediment (Kaiho, 1994). Moodley & Hess (1992) showed in experiments that infaunal foraminifera may migrate toward a source of oxygen, such as aerated artificial macrofaunal burrows. The mobility of foraminifera enables them to undergo substantial depth changes within a few days. But even if some species (e.g. *Ammonia beccarii*, *Elphidium excavatum* and *Quinqueloculina seminulum*) are well adapted to anoxia/dysoxia (Moodley et al., 1998), this does not explain why some other foraminiferal species escape by migration to the sediment surfaces. Possibly, the energy required for respiration may lead this foraminifera to the typically richer food zones in the surface layers.

Migration classes

Migration speed analyses of benthic foraminifera revealed, despite variability between species, a possible grouping into three migration classes, slow, fast and very fast, when speeds of taxa were analysed with a χ^2-test. These species groupings were consistent at 4 °, 10 ° and 15 °C. Obviously, within the benthic community a separation into different life styles can be expected. Different migration speeds may have considerable ecological importance in bathyal and other deep-sea settings where food availability is limited.

The formation of cysts or secondary tests (Cedhagen, 1996) did not deter the protozoans from searching for elevated or sheltered microhabitats (Lutze & Thiel, 1989). But in these cases, a moderate or slow rate of movement is sufficient. A hemisessile life strategy with low migration speeds seems to be of advantage for some species living under high predation pressure. For example, camouflage of calcareous species under detrital cysts, as seen in *Fontbotia wuellerstorfii*, was speculated to be a response to predation pressure by macrobenthic snails (Cedhagen, 1996).

Other more infaunal species are characterised by moderately fast migration speeds. The third group of very fast moving species, however, can be considered to be adapted to a hunting life style. Higher speeds might be of advantage for some species under prolonged oligotrophic conditions. Higher migration rates were suggested to be an adaptation to a carnivore life strategy in the epifaunal foraminifer *Fissurina* sp. (Kitazato, 1988). Such species are constantly in search for food, mainly on the surface of the sediment in order to cover their high energy demand caused by their active life style. Another requirement is to withstand competition and predation by other meiobenthic animals on the surface of the sediment. Rapid movement may help to avoid being eaten, or to quickly gain access to deeper sediment layers, as suggested by Gooday (1986), Corliss (1985) and Corliss & Silva (1993). The high migration activity, however, demonstrates the complex responses displayed by the foraminifera to varying environmental conditions.

Conclusion

Like most organisms, foraminifera must gather food and at the same time avoid predation or unfavourable environmental conditions. Horizontal and vertical migration from the food-rich surface into deeper layers is a possible mechanism to avoid predators. However, in deeper sediment layers, the amount of food is reduced and the biogeochemical environment (e.g. low dissolved oxygen content) may stress the foraminiferan cell. In such case, the ability to migrate out of the reduced conditions is advantageous. Active migration may also become necessary when bioturbation processes distribute individuals to deeper sediment layers.

All species examined here were able to migrate within the sediment layers. Escape mechanism in order to avoid oxygen depletion and search for obtaining food were documented. However, owing to the technical difficulties involved in in situ studies, we are still at the beginning of understanding the behavioural diversity of these protists.

Acknowledgements

I would like to thank the captain, crew and participants of the R.V. *'GEORGE PETIT'* and R.V. *'METEOR'* cruises for their assistance at sea. Thanks are due to Beate Breitinger, Christine Hock and Bodo Rödiger for the help in establishing and monitoring the laboratory work. Thanks are also due to Rolf Koppelmann, Ulf Luth, Andy Gooday, Andreas Mackensen and another anonymous referee for their linguistic improvements and critical comments to earlier versions of the manuscript. This work was part of a Ph D. thesis in the working group of Prof. Ch. Hemleben at the University of Tübingen supported by the Deutsche Forschungsgemeinschaft.

References

Alongi, D. M., 1992. Bathymetric patterns of deep-sea benthic communities from bathyal to abyssal depths in the western South Pacific (Solomon and Coral Seas). Deep-Sea Res. 39: 549–565.

Altenbach, A. V., 1992. Short term processes and patterns in the foraminiferal response to organic flux rates. Mar. Micropaleontol. 19: 119–129.

Anderson, O. R, J. J. Lee & W. W. Faber, 1991. Collection, maintenance and culture methods for the study of living foraminifera. In Lee, J. J. & O. R. Anderson (eds), Biology of Foraminifera. Academic Press, London: 335–358.

Alve, E. & J. M. Bernhard, 1995. Vertical migratory response of benthic foraminifera to controlled oxygen concentrations in an experimental mesocosm. Mar. Ecol. Progr. Ser. 116: 137–151.

Barnet, P. R. O., J. Watson & D. Conelly, 1984. A multiple corer for taking virtually undisturbed samples from shelf, bathyal and abyssal sediments. Oceanol. Acta 7: 399–408.

Bernhard, J. M., 1993. Experimental and field evidence of Antarctic foraminiferal tolerance to anoxia and hydrogen sulfide. Mar. Micropaleontol. 20: 203–213.

Bernhard, J. M., 1996. Microaerophilic and facultative anaerobic benthic foraminifera: a review of experimental and ultrastructural evidence. Rev. Paleobiol. 15: 261–275.

Bernhard, J. M. & E. Alve, 1996. Survival, ATP pool and ultrastructural characterization of benthic foraminifera from Drammensfjord (Norway): response to anoxia. Mar. Mircropal. 28: 5–17.

Bernhard, J. M. & S. S. Bowser, 1992. Bacterial biofilms as a trophic resource for certain benthic foraminifera. Mar. Ecol. Progr. Ser. 83: 263–272.

Bernstein, B. B. & J. P. Meador, 1979. Temporal persistence of biological patch structure in an abyssal benthic community. Mar. Biol. 51: 179–183.

Bornmalm, L., B. H. Corliss & K. Tedesco, 1997. Laboratory observations of rates and patterns of movement of continental margin benthic foraminifera. Mar. Micropal. 29: 175–184.

Boltovskoy, E., 1966. Depth at which Foraminifera can survive in sediments. Cushman Found. Foram. Res. Contr. 17: 43–45.

Cedhagen, T., 1996. Foraminiferans as food for cephalaspideans (Gastropoda: Opisthobranchia), with notes on secondary tests around calcareaous foraminiferans. Phuket Mar. Biol. Cent. Spec. Publ. 16: 279–290.

Chandler, G.T., Williams D. F. & G. Xiaodong, 1996. Sediment microhabitat effects on carbon isotopic signatures of mircrocosm-cultered benthic Foraminifera. Limnol. Oceanogr. 41: 680–688.

Corliss, B. H., 1985. Microhabitats of benthic foraminifera within deep-sea sediments. Nature 314: 435–438.

Corliss, B. H., 1991. Morphology and microhabitat preferences of benthic foraminifera from the northwest Atlantic Ocean. Mar. Micropaleontol. 17: 195–236.

Corliss, B. & K. A. Silva, 1993. Rapid growth of deep-sea benthic foraminifera. Geology 21: 991–994.

Gage, J. D., 1996. Why are there so many species in deep-sea sediments ?. J.exp. mar. Biol. Ecol. 200: 257–286.

Gooday, A. J., 1986. Meiofaunal foraminiferans from the bathyal Porcupine Seabight (northeast Atlantic): size structure, standing stock, taxonomic composition, species diversity and vertical distribution in the sediment. Deep-Sea Res. 33: 1345–1373.

Gooday, A. J., 1988. A response by benthic foraminifera to the deposition of phytodetritus in the deep sea. Nature 332: 70–73.

Gooday, A. J., 1993. Deep-sea benthic foraminiferal species which exploit phytodetritus: characteristic features and controls on distribution. Mar. Micropaleontol. 22: 187–205.

Gooday, A. J., 1994. The biology of deep-sea foraminifera: a review of some advances and their applications in paleoceanography. Palaios 9: 14–31.

Gooday, A. J. & P. J. D. Lambshead, 1989. Influence of seasonally deposited phytodetritus on benthic foraminiferal populations in the bathyal northeast Atlantic: the species response. Mar. Ecol. Prog. Ser. 58: 53–67.

Gooday, A. J. & C. M. Turley, 1990. Responses by benthic organisms to inputs of organic material to the ocean floor: a review. Phil. Trans. r. Soc., Lond. 331(119): 119–138.

Gooday, A. J., L. A. Levin, P. Linke & T. Heeger, 1992. The role of benthic foraminifera in the deep-sea food webs and carbon cycling. In Rowe, G. T. & V. Pariente (eds), Deep-Sea Food Chains and the Global Carbon Cycle. NATO Asi Ser. C., 360, Kluwer Academic Publishers, Dordrecht, The Netherlands: 63–91.

Graf, G. & P. Linke, 1992. Adenosine nucleotides as indicators of deep-sea benthic metabolism. In Rowe, G. T. & V. Pariente (eds), Deep-Sea Food Chains and the Global Carbon Cycle, NATO Asi Ser. C., 360, Kluwer Academic Publishers, Dordrecht, The Netherlands: 92–105.

Graf, G., S. A. Gerlach, P. Linke, W. Queisser, W. Ritzrau, A. Scheltz, L. Thompsen & U. Witte, 1995. Benthic–pelagic coupling in the Greenland-Norwegian Sea and its effect on the geological record. Geol. Rdsch. 84: 49–58.

Gross, O., 1998. Investigations on autecology, migration and bioturbation of living benthic deep-sea Foraminifera (Protozoa). Ber. Zentr. Meeres- Klimaforsch. Reihe E. Hydrobiol. Fischereiwiss. 15: 1–224.

Hannah, F. & R. Rogerson, 1997. The temporal and spatial distribution of foraminiferans in marine benthic sediments of the Clyde Sea area, Scotland. Estuar. coast. shelf. Sci. 44: 377–383.

Hannah, F., R. Rogerson & J. Laybourn-Parry, 1994. Respiration rates and biovolumes of common benthic foraminifera (Protozoa). J. mar. biol. Ass. U. K. 74: 301–312.

Hemleben, Ch. & H. Kitazato, 1995. Deep-sea foraminifera under long time observation in the laboratory. Deep-Sea Res. 42: 827–832.

Hieke, W., P. Halbach, M. Türkay & H. Weikart,1994. Mittelmeer 1993. Cruise No. 25. Meteor-Berichte: 94–3.

Jorissen, F. J., H. C. de Stigter & J. G. V. Widmark, 1995. A conceptual model explaining benthic foraminiferal microhabitats. Mar. Micropaleontol. 26: 3–15.

Kaiho, K., 1994. Benthic foraminiferal dissolved-oxygen index and dissolved-oxygen levels in the modern ocean. Geology 22: 719–722.

Kaminski, M. A., A. M. Al Hassawai & W. Kuhnt, 1994. Seasonality in microhabitats of rose bengal stained deep-sea benthic foraminifera from the New Jersey continental margin, Abstract, Forams 94, Paleobios Suppl.: 39.

Kitazato, H., 1981. Observation of behaviour and mode of life of benthic foraminifers in laboratory. Geosci. Rep. Shizuoka Univ.6: 61–71.

Kitazato, H., 1983. Microhabitats of benthic foraminifera and their application to fossil assemblages. In Oertli, H. J. (ed.), Benthos' 83: 2nd Int. Symp. Benthic Foraminifera, Elf Aquitaine, Esso REP, Total DFP. Pau and Bordeaux: 339–344.

Kitazato, H., 1988. Locomotion of some benthic foraminifera in and on sediments. J. Foram. Res. 18: 344–349.

Kitazato, H. & T. Ohga, 1995. Seasonal changes in deep-sea benthic foraminiferal populations: results of long-term observations at Sagami Bay, Japan. In Sakai, H. & Y. Nozaki (eds), Biochemical Processes and Ocean Flux in the Western Pacific, Terra Sci. Publ., Tokyo: 331–342.

Lampitt, R. S., 1990. Directly measured rapid growth of a deep-sea barnacle. Nature 345: 805–807.

Langer, M. R. & H. Bagi, 1994. Tubicolous polychaetes as substrates for epizoic foraminifera. J. Micropaleontol. 132.

Langer, M. R. & C. A. Gehring, 1993. Bacteria farming: a possible feeding strategy of some smaller, motile Foraminifera. J. Foram. Res. 23: 40–46.

Langer, M. R., L. Hottinger & B. Huber, 1989. Functional morphology in low-diverse benthic foraminiferal assemblages from tidal flats of the North Sea. Senckenberg. marit. 20: 81–99.

Lassere, P., 1976. Metabolic activities of benthic microfauna and meiofauna, In McCave, I. N. (ed.), The Benthic Boundary Layer. Plenum Press, New York, London: 95–142.

Lee, J. J., 1974. Towards understanding the niche of foraminifera. In Hedley, R. & C. G. Adams (eds), Foraminifera. Vol. 1. Academic Press, London: 201–260.

Levin, L. A., 1991. Interactions between metazoans and large, agglutinating protozoans: implications for the community structure of deep-sea benthos. Am. Zool. 31: 886–900.

Linke, P., 1992. Metabolic adaptations of deep-sea benthic foraminifera to seasonality varying food input. Mar. Ecol. Prog. Ser.: 51–63.

Linke, P. & G. F. Lutze, 1993. Microhabitat preferences of benthic foraminifera – a static concept or a dynamic adaptation to optimize food acquisition? Mar. Micropaleontol. 20: 215–234.

Linke, P., A. V. Altenbach, G. Graf & T. Heeger, 1995. Response of deep-sea benthic foraminifera to a simulated sedimentation event. J. Foram. Res. 25: 75–82.

Loubere, P., A. Gary & M. Lagoe, 1993. Sea-bed biogeochemistry and benthic foraminiferal bathymetric zonation on the slope of the Northwest Gulf of Mexico. Palaios 8: 439–449.

Loubere, P., P. Meyers & A. Gary, 1995. Benthic foraminiferal microhabitat selection, carbon isotope values and association with larger animals: a test with *Uvigerina peregrina*. J. Foram. Res. 25: 83–95.

Lutze, G. F. & H. Thiel, 1989. Epibenthic foraminifera from elevated microhabitats: *Cibicidoides wuellerstorfi* and *Planulina ariminensis*. J. Foram. Res. 19: 153–158.

Mackensen, A., H. W. Hubberten, T. Bickert, G. Fischer & D. K. Fütterer, 1993. The ^{13}C in benthic foraminiferal tests of *Fontbotia Wuellerstorfi* (Schwager) relative to the ^{13}C of dissolved inorganic carbon in southern ocean deep water: implications for glacial ocean circulation models. Paleoceanography 8: 587–610.

Meyers, M. B., E. N. Powell & H. Fossing, 1988. Movement of oxybiotic and thiobiotic meiofauna in response to changes in pore-water oxygen and sulfide gradients around macro-infaunal tubes. Mar. Biol. 98: 395–414.

Moodley, L. & C. Hess, 1992. Tolerance of infaunal benthic foraminifera for low and high oxygen concentrations. Biol. Bull. 183: 94–98.

Moodley, L., P. M. J. Herman, J. J. Middelburg &. C. H. R. Heip, 1996. Comment: subsurface activity of benthic foraminifera. Mar. Ecol. Prog. Ser. 145: 303–304.

Moodley, L., B. E. M. Schaub, G. J. Van der Zwaan & P. M. J. Herman, 1998. Tolerance of benthic foraminifera (Protista: Sarcodina) to hydrogen sulphide. Mar. Ecol. Prog. Ser. 169: 77–86.

Moodley, L., G. J. Van der Zwaan, P. M. J. Herman, L. Kempers & P. Von Breugel, 1997. Differential response of benthic meiofauna to anoxia with special reference to Foraminifera (Protista: Sarcodina). Mar. Ecol. Prog. Ser. 158: 151–163.

Pfannkuche, O., 1993. Benthic response to the sedimentation of particulate organic matter at the BIOTRANS station, 47 ° N, 20 ° W. Deep-Sea Res. 40: 135–149.

Pfannkuche, O., W. Balzer & F. Schott, 1994. Carbon-cycle and transport of water masses in the North Atlantic– the winter situation, Cruise No. 27. *'Meteor'*-Berichte: 94–6.

Poremba, K., 1994. Simulated degradation of phytodetritus in deep-sea sediments of the NE Atlantic (47 ° N, 19 ° W). Mar. Ecol. Progr. Ser. 105: 291–299.

Price, R. & R. M. Warwick, 1980. The effect of temperature on the respiration rate of meiofauna. Oecologia 44: 145–148.

Rex, M. A., 1997. An oblique slant on deep-sea biodiversity. Nature 385: 577–578.

Severin, K. P., 1987. Laboratory observations of the rate of subsurface movement of a small miliolid foraminifer. J. Foram. Res. 17: 110–116.

Severin, K. P. & M. G. Erskian, 1981. Laboratory experiments on the vertical movement of *Quinqueloculina impressa* Reuss through sand. J. Foram. Res. 11: 133–136.

Severin, K. P., S. J. Culver & C. Blanpied, 1982. Burrows and trails produced by *Quinqueloculina impressa* Reuss, a benthic foraminifer in fine-grained sediment. Sedimentology 29: 897–901.

Silva, K. A., B. H. Corliss, A. E. Rathburn & R. C. Thunell, 1996. Seasonality of living benthic foraminifera from the San Pedro Basin, California Borderland. J. Foram. Res. 26: 71–93.

Smith, C. R., 1994. Tempo and mode in deep-sea benthic ecology: punctuated equlibrium revisited. Palaios 9: 3–13.

Smith, K. L., R. S. Kaufmann & R. J. Baldwin, 1994. Coupling of near-bottom pelagic processes at abyssal depths in the eastern North Pacific Ocean. Limnol. Oceanogr. 39: 1101–1118.

Stigter, H. C., 1996. Recent and fossil benthic foraminifera in the Adriatic Sea: distributional patterns in relation to organic flux and oxygen concentration at the seabed. Ph D. Thesis. Univ. Utrecht, Geologica Ultraiectina 144: 1–254.

Thiel, H., 1983. Meiobenthos and nanobenthos of the deep-sea. In Rowe, G. T. (ed.), The Sea, Vol. 8. Deep-Sea Biology, John Wiley, New York: 157–230.

Thiel, H., O. Pfannkuche, G. Schriever, K. Lochte, A. J. Gooday, C. Hemleben, R. F. C. Montoura, C. M. Turley, J. W. Patching & F. Riemann, 1988. Phytodetritus on the deep-sea floor in a Central Oceanic Region of the Northeast Atlantic. Biol. Oceanogr. 6: 203–239.

Thistle, D., 1983. The role of biologically produced habitat heterogeneity in deep-sea diversity maintenance. Deep-Sea Res. 30: 1234–1245.

Travis, J. J. & S. S. Bowser, 1991. The motility of foraminifera. In Lee, J. J. & O. R. Anderson (eds), The Biology of Foraminifera. Academic Press, London: 91–155.

Turley, C. M. & M. Carstens, 1991. Pressure tolerance of oceanic flagellates: implications for remineralization of organic matter. Deep-Sea Res. 38: 403–413.

Turley, C. M., K. Lochte & D. J. Patterson, 1988. A barophilic flagellate isolated from 4500 m in the mid-North Atlantic. Deep-Sea Res. 35: 1079–1092.

Van Gemerden, H., C. S. Tughan, R. deWit & R. Herbert, 1989. Laminated microbial ecosystems on sheltered beaches in Scapa Flow, Orkney Islands. Microbiol. Ecol. 62: 87–102.

Weinberg, J. R., 1991. Rates of movement and sedimentary traces of deep-sea foraminifera and mollusca in the laboratory. J. Foram. Res. 3: 213–217.

Wetmore, K. L., 1988. Burrowing and sediment movement by benthic foraminifera, as shown by time-lapse cinematography. Rev. Paléobiol., Vol. Spéc. 2, Benthos 86: 921–927.

Hydrobiologia **426:** 139–156, 2000.
G. Liebezeit, S. Dittmann & I. Kröncke (eds), Life at Interfaces and Under Extreme Conditions.

Main characteristics of the boundary layer macrofauna in the English Channel

Jean-Claude Dauvin[1,*], Carole Vallet[2], Pascal Mouny[1,2] & Souaad Zouhiri[2]
[1]*Station Marine de Wimereux, UPRES A 8013 ELICO, Université des Sciences et Technologie de Lille, BP 80, 28 Avenue Foch, F-62930 Wimereux, France*
[2]*Muséum National d'Histoire Naturelle, Laboratoire de Biologie des Invertébrés Marins et Malacologie, UPRESA, 57 rue Cuvier, F-75231 Paris Cedex 05, France*
*E-mail: dauvin@pop.univ-lille1.fr (*author for correspondence)*

Key words: English Channel, benthic boundary layer, macrofauna, suprabenthic fauna, biomass, abundance

Abstract

Four hundred and thirty two suprabenthic hauls with a modified Macer-GIROQ sledge were taken over the period 1988–1996. Fifteen sites within the English Channel were sampled from the Roscoff and Plymouth waters to the Dover Strait, including the Bay of Morlaix, Bay of Saint Brieuc, the Bay of Seine and the Seine Estuary. The main characteristics of the Benthic Boundary Layer (BBL) macrofauna in the English Channel were elucidated from these data. BBL macrofauna was found over the whole water column sampled (10–150 m). Firstly, the number of species collected was very high (up to 116 species for five hauls) except in the Seine estuary. Numerically, mysids were the dominant group of organisms collected with the sledge, followed by amphipods, cumaceans, isopods and decapods. Abundance and biomass of the fauna were very high in the Seine estuary, especially in the mesohaline and oligohaline zones (maximum >216 000 ind. 100 m^{-3}; >200 g AFDW 100 m^{-3}) where the specialised mysids *Neomysis integer* and *Mesopodopsis slabberi* occurred in very high abundance. At other sites, the abundance and biomass were moderate (100–10 000 ind. 100 m^{-3}; 15–2000 mg AFDW 100 m^{-3}). Abundance reached its maximum when mysids or swarms of amphipods of the genus *Apherusa* were collected, and during the reproduction of the amphipods of the genus *Ampelisca*, as in June in the Bay of Morlaix. The number of species, abundance and biomass showed seasonal changes in all sites with, in general, lowest values in winter or spring, and highest values in autumn. For the open sea sites, two main geographical gradients were identified: i) in spring, a higher number of species in the western part of the Channel compared with the eastern part of the Channel and, in autumn, a higher number of species in the eastern part of the Channel compared with the western part, and ii) an increasing density gradient from west to east at spring and a decreasing gradient from west to east during the autumn. The suprabenthic fauna of the Channel can be considered as a unique community including three kinds of species: (i) good swimmer species which were common at all the time and all sites: e.g. *Anchialina agilis*, *Gastrosaccus* spp., *Haplostylus* spp. and *Apherusa* spp., (ii) species present at all sites only at night: e.g. *Atylus vedlomensis*, *Megaluropus agilis*, *Melphipidella macra*, *Orchomenella nana*, *Pseudocuma longicornis*, *Schistomysis ornata* and *Synchelidium maculatum*, and iii) species collected at one or a limited number of sites only at night: e.g. *Ampelisca* spp. at Pierre Noire and *Diastylis* spp. at site F. The Seine estuary supports a typical suprabenthic estuarine community. Elements of the suprabenthic fauna also showed significant daily rhythms which took several forms depending on the species under consideration.

Introduction

Many benthic species are known to swim actively for much or all their life cycle in the layer just above the sea bed. This fauna has generally been called suprabenthos, hyperbenthos, Benthic Boundary Layer macrofauna, demersal zooplankton or benthopelagic plankton (Brunel et al., 1978; Mees, 1994; Dauvin et al., 1994, 1995; Mees & Jones, 1997). In spite of differences in terminology used for this fauna, in recent years there have been considerable developments in research on this topic from the tropical seas to the Arctic Ocean, and from shallow waters to deep seas (Mees & Jones, 1997). As defined by Brunel et al. (1978), the suprabenthic fauna collected in the Benthic Boundary Layer includes all bottom-dependent animals that perform, with variable regularity, daily or seasonal vertical migrations. Crustaceans, especially decapods and peracarids, dominate in the water column near the bottom. Amongst the peracarids, amphipods and mysids are dominant in the English Channel as in other parts of the world ocean (Kaartvedt, 1986, 1989; Buhl-Jensen & Fosså, 1991; Vallet & Dauvin, 1995, 1998, 1999; Vallet, 1997; Zouhiri et al., 1998). The composition and swimming activity of amphipods of the English Channel have been described previously (Vallet & Dauvin, 1995; Dauvin & Zouhiri, 1996; Zouhiri & Dauvin, 1996). In the English Channel, mysids form important swarms in the Benthic Boundary Layer and in the water column in coastal areas (Zouhiri et al., 1998). They form a direct link between the benthos and the pelagos because of their daily vertical and horizontal migrations. In other areas, the suprabenthos is an important food source of nectobenthic fishes; decapods and peracarids constitute the main prey of many fish species (Sorbe, 1984; Mees & Jones, 1997).

This paper presents information on the species composition, densities and biomass of the suprabenthic fauna of the English Channel derived from 432 suprabenthic Macer-GIROQ hauls taken between 1988 and 1996. Fifteen sites in the English Channel from the Roscoff and Plymouth waters to the Dover Strait, as well as the Bay of Morlaix, Bay of Saint Brieuc and the Bay of Seine, and the Seine Estuary were sampled. The main characteristics of the Benthic Boundary Layer (BBL) macrofauna in the English Channel are described and compared with other temperate continental shelf BBL data.

Materials and methods

Study sites

Sampling sites were located in the English Channel and the Seine Estuary (Figure 1, Table 1). The sites chosen were representative of the range of hydrological and sedimentological conditions which are encountered in this area. The Channel water is characterised by the occurrence of generally strong tidal currents which control the sediment distribution. In areas of weak currents, muddy fine sand predominates, whereas in the open sea where currents are stronger, medium and coarse sands, pebble and gravel predominate.

At the seven western sites, the salinity remained around 35 PSU throughout the year; the temperature showed an annual cycle with the lowest temperature in winter (6–8 °C) and the highest at the end of the summer (15.5–20 °C). To the south of Plymouth, at site 2, a thermocline was present from June to September, separating surface and bottom waters at 25 m depth (Vallet, 1997). Three sites were in the circalittoral zone: 33–75 m (medium sand to coarse sand and pebble), while the four others were in the infralittoral zone: 5–25 m (fine sand, fine sand with shell to medium sand).

Of the five sites which were located in the open eastern Channel (BV, 4, 5, 6 and F), three were in the circalittoral zone: 25–29 m depth (medium sand to pebble), the BV site was in the infralittoral zone: 10 m on fine sand and site F was in the infralittoral zone: 5 m on muddy fine sand. Three sites (EO, EM and EP), all at ca. 5 m depth at low tide and ranging from muddy fine sand to coarse sand, were located in the Seine Estuary in an area with high variability in salinity and temperature (0.3–30 PSU; 3–23 °C): one in the oligohaline zone (0.3–5.0 PSU), one in the mesohaline zone (5.1–18.0 PSU) and one in the polyhaline zone (18.1–30.0 PSU). In the eastern part of the Channel (sites F, 4, 5 and 6), the salinity was affected by the freshwater input of the Seine River, and the salinity remained lower than 34 PSU throughout the year. At site F, sea bottom salinity may reach lower values than 30 PSU during Seine flooding (Vallet, 1997). Temperature varied seasonally with the lowest temperature in winter (4–6 °C) and the highest at the end of the summer (20–22 °C) (Vallet, 1997).

Sampling strategy

All the hauls were collected with a new version of the Macer-GIROQ sledge (Dauvin & Lorgeré, 1989; Dauvin et al., 1995). This sledge allowed simultaneous sampling of the fauna at four levels between 0.10 and 1.45 m (net 1: 0.10–0.40; net 2: 0.45–0.75; net 3: 0.80–1.10 and net 4: 1.15–1.45 m) above the bottom with four WP2 plankton nets (0.5 mm mesh size). Each net was equipped with a Tsurimi Seiki Ku flow meter to measure the volume of the sea water filtered. The hauls were sampled at different periods of the year, with spring and autumn sampling at nine sites,

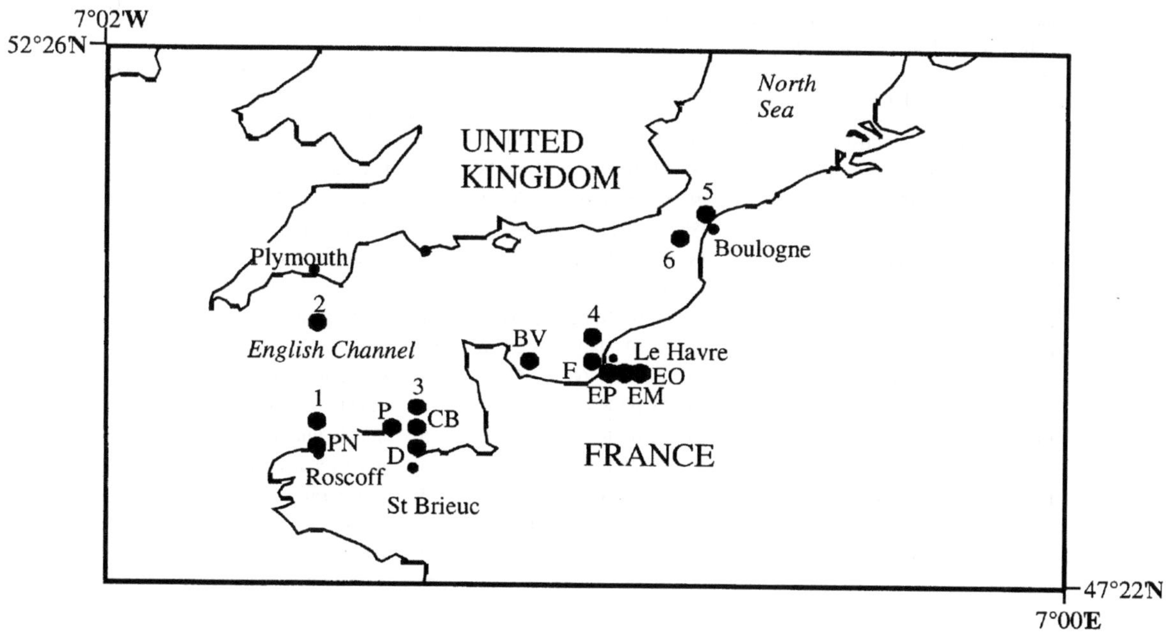

Figure 1. Location of the 15 sites in the English Channel.

with summer and autumn sampling at the BV site, and during an annual cycle at site 1 (Trezen Vraz) and site 3 (Norman-Breton Gulf) in the western Channel, and at site F and in the three zones in the Seine estuary (Table 2). A total of 432 hauls were taken (Table 2). Organisms were fixed with 10% neutralised formaldehyde, rinsed and transferred to 70% ethanol. All the organisms were sorted under a dissection microscope, counted and identified to species level. In this paper, we have counted only the suprabenthic organisms as defined by Brunel et al. (1978): decapods, peracarids, pycnogonids and leptostraceans, and two species of fish very common in the Seine estuary, considered as suprabenthic species (Mouny et al., 1996, 1998). The numbers of individuals were standardised to 100 m^{-3} volume.

Data analysis

A Kolmogorov-Smirnov test (Scherrer, 1984) was used to determine if there were significant differences between spring/autumn S (number of species per site), A (abundance standardised to 100 m^{-3} volume) and B (biomass mg AFDW standardised to 100 m^{-3} volume).

The affinity of assemblages both among sites and species was established from the sites × species matrix using correspondence analyses (CA) (Benzecri, 1973). Site and species groups were then distinguished by hierarchical classification (Jambu & Lebeaux, 1978; Bruynhooge's algorithm, 1978). Prior to this analysis, due to the large number of rare species collected, only those with a mean abundance higher than 1 ind. 100 m^{-3} were selected for the analyses. These comprised 63 amphipods, nine cumaceans, four isopods, 20 mysids, 12 decapods, one leptostracean and two fish (Table 3). Abundances were also transformed to the following classes: 1 (1–5 ind. 100 m^{-3}), 2 (5.1–20 ind. 100 m^{-3}), 3 (20.1–100 ind. 100 m^{-3}), 4 (100.1–500 ind. 100 m^{-3}) and 5 (>500 ind. 100 m^{-3}).

The swimming activity of the dominant species was measured using three coefficients: K1 = Nf2 / Nt, K2 = Nf3 / Nt, K3 = Nf4 / Nt. Nf2, Nf3, Nf4 were the densities per 100 m^3 in nets 2, 3 and 4 respectively, and Nt corresponded with the total density per 400 m^3 (Dauvin et al., 1994). The ash free dry weight (AFDW) was measured according to Vallet & Dauvin (1998).

Results

Species composition

A total of 261 suprabenthic species was identified at the 15 sampling sites. In rank order of dominance, these comprised amphipods (142 species, 54.4% of the total), decapods (46 species, 17.6%), mysids (27 species, 10.4%), isopods (22 species, 8.5%), cumaceans

Table 1. Main geographical and sedimentological characteristics of the 15 sampling sites in the English Channel and the Seine Estuary. *d*: depth [m]

Site	Site description	Location	*d*	Type of sediment
1	Trezen Vraz, offshore Roscoff	48° 51.20′ N–03° 53.20′ W	75	coarse sand
PN	Pierre Noire, Bay of Morlaix	48° 42.30′ N–03° 51.58′ W	17	fine sand
2	Offshore Plymouth	49° 53.75′ N–04° 13.00′ W	75	medium sand
3	Norman-Breton Gulf	48° 50.20′ N–02° 42.50′ W	33	coarse sand and pebble
P	Le Palus, Bay of Saint Brieuc	48° 35.90′ N–02° 37.50′ W	23	muddy fine sand with shell
D	Dahouët, Bay of Saint Brieuc	48° 41.80′ N–02° 50.50′ W	12	muddy fine sand with shell
CB	Center Bay of Saint Brieuc	48° 39.50′ N–02° 40.20′ W	24	medium sand
BV	Baie des Veys	49° 26.50′ N–01° 08.25′ W	10	fine sand
4	Offshore Bay of Seine	49° 37.60′ N–00° 12.57′ W	29	medium, coarse sand and gravel
F	Outer side of Seine Estuary	49° 26.26′ N–00° 01.30′ E	8	muddy fine sand
EP	Polyhaline Seine Estuary	49° 26.10′ N–00° 06.15′ E	5	muddy fine sand
EM	Mesohaline Seine Estuary	49° 26.08′ N–00° 15.00′ E	5	fine to coarse sand
EO	Oligohaline Seine Estuary	49° 26.80′ N–00° 25.00′ E	5	coarse sand
5	Dover Strait	50° 54.50′ N–01° 34.00′ E	25	pebble
6	Dover Strait	50° 48.00′ N–01° 19.50′ E	29	medium sand

(13 species, 5.0%), pycnogonids (8 species, 3.1%), fish and tanaids (2 species and 0.8% each) and leptostraceans (one species, 0.4%). The number of species ranged from 3 in the oligohaline zone of the Seine estuary to 102 at site P in spring, and from 6 in the mesohaline zone of the Seine estuary to 116 at site 5 in autumn (Table 4). Excluding the Seine estuary sites where the number of species was low in both seasons, the number of species varied from 44 at site F to 102 at site P in spring, and from 33 at site F to 116 at site 5 in autumn; however, no significant difference was found between spring and autumn values (KS test). In spring, there was a higher number of species in the western part of the Channel compared with the eastern part, but in autumn there was a higher number of species in the eastern part of the Channel compared with the western part. In both seasons, there was an increase in the number of species from the shallower site F to the deeper sites 1 and 2. The number of species was particularly high at site 5 in the Dover Strait on pebble during the autumn with 116 recorded from only five hauls.

Excluding the Seine estuary sites, the number of species of amphipods was of the same order of magnitude in both seasons and varied from 38 at site 5 to 60 at site 1 in spring, and from 38 at site D to 59 at sites 2 and 6 in autumn. For mysids, the number of species varied from eight at site 4 and 14 at sites P and PN in spring, and was slightly higher in autumn when it varied from 9 at sites D and 1, to 16 at sites P and 5. The number of species of the other zoological groups was lower than those of amphipods and mysids, and was of the same order of magnitude in both seasons.

Abundance

Abundance varied greatly from one site to another and between both seasons (Table 4), although there was no significant difference between spring and autumn values (KS test). For the sites outside the Seine estuary, there was a gradient of increasing density from west to east and, in autumn, an increase from the shallower site F to the deeper sites 1 and 2. Nevertheless, at site 2 during autumn, the mean abundance reached 7052 ind. 100 m^{-3} due to the sampling of large numbers of mysids (*Anchialina agilis*, *Erythrops elegans* and *Schistomysis ornata*) and the amphipod *Apherusa clevei*.

In spring, the density (Table 4) was particularly low (<650 ind. 100 m^{-3}) at 10 sites and did not increase beyond 200 ind. 100 m^{-3} at five sites, including four in the western part of the Channel. Values reached >2500 ind. 100 m^{-3} at only four sites: at PN where there was high density during nuptial swarming of the amphipod *Ampelisca* spp. and at three sites from the Seine estuary with high abundance of both mysids *Mesopodopsis slabberi* and *Neomysis integer*. During autumn (Table 4), the density remained lower than 650 ind. 100 m^{-3} at only six sites, including five in the

Table 2. Date of suprabenthic sampling and number of hauls (*N*) for each campaign at the 15 sampling sites in the English Channel and the Seine Estuary

Site	Date	*N*	Reference	Site	Date	*N*	Reference
1	03/10/88	1	Dauvin et al., 1994	EP	17/05/94	2	Mouny et al., 1996
	14–15/11/88	9			06/02/95	2	
	16/02/89	5			16/09/95	2	
	24/08/89	4			28/01/96	1	Mouny, 1998
	09/10/89	3			25/02/96	2	
	17–18/07/90	11	Zouhiri & Dauvin, 1996		27/03/96	1	
	05/06/93	5	Vallet, 1997		23/04/96	1	
PN	17/06/94	5	Dauvin & Zouhiri, 1996		28/05/96	2	
	15/11/95	5	Zouhiri, 1999		25/06/96	2	
					22/07/96	1	
					21/08/96	2	
2	10–13/06/93	5	Vallet, 1997		18/09/96	1	
	06–07/10/94	6			20/10/96	1	
					18/11/96	1	
					16/12/96	3	
3	03/06/93	5	Vallet, 1997	EM	17/05/94	13	Mouny et al., 1996
	15–16/02/94	11			06/02/95	12	
	21–22/04/94	11			16/09/95	15	
	16–17/06/94	10			28–29/01/96	5	Mouny, 1998
	30–31/08/94	7			25–27/02/96	5	
	01–03/10/94	5			27–29/03/96	6	
	14/11/95	5			23–24/04/96	5	
P	04/09/95	5	Vallet, 1997		28–29/05/96	6	
	11/05/96	5			24–25/06/96	5	
D	03/09/95	5	Vallet, 1997		22–23/07/96	5	
	10/05/96	5			20–21/08/96	5	
CB	05/09/95	5	Vallet, 1997		17–18/09/96	4	
	10/05/96	5			20–21/10/96	5	
BV	05/07/97	7	unpublished data		17–18/11/96	5	
	05–06/10/97	12			15–16/12/96	5	
4	28/05/93	5	Vallet, 1997	EQ	17/05/94	2	Mouny et al., 1996
	26–29/09/94	6			06/02/95	2	
	13/11/95	5	Zouhiri, 1999		16/09/95	2	
					28/01/96	1	Mouny, 1998
					25/02/96	2	
F	01–02/06/92	24	Wang & Dauvin, 1994		27/03/96	2	
	13/06/92	14			23/04/96	2	

Continued on p. 144

western part of the Channel, >1000 ind. 100 m^{-3}at five sites, and over 5000 ind. 100 m^{-3}at four other sites. The abundance was particularly high at site F (due to the cumaceans *Diastylis* spp.) and at site EM (due to the mysid *Neomysis integer*). Of the 246 species sampled in the 432 hauls, only 15 were able to form patches of abundance (>1000, >5000, >10 000 or >100 000 ind. 100 m^{-3}; Table 5). These comprised seven mysids, five amphipods (three *Ampelisca* and two *Apherusa*), one decapod, one cumacean and one

Table 2. Continued

Site	Date	*N*	Reference	Site	Date	*N*	Reference
F	12/11/95	5	Zouhiri, 1999		28/05/96	1	
	27/01/96	3			25/06/96	2	
	26/02/96	3			22/07/96	3	
	27–28/03/96	4			21/08/96	2	
	25/04/96	3			18/09/96	4	
	30/05/96	3			20/10/96	2	
	23/06/96	3			18/11/96	2	
	24/07/96	3			16/12/96	1	
	21/08/96	3		5	24–26/05/93	5	Vallet, 1997
	18/09/96	3			21–24/09/94	6	
	21/10/96	3		6	20–23/05/93	5	Vallet, 1997
	16/12/96	3			17–20/09/94	6	

fish; four of these were collected exclusively in the Seine estuary. Mysids were, therefore, the numerically dominant group of organisms collected with the sledge, followed by amphipods, cumaceans and decapods. *Palaemon longirostris* was abundant only in the Seine estuary. Seven of these 15 dominant species formed patches at only one site. The mysid *Anchialina agilis* showed high abundance at both sites 1 and 2 at the Atlantic entrance of the Channel, while the amphipod *Apherusa clevei* showed high abundance on one occasion at the western sites 1 and 2 and the eastern site 6, and the amphipod *A. bispinosa* at sites 2 and 3, in the western channel and site 4 in the Bay of Seine. Only three species reached abundances higher than >10 000 ind. 100 m^{-3}; of these, the estuarine mysid *Neomysis integer* reached a very high value of 216 000 ind. 100 m^{-3} in September in the mesohaline zone of the Seine estuary (Table 5).

Table 6 shows the monthly change in abundance of the suprabenthos at six sites. For both western sites 1 and 3, the abundance was greatest during the summer (July or August) and was very low during the winter and spring (often <100 ind. 100 m^{-3}). The mean annual abundance was very low at site 3 in the Norman-Breton Gulf, and just over 1000 ind. 100 m^{-3} at site 1 off the coast of Brittany. In the Seine Estuary, there was also a seasonal cycle of abundance with low values from November to April–May (often <1000 ind. 100 m^{-3}) and very high abundances from June to September. Highest abundance occurred earlier in the Seine estuary than the outer side of the Estuary at site F which, in the latter case, was due to nuptial swarming of the cumaceans *Diastylis* spp. in September.

Biomass

As with abundance, biomass varied greatly from one site to another and from one season to the next (Table 4) but, again, there was no significant difference between spring and autumn values (KS test). For the open sea sites, there was a gradient of increasing biomass from west to east and an increase from the shallower site F to the deeper sites 1 and 2 at autumn. In spring, the biomass reached 1 g AFDW 100 m^{-3} at site PN and in the Seine estuary, while in autumn the biomass reached 1 g AFDW 100 m^{-3} at sites 2, F and at the three Seine estuary sites. In the Seine estuary, high biomass (>75 g AFDW 100 m^{-3}) was due to a combination of the mysid *Neomysis integer* with low individual biomass but with high abundance, and high individual biomass both of the decapod *Palaemon longicornis* and the fish *Pomatoschistus microps.* It may be noted that the difference between the maximum and minimum values of biomass was one order of magnitude higher than the difference between maximum and minimum values of abundance (Table 4). This is mainly due to the high individual biomass of both estuarine species, the decapod *Palaemon longicornis* and the fish *Pomatoschistus microps*.

Table 3. List of the main species collected at the 15 sampling sites in the English Channel and the Seine Estuary. Abundances were transformed to abundance classes, as follows: '-' for abundance < 1 ind. 100 m^{-3}; '1': 1–5 ind. 100 m^{-3}; '2': 5.1–20 ind. 100 m^{-3}; '3': 20.1–100 ind. 100 m^{-3}; '4': 100.1–500 ind. 100 m^{-3}; '5': > 500 ind. 100 m^{-3}

Name	Site														
	1	PN	2	3	P	D	CB	BV	4	F	EP	EM	EO	5	6
Amphipoda															
Ampelisca armoricana	-	4	-	-	-	-	-	-	-	-	-	-	-	-	-
Ampelisca brevicornis	-	-	-	-	-	-	-	-	-	1	-	-	-	-	-
Ampelisca sarsi	-	5	-	-	-	-	-	-	-	-	-	-	-	-	-
Ampelisca spinipes	-	-	-	-	-	-	1	-	-	-	-	-	-	-	-
Ampelisca spooneri	-	-	2	-	-	-	2	-	4	-	-	-	-	-	-
Ampelisca tenuicornis	-	4	-	-	1	1	-	1	-	-	-	-	-	-	-
Ampelisca typica	-	2	-	-	1	-	2	-	2	-	-	-	-	-	-
Amphilochus neapolitanus	-	-	1	-	-	-	-	-	1	1	-	-	-	2	2
Aora typica	1	-	-	-	-	-	-	-	-	-	-	-	-	-	-
Apherusa bispinosa	4	1	5	4	1	1	3	1	5	-	-	-	-	3	4
Apherusa cirrus	3	-	-	-	-	-	-	-	-	-	-	-	-	-	-
Apherusa clevei	5	3	5	3	1	1	2	3	1	-	-	-	-	2	4
Apherusa ovalipes	2	2	1	1	-	-	-	3	-	-	-	-	-	-	3
Atylus falcatus	-	-	-	-	-	-	-	-	-	2	1	-	-	-	-
Atylus swammerdami	-	-	-	-	-	-	-	1	-	1	-	-	-	2	2
Atylus vedlomensis	1	1	2	1	1	1	2	-	3	1	-	-	-	-	-
Bathyporeia elegans	-	-	-	-	-	-	-	-	-	1	-	-	-	-	-
Bathyporeia nana	-	3	-	-	-	-	-	-	-	-	-	-	-	-	-
Bathyporeia tenuipes	-	2	-	-	-	-	-	-	-	-	-	-	-	-	-
Caprella linearis	-	-	-	-	-	-	-	-	-	-	-	-	-	2	-
Cheirocratus assimilis	1	-	1	-	-	-	-	-	-	-	-	-	-	-	-
Dexamine spinosa	-	-	-	-	-	1	-	-	-	-	-	-	-	-	-
Eusirus longipes	3	-	-	2	-	-	-	-	-	-	-	-	-	-	
Gammaropsis maculata	1	-	1	-	-	-	-	-	-	-	-	-	-	-	-
Gammarus crinicornis	-	-	-	-	1	-	-	-	-	-	-	-	-	-	-
Gammarus spp.	-	-	-	-	-	-	-	-	-	1	1	5	3	-	-
Gitana sarsi	1	-	-	-	-	-	-	-	-	-	-	-	-	1	
Gitanopsis bispinosa	-	-	1	-	-	-	-	-	-	-	-	-	-	-	-
Guernea coalita	1	1	1	-	-	-	-	-	1	-	-	-	-	-	-
Hippomedon denticulatus	-	-	1	-	-	-	-	-	-	-	-	-	-	-	-
Iphimedia obesa	1	-	-	-	-	-	-	-	-	-	-	-	-	-	-
Iphimedia perplexa	-	-	-	-	-	-	-	-	-	-	-	-	-	-	1
Iphimedia spatula	1	-	-	-	-	-	-	-	-	-	-	-	-	-	-
Ischyrocerus anguipes	-	-	-	-	-	-	-	-	-	1	-	-	-	1	-
Leucothoe incisa	-	1	1	-	-	-	-	-	-	1	-	-	-	-	-
Leucothoe spinicarpa	1	-	-	-	-	-	-	-	-	-	-	-	-	-	-
Megaluropus agilis	-	3	1	1	1	1	3	4	-	2	1	-	-	-	2
Megamphopus cornutus	1	-	1	1	-	-	-	-	1	-	-	-	-	-	-
Melita obtusata	-	-	1	-	-	-	-	-	-	1	-	-	-	1	-
Melphipidella macra	2	-	3	2	-	-	1	1	2	1	-	-	-	-	1
Metaphoxus fultoni	-	1	-	-	-	-	-	-	-	-	-	-	-	-	-
Metopa alderi	-	-	-	-	-	-	-	-	-	-	-	-	-	-	1
Metopa borealis	-	-	-	-	-	-	-	-	-	-	-	-	-	1	2
Microdeutopus versiculatus	-	-	-	-	-	1	-	-	-	-	-	-	-	-	-
Monoculodes carinatus	-	-	-	-	-	-	-	-	1	-	-	-	-	-	-
Monoculodes subnudus	-	-	-	-	1	-	-	-	-	-	-	-	-	-	-

Continued on p. 146

Table 3. Continued

Name	Site														
	1	PN	2	3	P	D	CB	BV	4	F	EP	EM	EO	5	6
Amphipoda															
Orchomene humilis	-	-	-	1	2	-	1	-	-	-	-	-	-	-	-
Orchomenella nana	1	1	2	1	3	1	3	3	3	1	-	-	-	2	1
Paramphilochoides odontonyx	-	-	1	-	-	-	-	-	-	-	-	-	-	-	-
Pariambus typicus	-	-	-	-	-	-	-	-	-	1	-	-	-	-	-
Perioculodes longimanus	-	4	-	-	-	1	-	4	-	1	-	-	-	-	1
Perrierella audouiniana	-	-	-	1	-	-	-	-	-	-	-	-	-	-	-
Phtisica marina	1	-	1	-	-	1	-	1	-	-	-	-	-	-	-
Pontocrates altamarinus	-	-	-	-	-	-	-	-	-	1	-	-	-	-	-
Pontocratus arenarius	-	-	2	1	-	-	-	-	1	-	-	-	-	-	-
Pseudoprotella phasma	1	-	-	-	-	-	-	-	-	-	-	-	-	-	-
Scopelocheirus hopei	-	-	1	-	-	-	-	-	-	-	-	-	-	-	-
Socarnes erythrophtalmus	-	-	-	-	-	-	-	-	-	-	-	-	-	-	1
Stenothoe marina	3	1	2	1	-	-	-	1	2	1	-	-	-	3	3
Synchelidium maculatum	1	3	1	2	3	2	3	-	2	-	-	-	-	-	1
Tmetonyx similis	-	-	-	-	-	-	1	-	-	-	-	-	-	-	-
Tritaeta gibbosa	1	-	-	1	2	-	-	-	-	-	-	-	-	-	-
Urothoe elegans	-	-	-	-	-	1	-	-	-	-	-	-	-	-	-
Cumacea															
Bodotria pulchella	-	3	-	-	-	-	1	4	1	-	-	-	-	-	-
Bodotria scorpioides	-	-	2	-	-	-	1	4	1	1	-	-	-	3	2
Campylapsis legendrei	-	-	-	1	-	-	-	-	3	-	-	-	-	-	-
Diastylis bradyi	-	-	-	-	-	-	-	1	-	5	1	-	-	-	-
Diastylis laevis	-	-	-	-	-	-	-	1	-	4	-	-	-	-	-
Diastylis rugosa	-	-	-	-	-	-	-	-	-	-	-	-	-	1	-
Eudorella emarginata	-	-	-	-	2	2	-	-	-	-	-	-	-	-	-
Pseudocuma longicornis	-	3	-	-	2	3	4	3	1	3	1	-	-	1	-
Pseudocuma similis	-	-	2	-	-	-	-	-	1	-	-	-	-	-	3
Isopoda															-
Astacilla longicornis	-	-	1	-	-	-	-	-	-	-	-	-	-	-	-
Eurydice pulchra	-	-	1	1	-	-	2	-	2	-	-	-	-	-	1
Gnathia oxyuraea	1	2	-	-	1	1	2	-	1	-	-	-	-	-	1
Sphaeroma rugicauda	-	-	-	-	-	-	-	-	1	-	-	-	-	-	-
Mysidacea															
Acanthomysis longicornis	-	1	3	1	-	-	-	-	1	-	-	-	-	2	3
Anchialina agilis	5	1	4	3	1	1	3	3	3	2	2	1	-	2	4
Erythrops elegans	-	-	4	-	-	-	-	-	-	-	-	-	-	-	-
Gastrossacus sanctus	-	1	-	-	-	-	-	-	-	-	-	-	-	-	-
Gastrosaccus spinifer	-	-	2	-	-	-	-	2	1	1	2	1	-	3	4
Haplostylus lobatus	1	1	1	2	2	1	5	-	-	-	-	-	-	-	1
Haplostylus normani	3	3	4	1	-	-	-	3	1	-	-	-	-	1	3
Leptomysis gracilis	2	-	1	-	-	-	-	1	-	-	-	-	-	-	-
Leptomysis lingvura	-	-	1	-	-	-	-	-	-	-	-	-	-	-	-
Mesopodopsis slabberi	-	-	-	-	-	-	-	-	-	2	2	5	5	-	-
Mysidopsis angusta	-	-	3	-	-	-	1	-	-	-	-	-	-	-	-

Continued on p. 147

Table 3. Continued

Name	Site														
	1	PN	2	3	P	D	CB	BV	4	F	EP	EM	EO	5	6
Mysidacea															
Mysidopsis gibbosa	-	-	1	1	1	1	-	-	2	-	-	-	-	1	2
Neomysis integer	-	-	-	-	-	-	-	-	-	-	5	5	5	-	-
Paramysis helleri	-	1	-	-	-	-	-	4	-	1	-	-	-	-	-
Paramysis nouveli	3	-	-	-	-	-	-	-	-	-	-	-	-	-	-
Schistomysis ornata	-	-	5	-	-	1	-	2	2	3	-	-	-	1	1
Siriella armata	-	-	-	-	1	-	1	-	-	-	-	-	-	-	-
Siriella clausii	-	-	-	2	3	2	2	-	-	-	-	-	-	1	1
Siriella jaltensis	2	-	-	2	-	-	-	-	-	-	-	-	-	-	-
Siriella norvegica	1	-	-	-	-	-	-	-	-	-	-	-	-	-	-
Decapoda	-	-	-	-	-	-	-	-	-	-	-	-	-	-	-
Crangon crangon	-	-	-	-	-	-	-	-	-	2	3	3	5	-	-
Galathea intermedia	-	1	1	1	-	-	-	-	-	-	-	-	-	-	-
Hippolyte varians	-	-	-	-	1	-	-	1	-	-	-	-	-	-	-
Liocarcinus spp.	1	-	-	-	-	1	-	-	-	-	-	-	-	-	-
Macropopia spp.	-	-	2	-	-	-	-	-	-	-	-	-	-	-	-
Palaemon longirostris	-	-	-	-	-	-	-	-	-	-	-	1	5	-	-
Pandalina brevirostris	2	-	1	-	1	-	-	-	1	-	-	-	-	1	-
Pandalus montagui	-	-	-	-	-	-	-	-	-	-	-	-	-	1	-
Philocheras bispinosus	-	-	2	-	-	-	-	3	-	-	-	-	-	-	
Philocheras sculptus	-	-	-	-	-	-	-	-	4	-	-	-	-	-	
Pisidia longicornis	-	2	-	-	-	-	-	-	-	-	-	-	-	-	
Processa spp.	-	-	-	-	-	-	-	3	-	-	-	-	-	-	
Leptostracea	-	-	-	-	-	-	-	-	-	-	-	-	-	-	
Nebalia bipes	-	1	-	-	1	-	2	-	-	-	-	-	-	-	
Fish															
Pomatoschistus microps	-	-	-	-	-	-	-	-	-	1	5	5	5	-	-
Pomatoschistus minutus	-	-	-	-	-	-	-	-	-	1	1	4	4	-	-

Spatial pattern in the English Channel and in the Seine estuary

A first CA was conducted on a matrix 15 sites × 111 species recorded with mean abundance >1 ind. 100 m^{-3}in the hauls (Table 3). Four assemblages were identified (Figure 2): 1. assemblage I comprised the oligohaline and the mesohaline sites of the Seine estuary; 2. assemblage II consisted of eastern sites together with site 2 (offshore Plymouth). Sub group IIa consisted of two sites under the influence of the Seine estuary (F and polyhaline zone), while sub group IIb consisted of the four other eastern site along with site 2 which is closely linked to site 4 (both characterised by a medium sand substratum); 3. assemblage III comprised western sites, with the subgroup IIIa containing the neighbouring Norman-Breton and Bay of Saint Brieuc sites, while sub-group IIIb was represented by Brittany site 1, and 4. assemblage IV represented by the single site PN. A second CA without the Seine estuary sites (matrix 12 sites × 109 species, two species *Palaemon longirostris* and *Neomysis integer* being strictly sampled in the Seine estuary) show similar assemblages for the open sea sites. Both analyses revealed a relatively good separation between the eastern sites together with site 2 (offshore Plymouth) which included many boreal species, and the western Brittany sites. Site PN possessed a distinctive

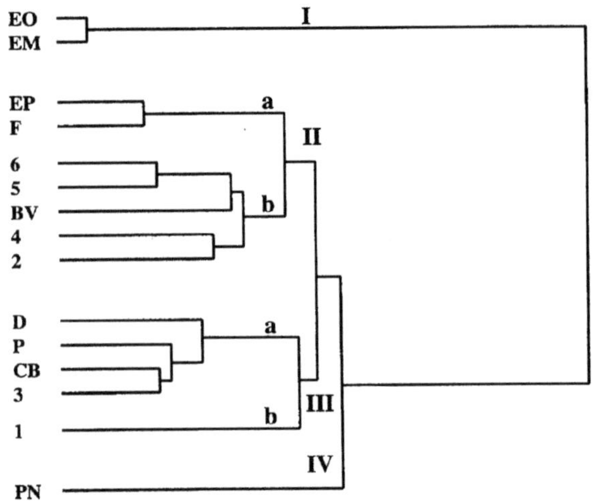

Figure 2. Hierarchical classification using CA showing the different groups of suprabenthic hauls (15 sites × 111 species); see Table 1 for the list of the sampling sites.

faunistic composition characterised by the dominance of *Ampelisca* species; sites F and the Seine estuary also possessed a distinctive fauna characterised by *Pomatoschistus minutus*, *Palaemon longirostris* and *Neomysis integer*.

Table 3 summarises the general features of the suprabenthic fauna in the Channel and the Seine estuary:

1. Forty three species were characteristic of only one site (among them, the amphipods *Gitana bispinosa*, *Paramphilochoides odontonyx*, *Metopa alderi* and the cumacean *Diastylis rugosa* were boreal species and collected only at the Dover Strait or Plymouth sites).
2. Eight species were characteristic of the sites under the influence of the Seine (among them *Palaemon longirostris*, *Mesopodopsis slabberi*, *Neomysis integer*, *Crangon crangon* and *Pomatoschistus* spp.).
3. Seven species were characteristic of the eastern part of the Channel (including *Diastylis laevis* and *D. bradyi*, and the boreal species *Metopa borealis* which was collected only at both northern Dover strait sites).
4. In contrast, 12 species were characteristic of the western English Channel (among them the amphipod *Eusirus longipes* which showed relatively high abundance at the Brittany sites 1 and 3, and the amphipod *Orchomene humilis* which was characteristic of the Norman-Breton Gulf and the bay of Saint Brieuc).
5. Twenty eight species showed a mean abundance higher than 100 ind. 100 m^{-3} and, therefore, represented the dominant species of the suprabenthic community in the Channel and the Seine estuary: 15 taxa (16 species, Table 5), and the mysids *Haplostylus normani* and *Paramysis helleri*, the amphipods *Ampelisca spooneri*, *Megaluropus agilis*, *Perioculodes longimanus* and *Gammarus* spp., the cumaceans *Bodotria scorpioides* and *B. pulchella*, *Pseudocuma longicornis*, the decapods *Crangon crangon* and *Philocheras sculptus*, and the fish *Pomatoschistus minutus*.
6. Only 12 species were dominant at 2/3 of the sites (excluding the three sites of the Seine estuary): the amphipod *Melphidipella macra*, the cumacean *Pseudocuma longicornis* and the mysids *Haplostylus lobatus* and *H. normani* were dominant at eight sites, the amphipods *Atylus vedlomensis*, *Megaluropus agilis*, *Stenothoe marina* and *Synchelidium maculatum* at nine sites, *Apherusa bispinosa* and *A. clevei* at 11 sites, and *Orchomenella nana* and *Anchialina agilis* were sampled in abundance at all the sites. Seven other species were characteristic of >50% of the sites: the amphipod *Apherusa ovalipes*, the mysids *Acanthomysis longicornis*, *Gastrosaccus spinifer*, *Mysidopsis gibbosa*, *Schistomysis ornata*, *Siriella clausii* and the isopod *Gnathia oxyuraea*.

Accordingly, using the criteria of numerical dominance at >50% of the sites and total abundance, a total of 37 (out of 246 species) characterised the suprabenthic communities of the English Channel and the Seine estuary.

Daily changes

Organisms showed active vertical migrations from sediments to the water column, or from the Benthic Boundary Layer (BBL) to the water column. The total abundance of the suprabenthos was generally low during the day hauls and increased rapidly just after sunset, reaching a maximum around midnight and decreasing thereafter. A second peak of abundance occasionally occurred at sunrise, abundance decreasing again in the morning. Nevertheless, at site 5 characterised by a pebble substratum, abundance during the night was lower than during the day.

Two main patterns of daily changes have been identified for the suprabenthic species: 1. species with high abundances in the BBL during the day and with low abundances during the night: e.g. mysids *Anchialina agilis* (Figure 3a) and *Haplostylus normani*. Nevertheless, both species showed two peaks

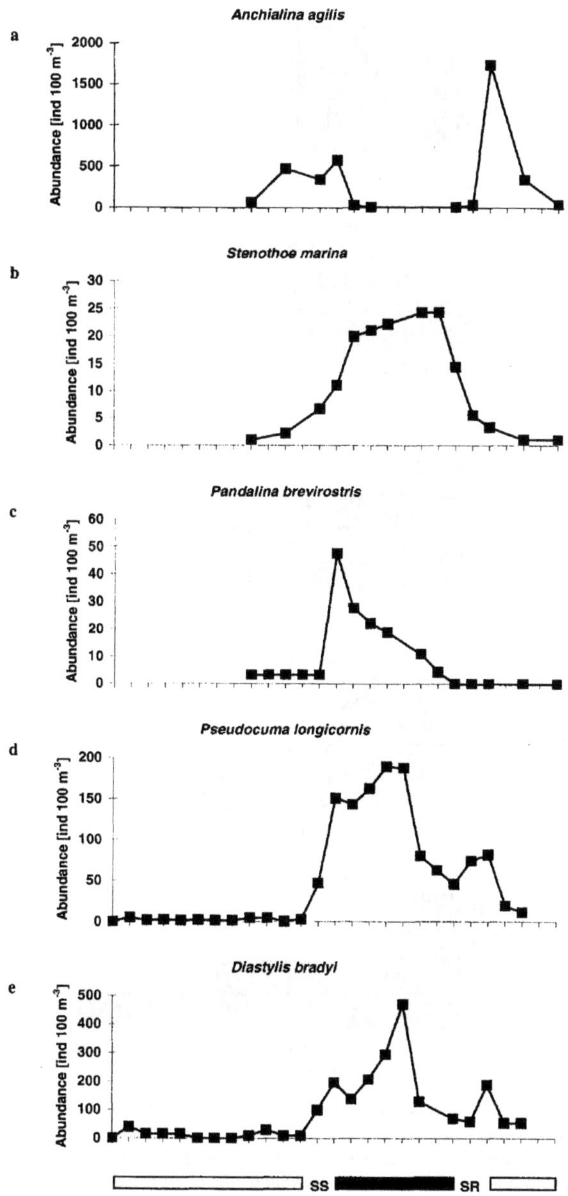

Figure 3. Diurnal migration (July) of *Anchialina agilis* (a) *Stenothoe marina* (b) and *Pandalina brevirostris* (c) at site 1 and of *Pseudocuma longicornis* (d) and *Diastylis bradyi* (e) at site F. SS: sunset, SR: sunrise.

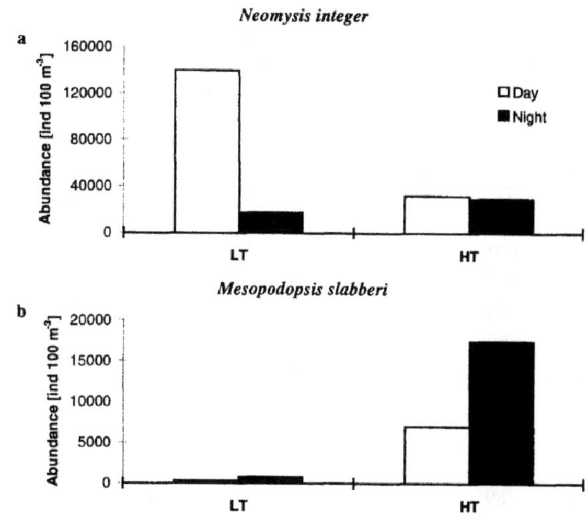

Figure 4. Temporal change (September) of abundance of *Neomysis integer* (a) and *Mesopodpsis slabberi* (b) during a 24 h time series in the Seine Estuary. LT: low tide; HT: high tide (from Zouhiri et al., 1998).

of abundance at sunset and at sunrise, and very low abundances at night. 2. species which performed an active migration and were present in the BBL only at night. In this group, several patterns could be identified: species which were in high abundance in the BBL throughout the night from: e.g. the amphipod *Stenothoe marina* (Figure 3b); species which were present in the BBL in high numbers just after sunset, after which abundance decreased regularly until sunrise: e.g. the decapod *Pandalina brevirostris* (Figure 3c); species which were present in the BBL at sunset with abundance increasing regularly until midnight then decreasing over the night with a second peak at sunrise; e.g. the cumaceans *Pseudocuma longicornis* (Figure 3d) and *Diastylis bradyi* (Figure 3e).

In the oligohaline estuary, daily changes of the suprabenthic fauna were investigated in September 1995 over a 24 h period. Temporal changes of the mysid *Neomysis integer* depended both on tidal advection (higher density during low tide) and daily changes. This species was sampled near the sea bottom in abundance only during the day; at night, it dispersed into the water column, where the abundance in the sledge was seven times lower than in day hauls (Figure 4a). *Mesopodopsis slabberi* was recorded only during high tide and was transported upstream by the tide (Figure 4b).

Swimming activity

Dominant species collected in benthic hauls were classified according to their swimming activity coefficient into five groups:

Group 1: suprabenthic species with strong swimming activity but occupying all the water column adjacent to the sea floor and $K1 = K2 = K3 \approx 0.25$;

Table 4. Number of species per site (*N*), Abundance (A) [ind. 100 m^{-3}], and biomass (B) [mg AFDW 100 m^{-3}] for the suprabenthos collected at the 15 sampling sites in the English Channel and the Seine Estuary. *biomass estimation, - Baie des Veys site, no available data

Site	Spring			Autumn		
	N	A	B	N	A	B
1	100	303	47	74	547	120*
PN	95	3957	1947*	61	352	77*
2	86	200	71	103	7 052	1611
3	88	144	25	87	537	58
P	102	94	25*	98	294	77
D	69	118	21*	69	217	39
CB	74	429	48*	71	1434	162
BV	–	–	–	–	1114	244*
4	89	640	147	90	2047	158
F	44	3223	706*	33	40 469	8863*
EP	7	2836	80 500*	11	8 527	77 000*
EM	10	12 910	116 000*	6	30 929	100 000
EO	3	293	112 500	9	4 217	202 820
5	71	169	289	116	379	267
6	91	379	136	101	2 285	424

Σ K1 + K2 + K3 ≈ 0.75: e.g. the mysid *Anchialina agilis* (Figure 5a).

Group 2: upper suprabenthic species with a very strong swimming activity and K1 < K2 < K3, K3 ≈ 0.30, Σ K1 + K2 + K3 ≥0.75: e.g. the amphipod *Atylus vedlomensis* (Figure 5b).

Group 3: transitional suprabenthic species which were abundant in the three lower nets with K1 = K2 ≈ 0.25 > K3; Σ K1 + K2 + K3 ≈ 0.70:e.g. the amphipod *Scopelocheirus hopei* (Figure 5c).

Group 4: lower suprabenthic species which were abundant in the two lower nets, but also present in the two upper nets, with K1 > K2 > K3, K3 >0.10: e.g. the amphipod *Stenothoe marina* (Figure 5d).

Group 5: lowermost suprabenthic species, which do not swim much further upward than net 4 or net 3, with K1 > K2 > K3, K3 <0.10: e.g.the amphipod *Orchomenella nana* (Figure 5e).

Discussion

In a recent paper, Mees & Jones (1997) reviewed the current state of knowledge of the hyperbenthos (named suprabenthos in this paper) from shallow waters to deep sea and from tropical and subtropical areas to cold temperate seas. The following categories of

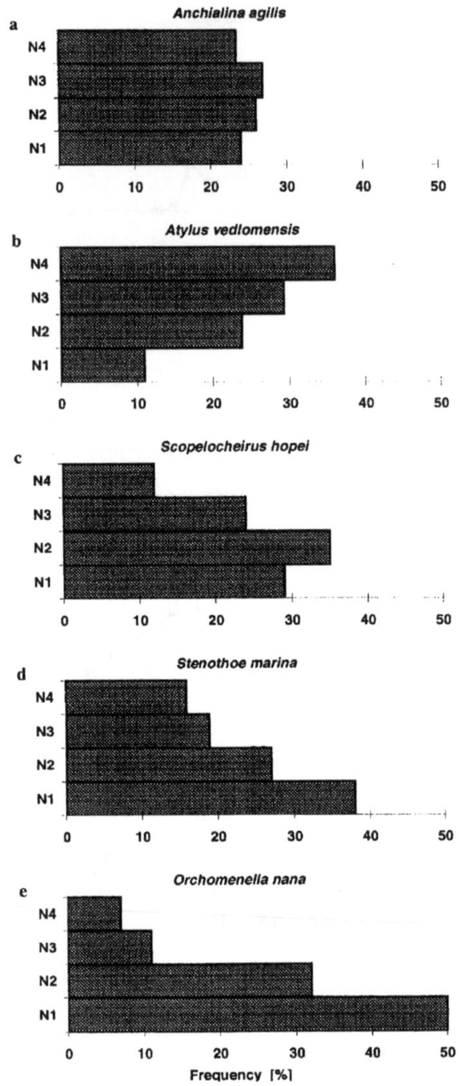

Figure 5. Frequency [%] in the four levels of the Macer-GIROQ sledge (N1, net 1; N2, net 2; N3, net 3; N4, net 4) of *Anchialina agilis* (a), *Atylus vedlomensis* (b), *Scopelocheirus hopei* (c), *Stenothoe marina* (d), and *Orchomenella nana* (e) in the English Channel sites.

suprabenthic research were identified:

1. taxonomic interest and diversity,
2. small-scale distribution and,
3. diurnal rhythms and vertical migrations, along with the role of the suprabenthos in marine food webs, in terms of feeding and predation.

The present discussion aims to complete this earlier synthesis, with special reference to the main characteristics of the suprabenthos in the English Channel.

The strong tidal currents in the English Channel have major consequences for the distribution and stability of sediment and, hence, the structure of benthic communities. Turbulence near the sea bottom also affects the vertical circulation of water and hence the distribution of the suprabenthic species in the Benthic Boundary Layer. In addition, the Channel is a zone of confluence between boreal and cold temperate fauna and temperate and warm temperate fauna. Finally, the Seine discharge to the Bay of Seine in the eastern English Channel is the most significant source of nutrients from a river system, resulting in enhanced primary production along the French coast of the eastern Channel. In response for these conditions, the suprabenthic macrofauna showed a number of distinctive features.

The total number of species (261) and amphipods (142) collected in the Channel revealed a very diversified fauna which was greater than that observed in other circalittoral sites in the Atlantic Ocean, in the Bay of Fundy, the Gullmarfjord or in the Baie des Chaleurs with a similar effort of sampling and similar sieve mesh: 0.5 mm (Buhl-Jensen & Fosså, 1991; Chevrier et al., 1991; Wildish et al., 1992). The present survey has added several species to the list of the marine fauna from Roscoff (23 species), Plymouth (17 species), Normandy coast (seven species) and Wimereux (52 species) (Dauvin & Vallet, 1997).

The estimated densities for the open sea sites overlying fine, medium and coarse sands and pebbles of the Channel are of the same order of magnitude of those reported by Mees & Jones (1997) for shallow coastal waters: 100–80 000 ind. 100 m^{-3}. For the Seine estuary, the maximum abundance observed for *Neomysis integer* (216 000 ind. 100 m^{-3}) is higher than those reported by Mees & Jones (1997) for other European estuaries. However, the maximum abundance of *Mesopodopsis slabberi* observed in the Seine estuary (34 300 ind. 100 m^{-3}) is lower than that observed in spring in the Tamar Estuary (>100 000 ind. 100 m^{-3}in summer) (Mees & Jones, 1997). The biomass of the suprabenthic fauna appeared especially high in the Seine estuary (>200 g AFDW 100 m^{-3}). At most of the Channel sites, the biomass was moderate in both seasons (25–425 mg AFDW ind. 100 m^{-3}) and >1500 mg AFDW 100 m^{-3}only when nuptial swarms of amphipods or mysids occurred as in spring at PN and in autumn at site 2. The trophic group composed of carnivores, scavengers and omnivores, which consists of motile predators with strong swimming activity, dominates the suprabenthos at all sites (>50% of the total biomass). Deposit feeders, which feed on detritus, benthic microphytobenthos and bacteria, is the second most dominant group in terms of biomass (>30%), while the suspension feeders, including the amphipod *Apherusa* spp. at some open sea sites, comprise <20% of the total biomass (Vallet & Dauvin, 1998).

Seasonal changes in number of species, abundance and biomass of the suprabenthic macrofauna, with maxima in summer or the beginning of autumn and minima in winter or the beginning of spring, are similar to these observed for the endobenthic and epibenthic macrofauna of the Channel and temperate waters (Dauvin, 1984). The fact that there is an increase in the number of species from shallow to deep water may be due to the more stable conditions which occur in the deeper western English Channel with relatively small changes in temperature during the year at the entrance of the Channel. However, this gradient is probably also a result of the increase in habitat heterogeneity from the muddy fine sand at the shallowest sites to the coarse sand with shells at the deeper sites which favour an increase in biodiversity. The higher abundance and biomass of the suprabenthic fauna of the eastern English Channel in autumn may be due to higher primary production in this part of the Channel (Vallet, 1997; Vallet & Dauvin, 1998).

The mysid distribution along the salinity gradient from site 4 in the Bay of Seine to the upper part of the estuary showed that the number of species declined with decreasing salinity. Nevertheless, maximum mysid density occurred at low salinities (3–10 PSU). *Schistomysis* spp. characterised the outer part of the estuary, *Mesopodopsis slabberi* characterised the mesohaline estuary (10–15 PSU), while *Neomysis integer* occurred in high abundance in the oligohaline estuary where the salinity was lower than 10 PSU. Upstream of 1 PSU, the density of *N. integer* decreased dramatically (Zouhiri et al., 1998).

The sampling of suprabenthos along a sedimentary gradient from muddy sand to pebble shows that some species occur only in one type of habitat: e.g. *Ampelisca armoricana*, *A. sarsi* and *A. tenuicornis* on fine sand at the Pierre Noire site; *Diastylis brayi* and *D. laevis* on the muddy fine sand site F in the outer part of the Seine estuary; *Ampelisca spooneri* and *Gastrosaccus spinifer* on the medium sand of sites 2, 4 and 6. Difference between western and eastern sites may also be due to biogeographical factors. Some temperate and warm temperate species (e.g. *Eusirus longipes* and *Orchomene humilis*) occur only along the Brittany coast in the western Channel, in contrast to boreal

Table 5. Maximal estimated abundance in a haul [ind. 100 m^{-3}] of suprabenthic species collected in the English Channel and the Seine Estuary. 'Mysi': mysid, 'Amph': amphipod, 'Deca': decapod and 'Cuma': cumacean

Abundance	Species	Group	Abundance	Site	Date
>1000	*Anchialina agilis*	Mysi	1523	2	07/10/94
	Apherusa clevei	Amph	1575	6	19/09/94
	Erythrops elegans	Mysi	1640	2	07/10/94
	Gastrosaccus spinifer	Mysi	1814	6	19/09/94
	Anchialina agilis	Amph	1874	1	09/10/89
	Pomatoschistus microps	Fish	1948	EM	18/11/96
	Haplostylus lobatus	Mysi	2216	CB	05/09/95
	Ampelisca armoricana	Amph	2404	PN	17/06/94
	Apherusa bispinosa	Amph	2470	3	30/08/94
	Apherusa clevei	Amph	2650	2	07/10/94
	Palaemon longirostris	Deca	2805	EM	18/11/96
	Apherusa bispinosa	Amph	2860	2	07/10/94
	Ampelisca tenuicornis	Amph	2970	PN	17/06/94
	Palaemon longirostris	Deca	4160	EO	21/08/96
	Apherusa bispinosa	Amph	4220	4	27/09/94
>5000	*Ampelisca sarsi*	Amph	5598	PN	17/06/94
	Schistomysis ornata	Mysi	8077	2	07/10/94
	Neomysis integer	Mysi	8149	EP	22/07/96
>10 000	*Apherusa clevei*	Amph	10 081	1	17/07/90
	Mesopodopsis slabberi	Mysi	13 172	EP	22/07/96
	Mesopodopsis slabberi	Mysi	15 168	EM	22/07/96
	Mesopodopsis slabberi	Mysi	18 400	F	18/09/96
	Diastylis spp.	Cuma	20 400	F	18/09/96
	Mesopodopsis slabberi	Mysi	34 329	EM	17/09/95
	Neomysis integer	Mysi	90 337	EO	21/08/96
>100 000	*Neomysis integer*	Mysi	216 000	EM	17/09/95

Table 6. Monthly abundance [ind. 100 m^{-3}] for the suprabenthos collected at six sampling sites in the English Channel and the Seine Estuary. Mean: mean annual abundance. '–': no data

Site	Jan.	Feb.	Mar.	Apr.	May	Jun.	Jul.	Aug.	Sep.	Oct.	Nov.	Dec.	Mean
1	–	410	–	–	–	303	4622	639	–	547	89		1050
3	–	58	–	43	–	79	–	1384	–	692	1234	–	582
F	706	402	449	385	1445	3223	4277	5965	40 469	1269	400	439	4 952
EP	2561	62	932	651	2836	7677	22 123	9681	8527	7139	309	561	5255
EM	1174	348	1105	5572	12 910	78 069	48 671	21 827	30 929	6 119	4 914	983	17 714
EO	2530	929	1275	370	293	1526	12 014	48 083	4217	513	900	294	6079

and cold temperate species which are collected only at site 2 in the northern part of the western Channel or at the eastern Dover Strait sites (e.g. *Gitana bispinosa, Metopa alderi, M. borealis, Paramphilochoides odontonyx, Diastylis rugosa*). Finally, brackish species (e.g. *Neomysis integer, Palaemon longirostris*) are found only in the Seine estuary. These three factors appear to account for the separation of the suprabenthic macrofauna into four assemblages, namely:

1. Seine estuary,
2. eastern Channel and Plymouth area,
3. western Channel along Brittany coast, and
4. fine sand Pierre Noire in the Bay of Morlaix.

Nevertheless, the suprabenthic fauna of the Channel appears relatively homogeneous at a meso-scale level (Vallet & Dauvin, 1998) and can be considered as a unique community including three kinds of species: 1. good swimmer species which are common at all times and at all sites: e.g. *Anchialina agilis, Gastrosaccus* spp., *Haplostylus* spp. and *Apherusa* spp., 2. species which are present at all sites only at night: e.g. *Atylus vedlomensis, Megaluropus agilis, Melphipidella macra, Orchomenella nana, Pseudocuma longicornis, Schistomysis ornata* and *Synchelidium maculatum* and 3. species occurring at one or a limited number of sites nocturnally: e.g. *Ampelisca* spp. at Pierre Noire, and *Diastylis* spp. at site F. The suprabenthic assemblage sampled in the Seine estuary is very different from the other three, but is comparable with that found in other northeastern Atlantic estuaries: e.g. Ems, Westerschelde and Gironde estuaries (Sorbe, 1981; Cattrijsse et al., 1993; Mees et al., 1993, 1995; Mees, 1994; Mouny et al., 1996, 1998).

Changes in light intensity seem to be the most important factor determining migratory activities of suprabenthic species (see e.g. Hesthagen, 1973; Sainte-Marie & Brunel, 1985; Kaartvedt, 1985, 1986, 1989; Chevrier et al., 1991; Macquart-Moulin & Ribera Maycas, 1995; Mees & Jones, 1997). In the English Channel, there are different patterns of nocturnal migration in the water column: some species migrate as early as dusk and show maximum abundance in the water column during the first part of the night, while other species migrating only at night are found in large numbers later in the night (Vallet et al., 1995; Dauvin & Zouhiri, 1996). A few species show a second peak of swimming activity in the water column at sunrise (Wang et al., 1995). Among mysids, some species show a concentration in the Benthic Boundary Layer during the day and migrate into the water column at night. Differences in swimming activity between sexes and development stages have also been reported for mysids in the English Channel by Zouhiri et al. (1998). Swimming activity also varies according to reproductive stage. Swimming ability of males is generally greater than that of females (Dauvin & Zouhiri, 1996). For example, the males of *Ampelisca sarsi* show an increase in abundance from the lower net to the upper net (Dauvin & Zouhiri, 1996); conversely, the females are concentrated near the bottom, especially in nets 1 and 3, and are collected in low abundance in the upper net 4. The swimming behaviour of the mysid *Gastrosaccus spinifer* varies according to hydrodynamic conditions. At site 5 in the Dover Strait, characterised by a pebble substratum and strong currents, the individuals are concentrated in the upper nets whereas at site 6, also in the Dover Strait, characterised by a medium sand substratum and weaker currents, individuals are collected mainly in the lower net (Vallet, 1997; Zouhiri et al., 1998). In the Bay of Biscay where the Macer-GIROQ sledge has been used from 400 to 1000 m (Elizalde et al., 1991, 1993; Dauvin & Sorbe, 1995), the suprabenthic fauna is concentrated near the sea bottom and, therefore, is collected mainly in the lower net of the sledge (Figure 6). Conversely, in the English Channel, the suprabenthic fauna occupies all the Benthic Boundary Layer and the suprabenthos is sampled in the same order of magnitude in all four nets of the sledge (Figure 6). This characteristic of the suprabenthos in the Channel is probably due to the high turbulence which occurs throughout the water column and especially near the sea bottom. Such conditions are likely to maintain particulate organic matter in suspension for prolonged periods, to the advantage of the suprabenthic fauna. Turbulent conditions would also favour species with a strong swimming capability.

In the English Channel, as in the Bay of Seine (Wang & Dauvin, 1994) and off Roscoff (Zouhiri & Dauvin, 1996), the post-larvae and juveniles of Clupeiidae and Gobiidae are present in the Benthic Boundary Layer only at sunrise and sunset, suggesting that the concentration of suprabenthic organisms near the sea floor at these times can represent an attractive food resource for fish. Euphausiids are present in the BBL only during the day and are dispersed in the water column during the night (Vallet et al., 1995; Zouhiri & Dauvin, 1996). This indicates that the BBL is an area with important fluxes of individuals and biomass between the benthos and the water column (Zouhiri & Dauvin, 1996).

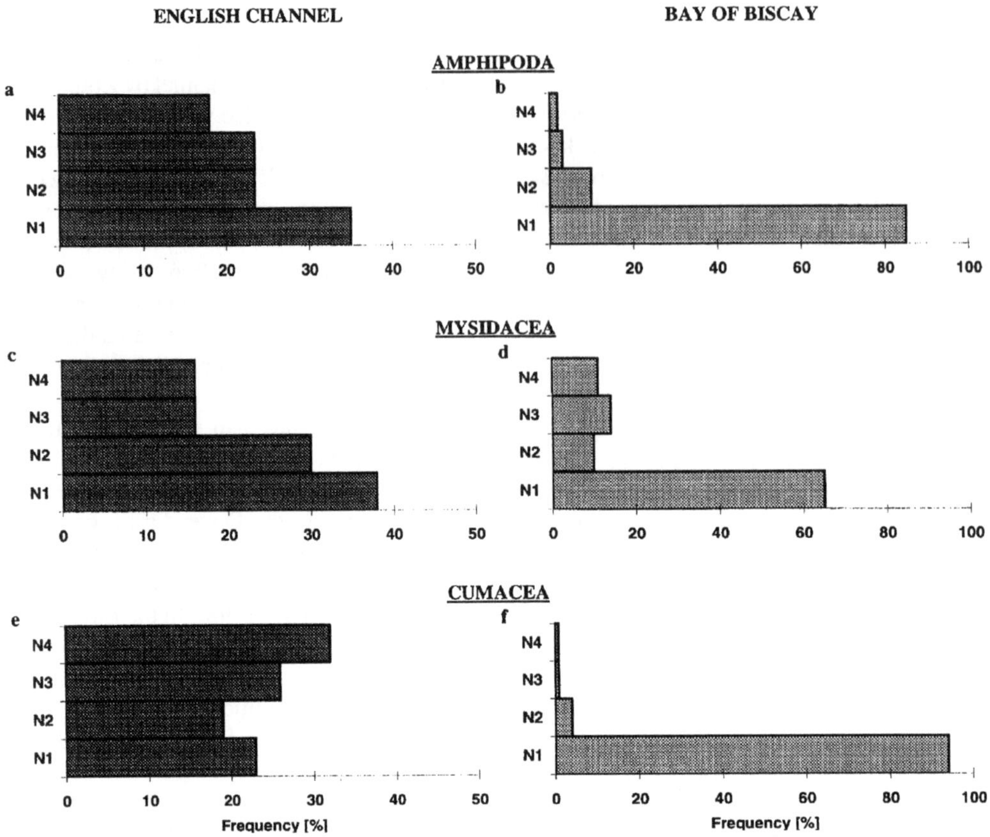

Figure 6. Frequency [%] in the four levels of the Macer-GIROQ sledge (N1, net 1; N2, net 2; N3, net 3; N4, net 4) of Amphipoda (a, b), Mysidacea (c, d), and Cumacea (e, f) sampled in the English Channel and in the Bay of Biscay (data of the Bay of Biscay: Elizalde et al., 1991, 1993; Dauvin & Sorbe, 1995).

In conclusion, in the Channel there is a very diversified suprabenthic fauna with high numbers of amphipods. Sediment and salinity gradients account for the presence of four distinct assemblages and two distinct communities: an open sea suprabenthic community with low abundances and biomass and an estuarine suprabenthic community with very high abundance and biomass. There are also distinct swimming activities which vary between species and are associated with diel cycles, seasonal cycles and reproductive periods. The turbulent nature of the Benthic Boundary Layer creates conditions which favour water column colonisation by the suprabenthic fauna.

In the English Channel, we consider that the descriptive phase is now complete, and the main priorities for future research on the suprabenthos should be: 1. the behavioural study of the vertical distribution of the species throughout the water column over 24 h periods in neap and spring tides, and 2. the study of the presence of successive phases of swimming and sinking of the species or a permanent presence in the BBL during the night. The study of the effects of the tidal currents on the horizontal transport of species during their pelagic phase and the role of suprabenthos in biomass exchanges between the water column, the BBL and sea bed sediments offer new perspectives for future research on suprabenthic fauna in the English Channel.

Acknowledgements

This study was carried out as a part of the "Programme National d'Océanographie Côtière (PNOC), chantier Manche (CNRS-INSU & IFREMER)", the 'Programme National Biodiversité', and the "Programme Seine Aval" coordinated by the 'Région Haute Normandie'. The authors thank the crews of the N. O. Pluteus II, Côte de Normandie, Côte d'Aquitaine, Le Suroît, Le Noroît and Le Thalia, J.-

C. Lorgeré and C. Conq for their technical supports during the cruises, K. Gherthos for help with the English, and Dr E. Flach and Dr H. L. Rees for their very useful remarks on the first version of the manuscript.

References

Benzecri, J. P., 1973. L'analyse des données. Tome 2. L'analyse des correspondances. Dunod, Paris: 619 pp.

Brunel, P., M. Besner, D. Messier, L. Poirier, D. Granger & M. Weinstein, 1978. Le traîneau Macer-GIROQ: appareil amélioré pour l'échantillonnage quantitatif de la petite faune nageuse au voisinage du fond. Int. Rev. ges. Hydrobiol. 63: 815–829.

Bruynhooge, M.,1978. Classification ascendante hiérarchique de grands ensembles de données: un algorithme rapide fondé sur la construction de voisinages réductibles. Cah. Anal. Don. 3: 7–33.

Buhl-Jensen, L. & J. H. Fosså, 1991. Hyperbenthic crustacean fauna of the Gullmarfjord area (western Sweden): species richness, seasonal variation and long-term changes. Mar. Biol. 109: 245–248.

Cattrijsse, A., J. Mees & O. Hamerlynck, 1993. The hyperbenthic Amphipoda and Isopoda of the Voordelta and the Westerschelde estuary. Cah. Biol. mar. 187–200.

Chevrier, A., P. Brunel & D. Wildish, 1991. Structure of a suprabenthic shelf sub-community of gammaridean Amphipoda in the bay of Fundy compared with similar sub-community in the Gulf of St Lawrence. Hydrobiologia 223: 81–104.

Dauvin, J. C., 1984. Dynamique d'écosytèmes macrobenthiques des fonds sédimentaires de la baie de Morlaix et leur perturbation par les hydrocarbures de l'Amoco Cadiz. Thèse de Doctorat d'Etat, Univ. P. & M. Curie, Paris VI: 468 pp.; annexes: 193 pp.

Dauvin, J. C. & J. C. Lorgeré, 1989. Modifications du traîneau Macer-GIROQ pour l'amélioration de l'échantillonnage quantitatif étagé de la faune suprabenthique. J. Rech. Océanogr. 14: 65–67.

Dauvin, J. C. & J. C. Sorbe, 1995. Suprabenthic amphipods from the southern margin of the Cape-Ferret Canyon (Bay of Biscay, Northeastern Atlantic Ocean): abundance and bathymetric distribution. Pol. Arch. Hydrobiol. 42: 441–460.

Dauvin, J. C. & C. Vallet, 1997. Apports d'échantillonnages suprabenthiques à la faunistique de la Manche et à la biogéographie du plateau continental nord-ouest européen. Crustacés et pycnogonides. Cah. Biol. mar. 38: 251–266.

Dauvin, J. C. & S. Zouhiri, 1996. Suprabenthic crustacea fauna of a dense *Ampelisca* community from the English Channel. J. mar. biol. Ass. U. K. 76: 909–929.

Dauvin, J. C., A. Iglesias & J. C. Lorgeré, 1994. Suprabenthic crustacean fauna of the circalittoral coarse sand community off Roscoff (English Channel): composition, swimming activity and seasonal variation. J. mar. biol. Ass. U. K. 74: 543–562.

Dauvin, J. C., J. C. Sorbe & J. C. Lorgeré, 1995. The Benthic Boundary Layer macrofauna from the upper continental slope and the Cap-Ferret Canyon (Bay of Biscay). Oceanol. Acta 18: 113–122.

Elizalde, M., J. C. Dauvin & J. C. Sorbe, 1991. Les mysidacés suprabenthiques de la marge sud du canyon du Cap-Ferret (Golfe de Gascogne): répartition bathymétrique et activité natatoire. Ann. Inst. Océanogr. 67: 129–144.

Elizalde, M., J. C. Dauvin & J. C. Sorbe, 1993. Las communidades suprabentonicos batiales del golfo de Vizcaya (margen sur del canon de Cap-Ferret): composicion faunistica y estructura. Publ. Espec. Inst. Esp. Oceanogr. 11: 247–258.

Hesthagen, I. H., 1973. Diurnal and seasonal variations in the near-bottom fauna – the hyperbenthos – in one of the deeper channels of the Kieler Bucht (Western baltic). Kieler Meeresforsch 29: 116–140.

Jambu, M. & M. O. Lebeaux, 1978. Classification automatique pour l'analyse des données. 2. Logiciels. Dunod, Paris: 399 pp.

Kaartvedt, S., 1985. Diel changes in small-scale vertical distribution of hyperbenthic mysids. Sarsia 70: 287–295.

Kaartvedt, S., 1986. Diel activity patterns in deep-living cumaceans and amphipods. Mar. Ecol. Progr. Ser. 30: 243–249.

Kaartvedt, S., 1989. Nocturnal swimming of gammaridean amphipod and cumacean Crustacea in Malsfjorden, Norway. Sarsia 74: 187–193.

Macquart-Moulin, C. & E. Ribera Maycas, 1995. Inshore and offshore diel migrations in European benthopelagic mysids, genera *Gastrosaccus*, *Anchialina* and *Haplostylus* (Crustacea, Mysidacea). J. Plankton Res. 17: 531–555.

Mees, J., 1994. The hyperbenthos of shallow coastal waters and estuaries: community structure and biology of the dominant species. PH. D. Thesis, Univ. Ghent: 212 pp.

Mees, J. & M. Jones, 1997. The hyperbenthos. Oceanogr. mar. Biol. Ann. Rev. 35: 221–255.

Mees, J., A. Dewicke & O. Hamerlynck, 1993. Seasonal and spatial distribution of hyperbenthic communities along estuarine gradients in the Westerschelde. Neth. J. Aquat. Ecol. 27: 359–376.

Mees, J., N. Fockedey & O. Hamerlynck, 1995. Comparative study of the hyperbenthos of three European estuaries. Hydrobiologia 311: 153–174.

Mouny, P. 1998. Structure spatio-temporelle du zooplancton et du suprabenthos de l'estuaire de la Seine. Dynamique et rôle des principales espèces dans la chaîne trophique pélagique. Thèse Doctorat du Muséum National d'Histoire Naturelle, 256 pp. + annexes.

Mouny, P., Z. Wang & J. C. Dauvin, 1996. Premières données sur la structure spatio-temporelle du mésozooplancton et du suprabenthos de l'estuaire de la Seine. J. Rech. Océanogr. 21: 100–108.

Mouny, P., J. C. Dauvin, C. Bessineton, B. Elkaim & S. Simon, 1998. Biological components from the Seine estuary: first results. Hydrobiologia 373/374 (Dev. Hydrobiol. 131): 333–347.

Sainte-Marie, B. & P. Brunel, 1985. Suprabenthic gradients of swimming activity by cold-water gammaridean amphipod Crustacea over a muddy shelf in the Gulf of Saint Lawrence. Mar. Ecol. Prog. Ser., 23: 57–69.

Scherrer, B., 1984. Biostatistique. Gäetan Morin Editeur: 850 pp.

Sorbe, J. C., 1981. La macrofaune vagile de l'estuaire de la Gironde. Distribution et migration des espèces. Modes de reproduction. Régimes alimentaires. Océanis 6: 579–592.

Sorbe, J. C., 1984. Contribution à la connaissance des peuplements suprabenthiques néritiques sud-Gascogne. Thèse Doctorat d'Etat, Univ. Bordeaux: 301 pp.

Vallet, C., 1997. Le compartiment suprabenthique des fonds circalittoraux de la Manche: composition faunistique et quantitative, et rôle dans les transferts benthos/pélagos. Thèse de Doctorat, Université de Paris VI: 160 pp.

Vallet, C. & J. C. Dauvin, 1995. Qualitative et quantitative composition of the hyperbenthic Amphipods from the English Channel. Polsk. Arch. Hydrobiol. 42: 461–481.

Vallet, C. & J. C. Dauvin, 1998. Composition and diversity of the Benthic Boundary Layer macrofauna from the English Channel. J. mar. biol. Ass. U.K. 78: 387–409.

Vallet, C. & J. C. Dauvin, 1999. Seasonal change of macrozooplankton and Benthic Boundary Layer macrofauna from the Bay of

Saint-Brieuc (western English Channel). J. Plankton. Res. 21: 35–49.
Vallet, C., S. Zouhiri, J. C. Dauvin & Z. Wang, 1995. Variations nycthémérales de l'abondance de la faune démersale en Manche. J. Rech. Océanogr. 20: 94–102.
Wang, Z. & J. C. Dauvin, 1994. The suprabenthic crustacean fauna of the infralittoral fine sand community in the Bay of Seine (eastern English Channel): composition, swimming activity and diurnal variations. Cah. Biol. mar. 35: 135–155.
Wang, Z., E. Thiébaut & J. C. Dauvin, 1995. Spring abundance and distribution of the ctenophore *Pleurobrachia pileus* in the Seine estuary: advective transport and diel vertical migration. Mar. Biol. 124: 313–324.
Wildish, D. J., A. J. Wilson & B. Frost, 1992. Benthic boundary layer macrofauna of Brown Banks, north-west Atlantic, as potential prey of juvenile benthic fish. Can. J. Fish. aquat. Sci. 49: 91–98.
Zouhiri, S., 1999. Composition qualitative de la faune démersale et structure trophique du suprabenthos en trois sites de l'Atlantique Nord-Est (Manche, Golfe de Gascogne et seamounts du sud des Açores). Thése Doctorat du Muséum National d'Histoire Naturelle, 109 pp. + annexes.
Zouhiri, S. & J. C. Dauvin, 1996. Diel changes of the Benthic Boundary Layer macrofauna over coarse sand sediment in the western English Channel. Oceanol. Acta 19: 141–153.
Zouhiri, S., C. Vallet, P. Mouny & J. C. Dauvin, 1998. Spatial distribution and biological rhythms of suprabenthic mysids of the English Channel. J. mar. biol. Ass. U.K. 78: 1181–1202.

Hydrobiologia **426:** 157–160, 2000.
G. Liebezeit, S. Dittmann & I. Kröncke (eds), Life at Interfaces and Under Extreme Conditions.

The physiological response of Northern krill *(Meganyctiphanes norvegica)* to temperature gradients in the Kattegat

Reinhard Saborowski*, Markus Salomon & Friedrich Buchholz
Biologische Anstalt Helgoland-AWI, Meeresstation, Postfach 180, D-27483 Helgoland, Germany
*E-mail: rsaborowski@awi-bremerhaven.de (*author for correspondence)*

Key words: Meganyctiphanes norvegica, respiration, temperature, vertical migration, Kattegat

Abstract

The Alkor-Deep (140 m), which forms part of a depression system in the northern Kattegat channel east of the island of Læsø (Denmark), is the location of a self sustaining population of Northern krill, *Meganyctiphanes norvegica* (Euphausiacea). This population is exposed to one of the most pronounced thermal gradients within the distributional range of this pelagic crustacean. During summer, the temperature of the water column ranges between 4 and 6 in the deep to 16 °C near the surface which results in the krill being exposed to temperature differences of 8–10 °C during diel vertical migration. Oxygen consumption rates were used to investigate the physiological adaptation of the animal to such gradients in temperature. The rates were found to increase exponentially from 31 μmol O_2 h^{-1} $g_{dw}{}^{-1}$ at 4 °C to 72 μmol O_2 h^{-1} $g_{dw}{}^{-1}$ at 16 °C, giving a Q_{10}-value of 2.0, and indicating that physiological adaptation to varying thermal conditions does not take place. Behavioural adaptations are discussed which may help the krill to cope with large temperature gradients in their environment.

Introduction

Northern krill, *Meganyctiphanes norvegica*, is an integral component of many zooplankton communities in the north and north-east Atlantic regions. It is abundant in an unusually wide area from the Sub-Arctic to the Mediterranean and is known to perform extensive diurnal vertical migrations (Mauchline & Fisher, 1969). Towards the Baltic Sea, krill is frequently found in the Skagerrak but is absent south of the Belt Sea and the Øresund. The Kattegat, where the marine waters of the Skagerrak and the brackish waters of the Baltic Sea meet, hosts an apparently self-sustaining krill population within a series of deeps east of the island of Læsø. The krill population of the Alkor-Deep was first described by Boysen & Buchholz (1984) and Buchholz & Boysen-Ennen (1988) and thereafter was the subject of several studies on e.g. population dynamics, biochemical composition, nutrition and biochemical adaptation. The diel vertical migration and distribution patterns were investigated by Buchholz et al. (1995) and Tarling et al. (1998). These studies showed that, during the day, krill remained within the basin at depths below 80 m and often concentrated at the north or south rim. Buchholz et al. (1995) found that krill did not migrate into the uppermost warm surface waters, but penetrated into the discontinuity layer. The authors proposed that the krill were prevented from migrating any higher because they had reached their upper thermal tolerance limit of 14–15 °C. Accordingly, krill were exposed to a maximum temperature difference of about 10 °C when migrating from the deep cold layers to the subsurface layers where their tolerance limit is met. Given the wide range of temperatures experienced by krill during a typical diurnal vertical migration in this region, this study investigates the capacity of krill to adapt physiologically to these conditions. Respiration rates served as a measure for the overall metabolic performance.

Materials and methods

Description of the study site

The Kattegat is a mostly shallow (<30 m) transition zone between the North Sea and the Baltic Sea. Five basins more than 100 m deep (Ulrich, 1983) are found in the northern part of the Kattegat, east of the island of Læsø. The Alkor Deep (11° 25.10 E, 57° 16.55 N), the current study site (Figure 1), is the largest of these basins with a span of 3.5 km × 1 km and a maximum depth of 140 m.

Hydrographically, the northern Kattegat is characterized by an outflow of brackish surface water from the Baltic Sea northward, and an opposing inflow of deep water from the North Sea. Accordingly, the water column is highly stratified into three main layers, a deep marine layer, an intermediate mixed layer and a brackish surface layer.

The vertical profiles of salinity were similar between summer and winter (Figure 2a) in the deeper part of the water column, where values around 33–34 PSU prevailed. In winter, salinity gradually decreased towards the surface, dropping below 31 PSU at 20 m and reaching a minimum of 26 PSU in the surface layer. The same gradual decrease was also found in summer, but salinities at the surface were lower than in winter (between 20 and 25 PSU).

The temperature profiles showed pronounced differences between seasons (Figure 2b). The bottom water was always cold between 4 and 6 °C and these temperatures were recorded throughout the whole water column during the winter situation. In summer, cold waters of 8 °C were found from the bottom up to 80 m. A slight increase in temperature towards 10 °C appeared between 70 and 50 m. In the mixed layer between 50 m to 15 m, there was a large temperature shift from 10 °C to 15 °C whilst in the surface layers, temperature was once again quite uniform, around 16 °C.

Respiration measurements

Respiration measurements were carried out during a winter and a summer cruise (4–18 March 1997 and 22 July to 5 August 1998, respectively) on board the research vessel FS Heincke. Samples were collected during short night hauls with a multiple opening and closing net and environmental sensing system (MOCNESS, Wiebe, 1985) which was towed at 2.5 knots in the 20–50 m depth layer. The krill were immediately transferred into large basins (200 l) with aerated water and were allowed to recover for a few hours. Then, individuals were transferred into new aquaria (5 l) which were adjusted to the experimental temperature of 4, 8, 12 and 16 °C, respectively. In order to minimize the effects of different maturation stages and body size, only males of an average length of 25–30 mm were selected. The respiration measurements were carried out in each of six chambers of 1.64 l volume. Five chambers were equipped with 2–5 krill, while one chamber served as a control. Measurements were carried out for several hours in the dark at constant temperature. The chambers were equipped with an internal current system as described in detail by Saborowski & Buchholz (1998) to encourage natural swimming behaviour. Oxygen consumption rates as measured with Clarke-type electrodes (Strathkelvin Instruments, model 928), were normalized to dry weight and expressed as μmol $O_2\ h^{-1}\ g_{dw}{}^{-1}$.

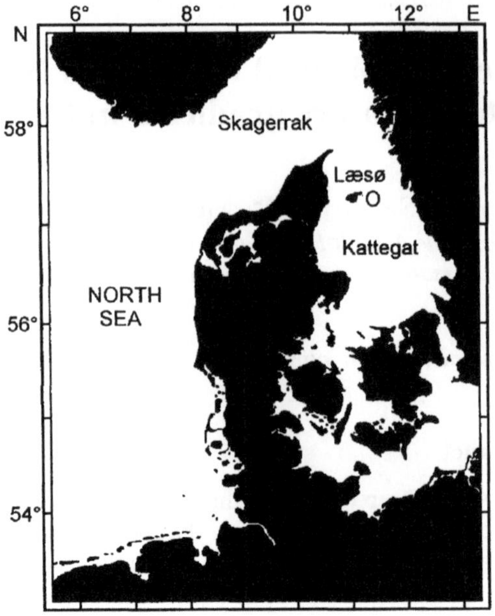

Figure 1. Sampling site east of the island Læsø in the Kattegat Channel. The position of the Alkor Deep (140 m) is indicated by the circle.

The Q_{10}-value, which provides a measure of the influence of temperature on metabolic processes, was calculated according to van t'Hoff's law:

$$Q_{10} = \left(\frac{R_2}{R_1}\right)^{10/(T_2-T_1)},$$

where R_2 and R_1 are the respiration rates at the temperatures T_2 and T_1.

Results and discussion

The respiration rates were not significantly different between summer and winter, and accordingly, data from both seasons were combined in the following. Oxygen consumption rates were lowest at 4 °C

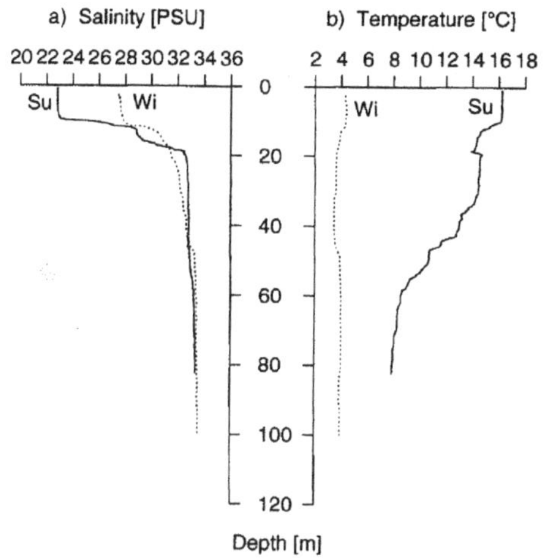

Figure 2. Vertical profiles of (a) salinity and (b) temperature in the Alkor-Deep during winter (early March 1997) and summer (August 1998).

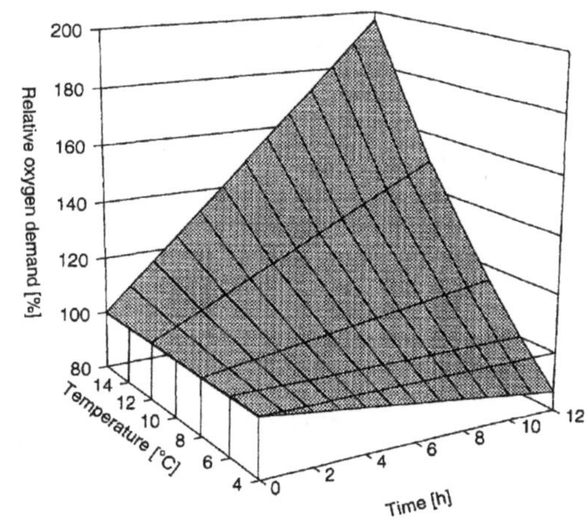

Figure 4. Relative oxygen demand in relation to temperature and duration of exposure. The reference value of 100% corresponds to averaged deep water temperatures (6 °C).

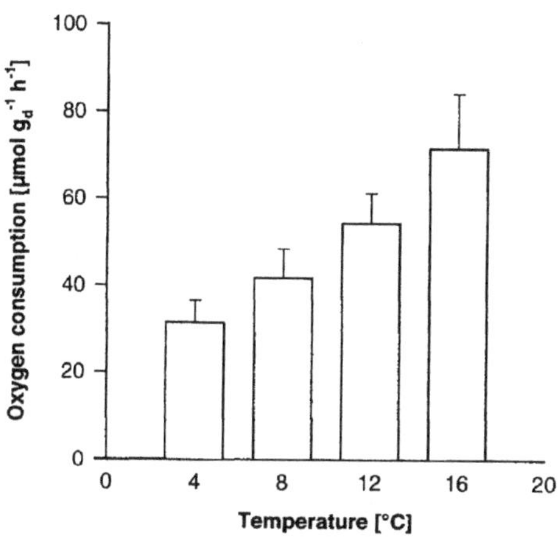

Figure 3. Respiration rates in relation to temperature of male krill. (Means ± s.d., n = 12–15).

amounting to 31.4 μmol O_2 h^{-1} $g_{dw}{}^{-1}$. At 8 °C, respiration increased to 41.8 μmol O_2 h^{-1} $g_{dw}{}^{-1}$ and reached 72 μmol O_2 h^{-1} $g_{dw}{}^{-1}$ at 16 °C (Figure 3). The increase of respiration rates with temperature followed an almost perfect exponential curve ($y = 24.06$ $e^{(0.06817 \cdot x)}$, $r^2 = 0.9998$). Accordingly, over the short term krill are not able to compensate for temperature effects by adaptive physiological mechanisms. This is also supported by the Q_{10}-value which was calculated from the respiration rates obtained over the entire range of experimental temperatures (4 °C–16 °C). The resulting Q_{10}-value of 2.0 is not exceptionally low as would be suggested in the case of metabolic adaptation. These results are in agreement with previous work by Hirche (1984) on krill from the Swedish Gullmarsfjord. Accordingly, krill had to spend more energy on basic metabolic processes in warm, rather than in cold, waters. As the temperature in the Kattegat gradually increases during summer over a comparatively large vertical distance, krill are subjected to a concomitant increase of metabolic costs during the course of their upward migration (Figure 4). On the assumption that 6 °C is the lowest summer temperature of the deep water, an average sized krill (50 mg dry weight, approx. 30 mm of length) would consume 1.81 μmol oxygen per hour which equals 43.46 μmol per day. If this krill moves up and enters water of 12 °C the oxygen consumption would increase to 2.73 μmol per hour and 65.43 μmoles per day. Nevertheless, since krill do not remain in upper water layers during the entire day but only during the night, a more realistic figure would be obtained through assuming they stayed 6 h up and 18 h down as observed in the summer, which gives an average value of 51.7 μmoles per day. This would increase the metabolic cost in a krill to 12% more than if it remained in the deep cool water over the entire 24 h period, and this does not include the potential extra cost of making a vertical ascent. However, a more significant effect than on the integrated daily energy consumption is given when concerning immediate metabolic changes. In this respect, a variation of temperature from 6 °C to 12 °C would increase metabolic rates by 50% which can be considered to cause stress on any physiological level.

In order to demonstrate a typical distribution pattern of high vertical resolution at day and night, we here extracted data from catches at noon and midnight

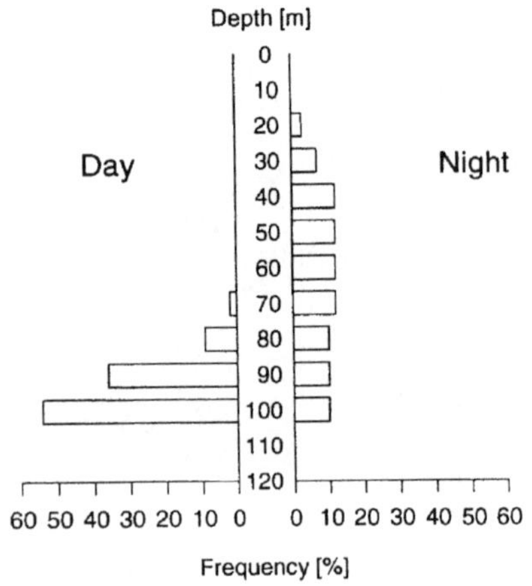

Figure 5. Vertical distribution of krill during day and night (based on data from Buchholz et al., 1995).

obtained in August 1989 (Buchholz et al., 1995) The pattern encountered in 1998 were largely identical. During the day, the entire krill population was found near the bottom layers of the Alkor Deep (Figure 5). Only very few animals were caught above 80 m. More than 80% of the total catch was obtained from 80 m and deeper. During the night, krill migrated into the upper water layers but simultaneously dispersed within the water column. While no krill was caught at the surface, numbers increased at 20 and 30 m. A comparatively low but uniform krill density was present from 40 m to 100 m. This dispersed nocturnal vertical distribution pattern indicates that not the entire krill population uniformly moved towards upper water layers and stayed there for the whole night feeding e.g. on phytoplankton. The time an individual spent in warmer layers appeared to be shorter than potentially possible presuming that this behaviour would be physiologically advantageous. It would result in a reduction of negative thermal effects by variation of duration or frequency of the ascent. Although metabolic extra costs in warmer waters should be covered by additional feeding, potential disadvantages, e.g. imbalances between anabolic and catabolic processes, can be minimized or avoided in this way.

In conclusion, it seems that krill are not able to adapt metabolically to the varying thermal conditions they encounter during the course of diurnal vertical migrations. However, they may compensate thermal effects by their vertical migration behaviour. Detailed fine scale studies on the distribution pattern including a comparison with regions where krill are exposed to isothermic conditions would help to clarify this suggestion.

Acknowledgements

This study is part of the project "Impact of a climatic gradient on the Physiological Ecology of a Pelagic crustacean" (PEP) of the EU MAST III program, contract no. MAS3-CT95-0013. We are grateful to the PEP-team for help on board.

References

Boysen, E. & F. Buchholz, 1984. *Meganyctiphanes norvegica* in the Kattegat: studies on the annual development of a pelagic population. Mar. Biol. 79: 195–207.

Buchholz, F. & E. Boysen-Ennen, 1988. *Meganyctiphanes norvegica* in the Kattegat: studies on the horizontal distribution in relation to hydrography and zooplankton. Ophelia 29: 71–82.

Buchholz, F., C. Buchholz, J. Reppin & J. Fischer, 1995. Diel vertical migrations of *Meganyctiphanes norvegica* in the Kattegat: comparison of net catches and measurements with acoustic doppler current profilers. Helgoländer Meeresunters. 49: 849–866.

Hirche, H.-J., 1984. Temperature and metabolism of plankton - 1. Respiration of Antarctic zooplankton at different temperatures with a comparison of Antarctic and Nordic krill. Comp. Biochem. Physiol. 77A: 361–368.

Mauchline, J. & L. R. Fisher, 1969. The Biology of Euphausiids. Adv. Mar Biol. 7: 1–545.

Saborowski, R. & F. Buchholz, 1998. Internal current generation in respiration chambers. Helgoländer Meeresunters. 52: 103–109.

Tarling, G. A., J. B. L. Matthews, R. Saborowski & F. Buchholz, 1998. Vertical migratory behaviour of the euphausiid, *Meganyctiphanes norvegica*, and its dispersion in the Kattegat Channel. Hydrobiologia 375/376 (Dev. Hydrobiol. 132): 331–341.

Ulrich, V. J., 1983. Zur Bathymetrie und Topographie der nördlichen Kattegat-Rinne. Meeresforsch. 30: 61–68.

Wiebe, P. H., A. W. Morton, A. M. Bradley, R. H. Backus, J. E. Craddock, V. Barber, T. J. Cowles & G. R. Flierl, 1985. New developments in the MOCNESS, an apparatus for sampling zooplankton and micronekton. Mar. Biol. 87: 313–323.

Hydrobiologia **426:** 161–168, 2000.
G. Liebezeit, S. Dittmann & I. Kröncke (eds), *Life at Interfaces and Under Extreme Conditions.*

Effects of salinity on larval and early juvenile growth of an extremely euryhaline crab species, *Armases miersii* (Decapoda: Grapsidae)

Klaus Anger*, Kim Riesebeck & Cornelia Püschel
*Alfred-Wegener-Institut für Polar- und Meeresforschung; Biologische Anstalt Helgoland, Meeresstation, D-27498 Helgoland, Germany (*author for corespondence)*

Key words: decapod larvae, salinity tolerance, larval growth, carbon, osmoregulation

Abstract

The neotropical crab *Armases miersii* (Rathbun, 1897) breeds in supratidal rock pools, where great salinity variations occur. In laboratory experiments, all larval stages and the first juveniles were reared at six different salinities (5–55 PSU, intervals of 10 PSU). In five series of experiments, exposure to these conditions began either from hatching (Zoea I) or from the onset of successively later stages (Zoea II, III, Megalopa, Crab I). Growth was measured in terms of dry weight, carbon, nitrogen and hydrogen content. At osmotically extreme conditions (5 and 55 PSU, resp.), all stages showed minimum biomass accumulation; this was consistent with maximum mortality and longest duration of development (data presented in a separate paper). Successively later exposure to these salinities tended to reduce these effects. Lowest mortality and shortest time of development occurred generally at 15–25 PSU, indicating an optimum at moderately reduced salinities. This response pattern, however, was not congruent with that observed in growth. Biomass accumulation was initially maximum within a wide range of salinities (15–45 PSU), but in the Zoea II and III stages, this range tended to narrow and to shift towards higher salinities (35–45 PSU). These trends reversed in the Megalopa and Crab I, where maximum growth occurred again in a wider range and at lower salinities (15–35 PSU). The reduction of zoeal growth in moderately dilute media (15–25 PSU), which were optimal for survival and development, is interpreted as an energetic cost of hyper-osmoregulation, which begins already at hatching. Five PSU caused hypo-osmotic stress, exceeding in the long term the larval capacity for hyper-regulation. Poor zoeal survival and growth at 55 PSU are interpreted as effects of hyper-osmotic stress. In the Megalopa and Crab I, reduced growth at salinities $\geq$35 PSU may reflect the energetic costs of hypo-osmoreguation beginning in these stages. Our data suggest that the physiological adaptations of larval and early juvenile *A. miersii* allowing for survival and development in a physically harsh and unpredictable habitat imply a trade-off with reduced growth, due to energetic costs of osmoregulation.

Introduction

Among the marine Brachyura, the family Grapsidae is known for its pronounced tendency to adapt to non-marine conditions. Numerous grapsid crab species live in estuarine and intertidal habitats, and several have become fully terrestrial or limnic (Hartnoll, 1988; Anger, 1995a). A neotropical marine species, *Armases miersii* (Rathbun, 1897), has only recently been recognized as a particularly successful invader in boundary habitats between marine, freshwater and terrestrial environments, namely in the coastal mangrove fringe of Jamaica and, probably, other Caribbean islands (Anger, 1995b, c, 1996; Anger & Schultze, 1995; Schuh & Diesel, 1995a, b). On the limestone coasts of northern Jamaica, *A. miersii* passes its larval development in ephemeral supratidal rock pools, where highly variable and sometimes extreme conditions of salinity, temperature, pH, oxygen concentration and food availability occur (Anger, 1995b, c, Schuh & Diesel, 1995a). Hence, the larval stages of this species have evolved an array of physiological adaptations to typically unpredictable environmental variations in their 'nursery' habitat. For instance, *A. miersii* larvae are facultatively lecithotrophic, ensuring their survival and development in a generally food-limited environment (Anger, 1995b, c). As an adaptation to salinity variations occurring in supratidal rock pools, the larvae show an extremely euryhaline response, tolerating both brackish and hypersaline conditions (Anger, 1995c, 1996; Schuh & Diesel, 1995b). This tolerance is based upon strong osmoregulatory capabilities in all postembryonic life-history stages (Charmantier et al., 1998).

The present paper describes ontogenetic patterns in larval and early juvenile growth of *A. miersii* (measured as rates of increase in the biomass of successive developmental instars) during an experimental exposure to various salinity conditions that occur also in their natural habitat (5–55 PSU; Anger, 1995b; Schuh & Diesel, 1995a). These results are compared with recent observations of survival and development under identical conditions (Anger, 1996) and with the ontogeny of osmoregulation in this species (Charmantier et al., 1998). In order to differentiate between long-term (chronic) and short-term (acute) effects of osmotic stress, we exposed the larval stages (Zoea I, II, III, Megalopa) and the first juvenile (Crab I) not only continually (from hatching), but also separately (after rearing in seawater) to the different test salinities.

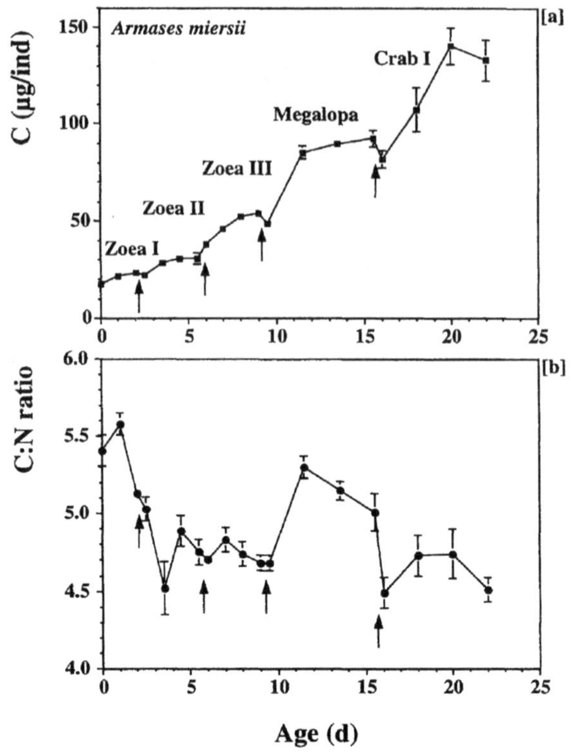

Figure 1. Growth and elemental composition of larval and early juvenile *Armases miersii* during development (days from hatching) at constant 35 PSU. (a) Carbon content (C) (μg ind^{-1}), (b) carbon:nitrogen (C:N) weight ratio. $\bar{x} \pm$ SD (n = 5; where no error bars are visible, SD smaller than plot symbol); arrows indicate the time of moulting to subsequent stages.

Materials and methods

Larvae of *Armases miersii* were obtained from laboratory-cultured ovigerous females that were kept in aquaria with seawater (32–33 PSU; 25 °C) under a 12:12 h light:dark regime. The parent generation originated from animals that had been collected in 1993 near the Discovery Bay Marine Laboratory, Jamaica (for further details see Anger, 1996). The larvae were mass-reared in ca. 400 ml glass bowls with freshly hatched *Artemia* spec. nauplii (ca. 10 ml^{-1}) given daily as food. Maximum rearing density was 50 individuals per bowl in the Zoea I and II stages, 30 in the Zoea III and 20 in the Megalopa and Crab I. The Megalopa and first juvenile crab instars were given a piece of nylon gauze (200 μm mesh size) as a substrate.

In five subsequent series of laboratory experiments, larval and early juvenile crabs were reared at six salinities (5, 15, 25, 35, 45, 55 PSU), and rates of growth were measured in short intervals; in parallel experiments with the same material, effects of salinity on larval survival and development duration were studied (see Anger, 1996). The present paper describes effects of salinity on growth in terms of biomass increase (see below for details). In one experimental series, the exposure to the different test salinities began within 12 h after hatching, and it continued through all four larval stages and the first juvenile crab instar. In the other four series, larvae were reared in seawater (32–33 PSU), where also hatching had taken place, and exposed only in later developmental stages (Zoea II, Zoea III, Megalopa or Crab I) to the various test salinities. Thus, this study comprised five sets of experiments beginning from five different stages, with six treatments each. Since this experimental design required a large amount of material, the number of larvae produced by a single female did not suffice for carrying out the entire study and a different hatch had to be used in each series. Only the results of the first experimental series (exposure from hatching) are documented (Table 1), because this protocol allows for a safe separation of maternal and environmental effects; the results from the other series are briefly described within graphs (Figures 2 and 4) and in the text.

Salinities <32 PSU were obtained by dilution of filtered (1 μm) seawater with desalinated water, higher

Table 1. Biomass and elemental composition of larval and early juvenile *Armases miersii* in relation to age within successive instars, during continual exposure (from hatching) to different salinities, dry weight (W), carbon (C), nitrogen (N), hydrogen (H) in μg per individual; C, N, H in% of weight; C:N weight ratio; Zoea II - Crab I: only final biomass (on day of moulting) shown; arithmetic mean ($\bar{x}$) n = 5 replicate analyses

Stage	Age	Salinity	W		C		N		H		C		N		H		C:N	
	[d]	[PSU]	[μg ind^{-1}]		[μg ind^{-1}]		[μg ind^{-1}]		[μg ind^{-1}]		[%]		[%]		[%]			
			$\bar{x}$	std.dev.	$\bar{x}$	std.dev.	$\bar{x}$	std.dev.	$\bar{x}$	std.dev.	$\bar{x}$	std.dev.	$\bar{x}$	std.dev.	$\bar{x}$	std.dev.	$\bar{x}$	std.dev.
Zoea I	0	32	38.3	1.3	17.5	0.7	3.2	0.1	2.4	0.1	45.7	0.6	8.5	0.1	6.4	0.2	5.41	0.10
	1	5	39.4	1.5	18.9	0.7	3.4	0.1	2.6	0.1	48.0	0.4	9.0	0.1	6.7	0.2	5.53	0.06
	1	15	44.4	1.4	21.3	0.7	3.9	0.1	2.9	0.1	47.6	0.3	8.7	0.1	6.6	0.2	5.49	0.05
	1	25	47.4	1.5	21.0	0.8	3.7	0.1	3.0	0.2	44.2	0.5	7.9	0.1	6.4	0.3	5.60	0.07
	1	35	49.9	2.4	21.9	1.1	3.9	0.2	3.1	0.2	43.9	0.2	7.9	0.1	6.3	0.1	5.58	0.07
	1	45	52.9	2.3	22.0	1.4	4.0	0.3	3.2	0.2	41.6	0.9	7.6	0.2	6.0	0.1	5.46	0.10
	1	55	50.7	2.2	19.7	1.4	4.0	0.2	2.8	0.2	38.8	1.2	7.9	0.3	5.6	0.2	4.91	0.28
	2	5	49.4	1.6	21.3	0.8	3.9	0.7	3.0	0.1	43.2	0.4	7.9	1.6	6.1	0.1	5.64	0.01
	2	15	45.5	1.6	22.1	0.9	4.3	0.1	3.2	0.2	48.5	0.5	9.5	0.1	6.9	0.2	5.13	0.07
	2	25	49.7	2.8	23.2	0.9	4.6	0.2	3.3	0.2	46.7	1.4	9.2	0.2	6.7	0.2	5.07	0.05
	2	35	50.9	1.3	23.5	0.3	4.6	0.1	3.4	0.1	46.1	0.9	9.0	0.2	6.7	0.1	5.13	0.04
	2	45	51.9	2.4	23.7	0.7	4.6	0.1	3.4	0.1	45.7	2.0	8.8	0.4	6.6	0.2	5.20	0.04
	2	55	50.8	2.3	21.8	1.0	4.2	0.2	3.2	0.2	43.0	0.6	8.2	0.2	6.3	0.2	5.26	0.10
Zoea II	3	5	85.9	1.5	35.3	0.9	7.4	0.2	5.0	0.2	41.0	0.5	8.6	0.0	5.8	0.1	4.76	0.04
	2	15	95.8	4.2	38.3	1.6	8.0	0.3	5.7	0.3	40.0	0.4	8.4	0.1	5.9	0.1	4.76	0.05
	2	25	105.2	4.7	41.7	2.3	8.8	0.5	6.3	0.4	39.6	0.5	8.4	0.1	5.9	0.1	4.73	0.01
	2	35	103.5	4.1	44.8	1.3	9.3	0.3	6.5	0.2	43.3	0.6	9.0	0.2	6.3	0.1	4.82	0.05
	3	45	111.8	2.3	46.0	1.3	9.4	0.3	6.9	0.2	41.2	1.1	8.4	0.2	6.1	0.2	4.92	0.07
	3	55	79.3	8.3	32.3	1.1	6.5	0.2	4.7	0.2	41.0	4.3	8.3	0.9	5.9	0.7	4.97	0.08
Zoea III	4	5	95.8	2.1	42.0	0.6	9.8	0.2	6.3	0.1	43.8	0.5	10.2	0.2	6.6	0.1	4.29	0.07
	3	15	118.7	4.7	51.4	2.3	11.2	0.4	7.5	0.4	43.3	0.4	9.4	0.2	6.3	0.1	4.59	0.08
	3	25	125.5	2.5	56.2	1.4	12.2	0.2	8.7	0.3	44.8	0.3	9.8	0.0	6.9	0.1	4.60	0.05
	3	35	129.2	2.4	57.6	1.8	12.5	0.3	9.0	0.3	44.6	0.9	9.6	0.1	7.0	0.2	4.62	0.07
	4	45	137.6	4.0	60.4	2.1	12.9	0.3	9.4	0.4	43.9	0.4	9.4	0.1	6.9	0.1	4.67	0.06
	5	55	124.3	1.9	53.0	0.5	11.2	0.3	8.3	0.1	42.7	0.5	9.0	0.1	6.7	0.1	4.72	0.06
Megalopa	6	15	221.4	8.8	89.0	3.3	18.5	0.5	12.9	0.5	40.2	0.6	8.3	0.2	5.8	0.1	4.82	0.06
	6	25	236.2	9.9	95.9	3.2	19.3	0.5	14.1	0.5	40.6	0.7	8.2	0.1	6.0	0.1	4.96	0.09
	6	35	246.2	5.7	98.4	2.6	19.7	0.4	14.6	0.5	40.0	1.3	8.0	0.2	5.9	0.2	5.01	0.04
	7	45	232.0	11.2	88.6	4.0	18.9	0.8	12.9	0.6	38.2	1.0	8.2	0.3	5.6	0.1	4.68	0.12
Crab I	5	15	431.9	25.4	155.9	10.0	31.8	2.1	21.2	1.8	36.1	0.6	7.4	0.1	4.9	0.2	4.90	0.06
	5	25	406.2	32.7	145.0	10.4	29.4	2.1	21.6	1.5	35.6	0.7	7.5	0.3	5.2	0.2	4.94	0.05
	5	35	412.0	18.5	144.9	6.7	30.0	1.5	21.8	1.2	35.2	0.3	7.3	0.1	5.3	0.1	4.83	0.08
	5	45	390.4	19.5	134.4	8.0	28.3	1.6	20.1	1.3	34.4	0.6	7.2	0.1	5.1	0.1	4.75	0.04

concentrations by freezing seawater at −18 °C. Salinities were checked with a temperature-compensated electric probe (WTW, Germany) or a hand refractometer (Krüss Salinometer S-28) to the nearest 0.1 or 1 PSU, respectively. Transfer of larvae or juveniles to 25 and 45 PSU was direct, whereas that to higher or lower salinities allowed for a stepwise adaptation (3 h per interval). Charmantier et al. (1998) showed that an adaptation time of ca. 2 h is sufficient for the early developmental stages of *A. miersii* to reach a constant haemolymph concentration. During daily water change and feeding, larvae and juveniles were checked for moults or mortality. Freshly moulted individuals were transferred to new marked bowls, so that each bowl contained exclusively individuals with an identical age within the moulting cycle of a given stage. Growth in the successive postembryonic instars was measured in terms of dry weight (W), carbon (C), nitrogen (N) and hydrogen (H). W was measured to the nearest 0.1 μg on a Mettler UM3 microbalance, and C, N and H with a model 1108 Carlo Erba CHN Analyser applying standard techniques (Anger & Harms, 1990). After exposure of a given developmental stage to the different test salinities, five rep-

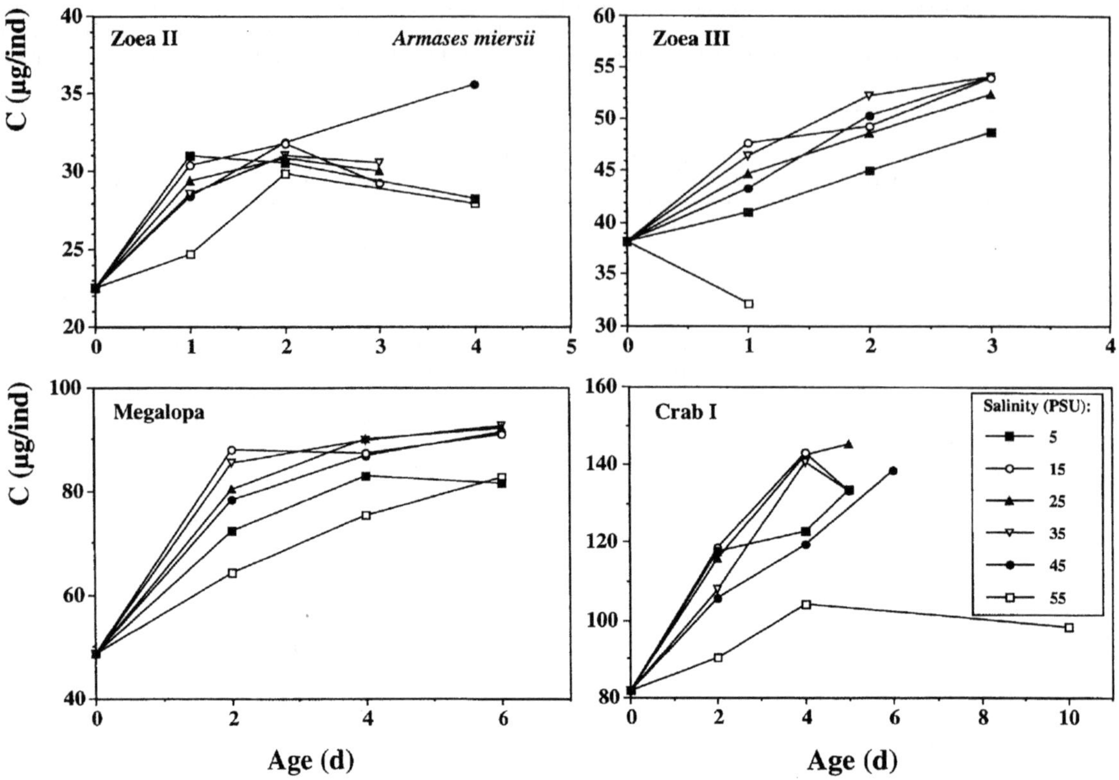

Figure 2. Growth (expressed as carbon content, C (μg ind^{-1}) of larval and early juvenile *Armases miersii* after acute exposure to different salinities (5–55 PSU; preceding development to each stage in seawater, 32–33 PSU); age (days) within the moulting cycle of a given stage; no error bars given for sake of clarity.

licate samples with 4–6 individuals each were taken daily (zoeal stages) or every other day (Megalopa, Crab I) and stored frozen at −20 °C until analysis. In the experimental series with exposure from hatching, growth was followed also through subsequent stages (with samples of final biomass only). When moulting began in a bowl, it was assumed that all individuals in the earlier stage were approaching the end of the moulting cycle and their biomass thus represented the final value in a given stage.

All statistical analyses followed standard techniques (Sokal & Rohlf, 1995). When data deviated significantly from a normal distribution (Kolmogorov-Smirnov test) or when variances deviated significantly from homogeneity (Levene's median test), the non-parametric Mann-Whitney rank sum test was used for comparisons of mean values, otherwise Student's *t*-test.

Results

When freshly hatched *Armases miersii* larvae were transferred from seawater (32 PSU) to the different rearing salinities (5–55 PSU), they exhibited in the full range of conditions a significant biomass accumulation, although this was consistently and significantly depressed at 5 and 55 PSU. The patterns of growth within individual developmental stages were similar in all experimental series and treatments (except for 55 PSU, where also losses of biomass occurred), regardless of the measure of biomass considered (W, C, N or H; as shown in Table 1).

This general pattern in biomass accumulation is exemplified with the carbon (C) content per individual during development at constant seawater conditions (35 PSU; Figure 1a). Growth in later stages, after acute exposure to different salinities, is shown in Figure 2. In general, a steep biomass increase occurred in the first half of the moulting cycle, followed by weaker increments, and eventually, often by constant or slightly decreasing C values when moulting was approached. During development at 35 PSU from the Zoea I (hatching) to the end of the Megalopa stage (metamorphosis), larval dry weight increased by a

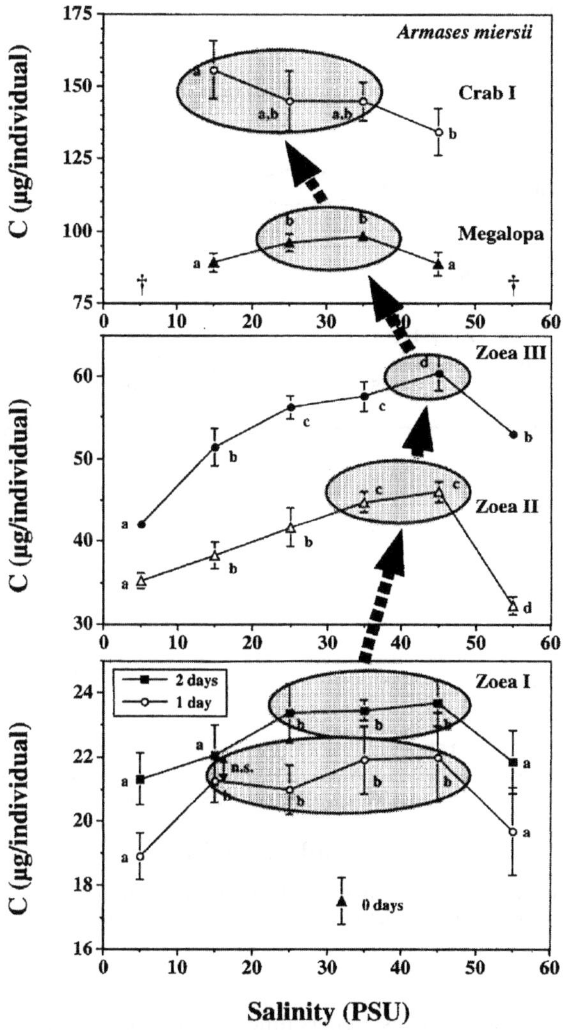

Figure 3. Growth (expressed as carbon content, C (μg ind^{-1}) of larval and early juvenile *Armases miersii* during development at different salinities (5–55 PSU; exposure from hatching); $\bar{x}\pm$ SD (n = 5); different letters indicate significant differences between salinities; †: complete mortality in indicated stage. Salinity range allowing for maximum growth and ontogenetic shifts in the apparent optimum for growth marked with shaded areas and arrows, respectively; maximum growth defined by maximum C content and values that were not significantly different from this.

factor of 6.0, C by 5.3, N by 5.8 and H by 5.6. Similar cumulative increments were measured during development at 25 PSU, while a slight but statistically significant depression occurred at 15 and 45 PSU.

At the two extreme salinities (5 and 55 PSU, resp.), development was generally delayed, so that the final biomass of successive instars was reached later than in the other treatments (see Zoea II, III, Megalopa in Table 1). In spite of the lengthened time available for growth, developmentally delayed individuals showed in most cases a significantly lower final biomass than siblings reared under more favourable conditions (exception: Zoea II at 45 PSU; Figure 2). Compared with 25 and 35 PSU, the duration of development in the Crab I stage was doubled at 55 PSU, and the increment in C was reduced to one half. Similar although weaker stress effects occurred at 5 and 45 PSU.

Similar patterns as in the C content were observed also in N and H per individual (Table 1). This indicates that the relative chemical composition of biomass remained fairly constant in most treatments. Significantly reduced percentage values of C, N and H, however, occurred in most stages under the exposure to very high salinities (55 PSU, sometimes also 45 PSU; see Table 1), indicating a stress-induced reduction of organic matter in relation to inorganic compounds. No consistent salinity effects were observed in the C:N weight ratio, which is generally considered as an indicator of changes in the lipid:protein relation.

Independent of the salinity, there were clear ontogenetic changes in the C:N ratio (Figure 1b; Table 1). This quotient decreased gradually throughout the zoeal phase, indicating a preferential degradation of lipid stores and/or a proportionally faster accumulation of protein. This decrease was followed by an increase in the Megalopa stage (again in all treatments), and low values after metamorphosis to the first juvenile crab.

Effects of chronic exposure to different salinities (from hatching) on larval and early juvenile growth are illustrated in Figure 3, using again the C content per individual as an example. In the first zoeal stage, initial growth was only at the two extreme salinites (5, 55 PSU) significantly reduced, while maximum C increments occurred in a wide intermediate range (15–45 PSU). Subsequently, however, the range allowing for maximum C accumulation tended to narrow and the optimum for growth (shaded areas in Figure 3) appeared to shift towards higher concentrations (arrows). These tendencies continued throughout the zoeal phase (Zoea I–III), but reversed thereafter. In the Megalopa and Crab I, maximum final biomass values were measured again in a wider range, including seawater and moderately brackish conditions (15–35 PSU). In order to allow for direct comparisons of growth data from different stages, treatments or experimental series (the latter measured in larvae from different females), we converted the biomass increments (in μg C per instar) to percentage values, defining the increments in the seawater (35 PSU) treatment as 'control' (=100%) values. In this presentation,

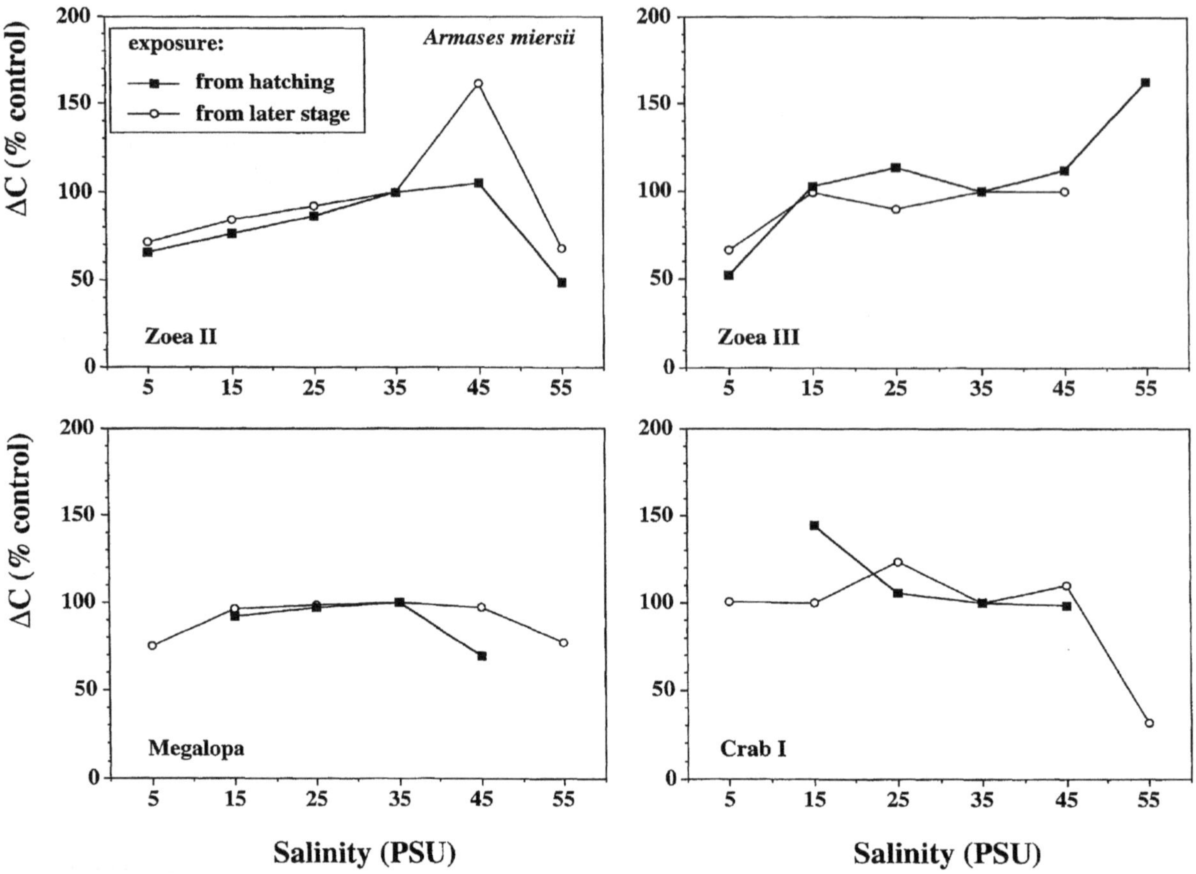

Figure 4. Relative carbon growth increments (ΔC [%]) in larval and early juvenile *Armases miersii* during development at different salinities (5–55 PSU), after exposure either from hatching (filled squares) or only in a later stage (open circles); ΔC expressed as a percentage of the C increment measured in a control condition (35 PSU).

mostly similar response patterns were found in experiments with either continuous or acute exposure to salinities ranging from 15 to 25 PSU, suggesting only insignificant effects of the timing of exposure (Figures 2 and 4). However, extreme conditions (5 and 55 PSU, resp., to some degree also 45 PSU) tended to affect growth stronger in experiments with chronic as compared with later beginning exposure (Zoea II, Megalopa in Figure 4).

Discussion

The present investigation confirms that the early life-history stages of *Armases miersii* are physiologically well adapted for survival and development in a physically harsh and unpredictable environment. Substantial biomass accumulation occurred in a wide range of salinities, comprising highly dilute brackish water (5 PSU), as well as hypersaline conditions (55 PSU). From a recent study of the ontogeny of osmoregulation in this species (Charmantier et al., 1998), we know that the extreme euryhalinity of the larval and early juvenile stages is based upon an unusually early appearance of osmoregulatory capabilities, including hyper-regulation from hatching of the Zoea I and hypo-regulation from the Megalopa stage.

Except for the most unfavourable condition (55 PSU), salinity did not greatly affect the patterns of growth within individual moulting cycles (Figure 2) or the relative chemical composition of biomass (Table 1). The parabola-shaped growth curves within successive developmental stages (Figure 1a) are typical of planktotrophic decapod crustacean larvae (Anger, 1998), although the overall level of biomass increase per moulting cycle appears to be exceptionally low in *A. miersii*. This was noted already in a preliminary study (Anger & Schultze, 1995) and interpreted as a consequence of partial utilization of enhanced internal energy reserves remaining from the egg yolk. These reserves enable the early larvae of this species to survive and develop, when necessary, also with limited or lacking planktonic food sources (Anger, 1995b). A relationship between particularly low zoeal growth and a high endotrophic potential was observed also in other decapod species that have partially emancipated from the sea (Anger, 1995a).

The present results corroborate this generalization. A decreasing C:N ratio during the early zoeal phase (Figure 1b) suggests a partial utilization of lipid reserves (predominantly stored in the hepatopancreas) and/or a concomitant synthesis of proteins (probably in nervous, muscular and epidermal tissues; Harzsch & Dawirs, 1993). A rapid initial increase in the C:N quotient after the moult to the Megalopa, by contrast, may reflect an increasing feeding activity in this stage (Anger, 1995b), and hence, an increasing lipid accumulation. Additionaly, however, there is a rapid initial cuticle calcification in the Megalopa, which should enhance the proportion of inorganic C within total dry weight, and thus, of total C in relation to N (Anger, 1998).

The biomass increments in successive stages were, in general, relatively little influenced by salinity, showing a conspicuous depression only at the most extreme conditions (5 and 55 PSU, resp.). Also the timing of exposure was not a very important factor in the salinity tolerance pattern of *A. miersii* larvae (Figure 4). As an overall tendency, later (acute) exposure to unfavourable salinites appeared to affect growth to a slightly lesser degree than long-term exposure. This corresponds with our observations of survival and development duration (Anger, 1996), where consistently a wider salinity range was tolerated in experiments with short-term as compared with chronic exposure. This response pattern indicates cumulative effects of continued osmotic stress. When the salinity range allowing for maximum growth (shaded areas in Figure 3) is compared with the optimum for survival and development (Anger, 1996), there appears to be a striking difference in the response patterns. The rates of survival and development were generally maximal in moderately dilute media (15–25 PSU), which corresponds with the prevailing conditions in natural breeding rock pools in spring on Jamaica (Anger, 1995b; Schuh & Diesel, 1995a, b). The salinity conditions allowing for maximum growth, by contrast, showed conspicuous ontogenetic changes. During development at 15–25 PSU, zoeal biomass accumulation was lower than in higher salt concentrations (35–45 PSU), and the initially wide salinity range with maximum growth tended to narrow throughout the zoeal phase (Figure 3). We interpret this as a consequence of cumulative energetic costs for hyper-osmoregulation in dilute media (Charmantier et al., 1998). At higher salinities, by contrast, the zoeal stages behave as osmoconformers, and hence, they do not invest significant amounts of energy in osmoregulation. We propose that this energy saving effect allows for maximum growth at 35–45 PSU, although other physiological processes associated with development and maintenance may not function at their optimum.

In the Megalopa and Crab I, the range of salinities allowing for maximum C accumulation appears to widen again, shifting back towards lower salt concentrations (Figure 3). This may be explained with two major ontogenetic changes. 1. The efficiency of hyper-osmoregulation in dilute media ($\leq$25 PSU) should significantly increase, due to the appearance of functional gills and an enlargement of other transport epithelia (Charmantier, 1998). As a consequence of a reduction in the energetic costs for hyper-osmoregulation, more energy should become available for growth in brackish water. 2. In the Megalopa stage, *A. miersii* attains also the capability of hypo-regulation in concentrated media. Reduced growth in the Megalopa and Crab I at high salinities ($\geq$35 PSU) may thus reflect an increasing energy demand for hypo-osmoregulation. When these late stages were exposed, without prior acclimatization, to enhanced salinities, also their mortality and development rates indicated particularly strong osmotic stress (Anger, 1996). The results of our studies of salinity tolerance and growth in the early life-history stages of *A. miersii* show consistently that this species is very well adapted to pass its larval and early juvenile development in supratidal rock pools, where great salinity variations occur. However, partially incongruous response patterns in survival, development and growth suggest that the adaptation of this semiterrestrial crab to non-marine conditions is not perfect. Reduced growth under otherwise optimum salinity conditions, presumably due to the energetic costs of osmoregulation, may indicate a trade-off between different physiological adaptations. Similar as in the partial independence from planktonic food sources (Anger, 1995a, b), the salinity tolerance of larval *A. miersii* appears to reflect a transitional stage in the evolutionary process of a crab species that is about to invade limnic and terrestrial environments.

References

Anger, K., 1995a. The conquest of freshwater and land by marine crabs: adaptations in life-history patterns and larval bioenergetics. J. exp. mar. Biol. Ecol. 193: 119–145.

Anger, K., 1995b. Developmental biology of *Armases miersii* (Grapsidae), a crab breeding in supratidal rock pools. I. Facultative lecithotrophy of larval stages. Mar. Ecol. Prog. Ser. 117: 75–81.

Anger, K., 1995c. Developmental biology of *Armases miersii* (Grapsidae), a crab breeding in supratidal rock pools. II. Food limitation in the nursery habitat and larval cannibalism. Mar. Ecol. Prog. Ser. 117: 83–89.

Anger, K., 1996. Salinity tolerance of the larvae and first juveniles of a semiterrestrial grapsid crab, *Armases miersii* (Rathbun). J. exp. mar. Biol. Ecol. 202: 205–223.

Anger, K., 1998. Patterns of growth and chemical composition in decapod crustacean larvae. Invert. Reprod. Develop. 33: 159–176.

Anger, K. & J. Harms, 1990. Elemental (CHN) and proximate biochemical composition of decapod crustacean larvae. Comp. Biochem. Physiol. 97B: 69–80.

Anger, K. & K. Schultze, 1995. Elemental composition (CHN), growth, and exuvial loss in the larval stages of two semiterrestrial crabs, *Sesarma curacaoense* and *Armases miersii* (Decapoda: Grapsidae). Comp. Biochem. Physiol. 111A: 615–623.

Charmantier, G., 1998. Ontogeny of osmoregulation in crustaceans: a review. Invert. Reprod. Develop 33: 177–190.

Charmantier, G., M. Charmantier-Daures & K. Anger, 1998. Ontogeny of osmoregulation in the grapsid crab *Armases miersii* (Crustacea, Decapoda). Mar. Ecol. Prog. Ser. 164: 285–292.

Hartnoll, R. G., 1988. Evolution, systematics, and geographical distribution. In Burggren, W. W. & B. R. McMahon (eds), Biology of the Land Crabs. Cambridge University Press, New York: 6–54.

Harzsch, S. & R. R. Dawirs, 1993. On the morphology of the central nervous system in larval stages of *Carcinus maenas* L. (Decapoda, Brachyura). Helgoländer wiss. Meeresunters. 47: 61–79.

Schuh, M. & R. Diesel, 1995a. Breeding in a rock pool: larvae of the semiterrestrial crab *Armases (=Sesarma)miersii* (Decapoda: Grapsidae) develop in a highly variable environment. J. exp. mar. Biol. Ecol. 185: 109–129.

Schuh, M. & R. Diesel, 1995b. Effects of salinity, temperature and starvation on the larval development of *Armases (=Sesarma) miersii* (Rathbun, 1897), a semiterrestrial crab with abbreviated development (Decapoda: Grapsidae). J. Crust. Biol. 15: 205–213.

Sokal, R. R. & F. J. Rohlf, 1995. Biometry. The Principles and Practice of Statistics in Biological Research. W. H. Freeman & Co., 3rd edn, San Francisco: 887 pp.

Hydrobiologia **426:** 169–172, 2000.
G. Liebezeit, S. Dittmann & I. Kröncke (eds), Life at Interfaces and Under Extreme Conditions.

Does fluctuating salinity induce branching of *Fucus vesiculosus*?

Ari Ruuskanen[1,*] & Mikko Kiirikki[2]
[1]*Department of Ecology and Systematics, Division of Hydrobiology, University of Helsinki P.O. Box 17, Finland*
*Tel. +358-9-1917359. Fax +358-9-1917301. E-mail: ari.ruuskanen@helsinki.fi (*author for correspondence)*
[2]*Finnish Environment Institute, P.O. Box 140, FIN-00251 Helsinki, Finland*
Tel. +358-9-40300214. Fax +358-9-40300291. E-mail: mikko.kiirikki@iki.fi

Key words: Bothnian Sea, branching, fluctuating, *Fucus vesiculosus*, proliferation salinity, SCUBA

Abstract

During the field surveys in the Bothnian Sea, i.e. towards low salinity areas (approx. 4‰), present authors observed that the frequency of irregularly branched *Fucus vesiculosus* plants increased. Salinity is known to decrease gradually towards the northern parts of the Baltic Sea. However, salinity is not steady but may fluctuate greatly on an annual and even a daily scale, and salinity can drop to zero for short periods. In order to demonstrate whether the fluctuating salinity induces irregular branching of *F. vesiculosusus,* a experiment was carried out in the Tvärminne archipelago, on the south coast of Finland in May–September 1997. First, plants were collected and then put into two containers both of which had a fresh water and a sea water flow throught. After 48 h of treatment, the plants were removed to the same place where they had been orginally collected. After the growing season, the plants were collected again, and the number of irregularly and normally branched tips were measured. The results shows that plants with the fresh water treatment have branched irregularly. In constrast, the control plants had only a few irregularly branched tips. This experiment brings us to conclude that low salinity during the critical growing season induces irregular branching.

Introduction

Fucus vesiculosus is the dominant perennial large submerget fucoid on the coast of Finland. The growth period is from May to July, and reproduction peak in June (Bäck et al., 1991). It has been reported in the areas of low salinity that the morphology of *Fucus vesiculosus* changes making the plants smaller and thinner (Luther, 1981; Raven & Samuelsson, 1988; Bäck et al., 1993). However, there are characteristics that can not be measured from every plant, but are only evident in some plants. One of these is irregular branching, called proliferation, which means that the tip of the *F. vesiculosus* frond has branched several times where normally it only branches once. Figures 1 A–B is a herbarium specimen plant collected from the exposed shore at a depth of three metres near Rauma city, in the Bothnian Sea, in 1995 (Figure 2.).

Low salinity areas occur on both large and small geographic scales. On a large scale, the salinity gradually decreases towards the northern and eastern parts of the Baltic Sea. In the Quark, in the Bothnian Sea, the mean salinity is approx. 5–4‰. Such a low concentration prevents the reproduction of *Fucus vesiculosus* and also forms a distribution barrier (Sérrao et al., 1996). On a small scale, low salinity areas occurs at the river end where fresh water meets sea water.

The authors have observed, in SCUBA dive field work, that the frequency of irregularly branched plants increases towards the low salinity areas. In some years, plants with irregularly branched tips can be found in great amounts, whereas in others years, it is almost impossible to find plants with irregularly branched tips. These field observations lead us to the assumption that in some years, there are environmental factors which induce branching, and in other years not.

The aim of this study was to demonstrate whether short pulses of low salinity induces irregular branching of *Fucus vesiculosus.*

Description of sites studied

The Baltic Sea is a brackish and tideless enclosed

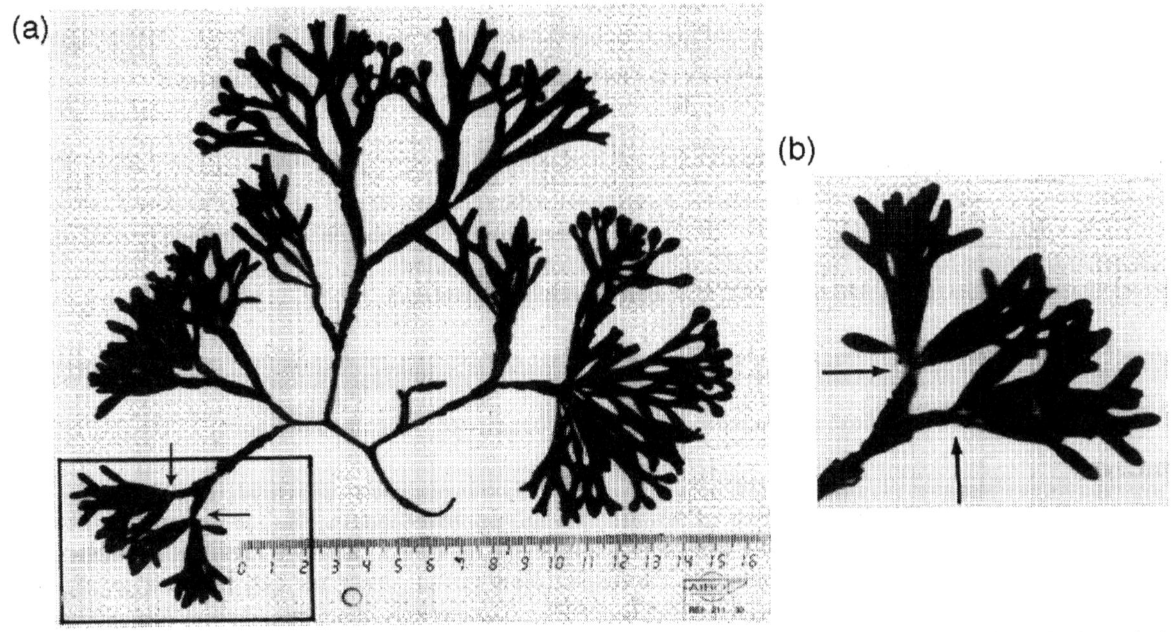

Figure 1. (A) A typical *Fucus veciculosus* growing in the low salinity areas in the Bothnian Sea (see Figure 2). Notice the 'year ring' of irregularly branched tips. (B) A close up picture of irregularly branched tips in the *Fucus vesiculosus*. Scale in mm.

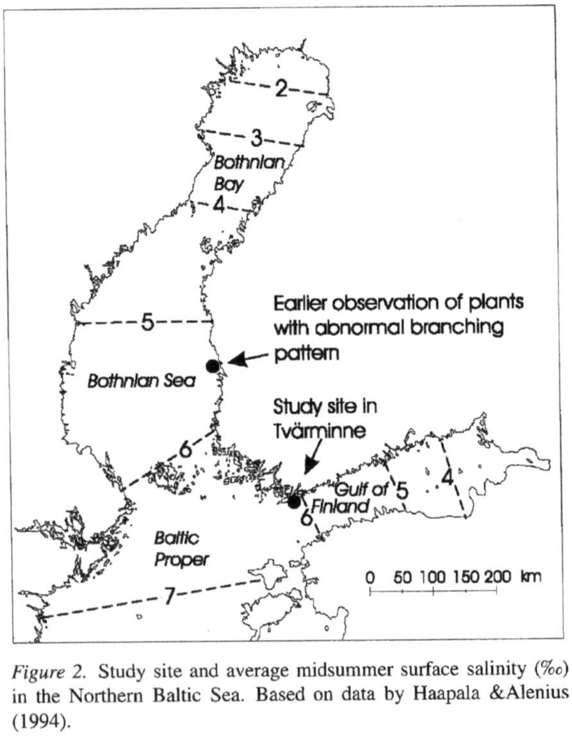

Figure 2. Study site and average midsummer surface salinity (‰) in the Northern Baltic Sea. Based on data by Haapala &Alenius (1994).

sea. On the coast of Finland salinity varies between 7 ppt and almost fresh water (Figure 2.). Salinity is generally known to decrease towards the northern and eastern parts of the Baltic Sea. There are permanent sampling stations run by environmental institutes on the open sea, which take samples once or twice a month. These don't give enough data to indicate fluctuations occurring on a daily scale. Moreover, long interval sampling gives the impression of steady salinity, which may not represent the true conditions (Figure 3 B). Continuous salinity measurement by an YSI 6000 sond revealed that even in the present study site, Tvärminne archipelago, short pulses of low salinity can be detected (Figure 3 A).

Materials and methods

The experiment took place in May 1997, in the Tvärminne archipelago, on the south coast of Finland (Figure 2). At the study site, the mean salinity was approx. 6‰. Forty *Fucus vesiculosus* plants were collected at a 2 m depth by SCUBA diving. All the plants were tied by the stipe onto a bricks. Twenty plants were put into a container (volume 340 l) with a fresh water flow through (approx. 10 l/min). The other 20 plants were put into a similar container with a sea water flow throughout. During the whole experiment,

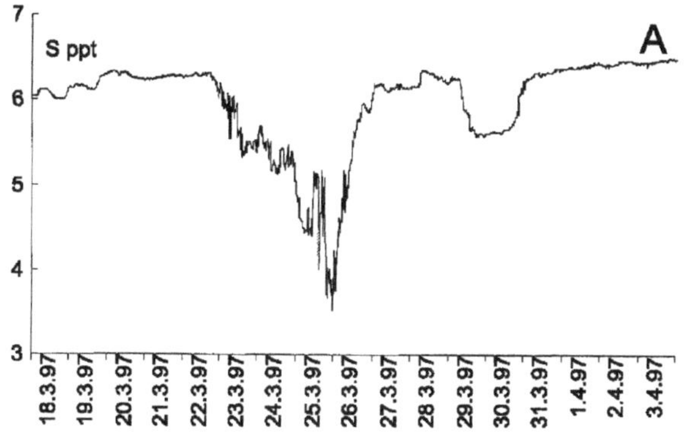

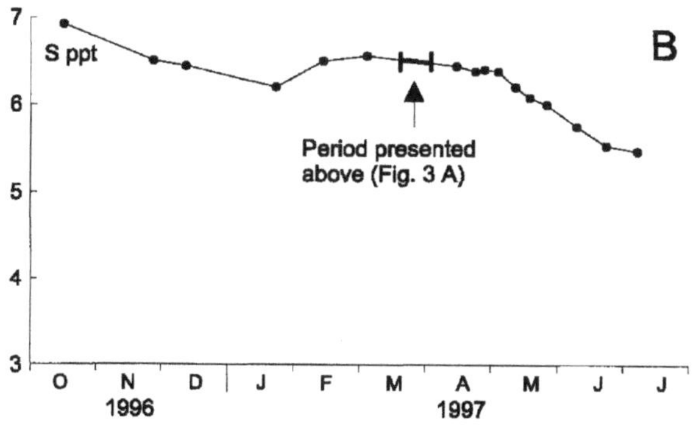

Figure 3. Surface salinity in the Tvärminne archipelago, SW Finland. (A) Continuous salinity measurement by YSI 6000 sond revealed a three-day-long low salinity water pulse. (B) Standard salinity measurements at a nearby intensive monitoring site gave an impression of stable salinity conditions.

temperature and pH were measured, and they stayed on the same level in the containers as in the sea. After 48 h, the plants were moved to the same place where they were orginally collected.

In September, after the growing season, the plants were recollected. The number of total tips, and the number of irregularly branched tips were measured.

Results

The results showed that all 20 plants with a fresh water treatment had branched irregularly. In constrast, the 20 plants with a sea water treatment only had a few irregularly branched plants (Table 1).

In Figure 4 it can be seen that the branching pattern of the plant treated with a fresh water pulse is similar to a irregularly branched plant collected in the field (Figure 1).

Figure 4. A close up picture of irregularly branched tips after fresh water through flow treatment and a five month growing period in the Tvärminne archipelago (see Figure 2). Scale in mm. Notice the similar branching model with 'natural irregularities' on a branched plant in Figure 1B.

Discussion

This experiment gives us a reason to conclude that low salinities during the critical growth period is one factor affecting irregular branching. Because salinity does not become too low every year, there may be years when *Fucus* has no irregular branching at all.

Irregular branching, or proliferation, can be caused by the lost of apical dominance (Moss, 1964). In this case, the fresh water pulse caused an osmotic shock which probably destroyed the thin walled apical meristem cells.

Even though continuous measurements showed a low salinity pulse in the Tvärminne archipelago, irregularly branched plants could not be found in any significant quantities. One explanation may be that in the Bothnian Sea, the salinity is lower and plants living near their salinity tolerance limit are more vulnerable to low salinity pulses.

Low salinity may only be one factor affecting the irregular branhing formation in the Baltic Sea. Rönnberg (1988) and Kiirikki & Ruuskanen (1996) found that the effect of ice may also affect in the condition of *Fucus,* and induce vegetastive regrowth of the plant. The grazing of crustaces, e.g. *Idotea* ssp., may induce the *Fucus* to branch irregularly.

Acknowledgements

The authors wish to thank both the Tvärminne Zoological Station and MSc Anne Heinänen for her help in laboratory. This study was financed by the Walter and Andrée de Nottbeck's Foundation.

Table 1. A mean (±S.E.) number of irregularly branched and normally branched tips of the*Fucus vesiculosus* plant treated with fresh and sea water flow throughout. (*n*=20)

	Fresh water flow through	Sea water water flow through
No of irregularly branched tips	17.6 (3.1)	1.7 (0.4)
No of regularly branched tips	63 (10)	72 (11)

References

Bäck, S., J. C. Collins & G. Russell, 1991. Aspects of reproductive biology of *Fucus vesiculosus* from the coast of SW Finland. Ophelia 34: 129–141.

Bäck, S., 1993. Morphological variation of the northern Baltic *Fucus vesiculosus* along the exposure gradient. Ann. bot. fenn. 30: 275–283.

Haapala, J. & P. Alenius, 1994. Temperature and salinity statistics for the Northern Baltic Sea 1961–1991. Finnish Mar. Res. 262: 51–121.

Kiirikki, M. & A. Ruuskanen, 1996. How does *Fucus vesiculosus* survive ice scraping? Bot. Mar. 39: 133–139.

Luther, H., 1981. Occurrence and ecological requirements of *Fucus vesiculosus* in semi-enclosed inlets of the Archipelago Sea, SW Finland. Ann. bot. fenn. 18: 187–200.

Moss, B., 1964. Apical dominance in *Fucus vesiculosus*. New Phytol. 64: 387–392.

Raven, J. A. & G. Samuelsson, 1988. Ecophysiology of *Fucus vesiculosus* L. close to its northern limit in the Gulf of Bothnia. Bot. mar. 31: 399–410.

Rönnberg, O. & I. Haahtela, 1988. Does anchor ice contribute to the decline of *Fucus* in the Baltic? Mar. Poll. Bull. 19: 388–389.

Serrão, E. A., L. Kautsky & S. H. Brawley, 1996. Distributional success of the marine seaweed *Fucus vesiculosus* L. in the brackish Baltic Sea correlates with osmotic capabilities of Baltic gametes. Oecologia 107: 1–12.

Hydrobiologia **426**: 173–183, 2000.
G. Liebezeit, S. Dittmann & I. Kröncke (eds), Life at Interfaces and Under Extreme Conditions.

Fungal degradation of the thermoplastic polymer poly-β-hydroxybutyric acid (PHB) under simulated deep sea pressure

K. E. Gonda[1], D. Jendrossek[2] & H. P. Molitoris[1,*]
[1]*Institut für Botanik, Pilzphysiologie, Universität Regensburg, D-93040 Regensburg, Germany*
[2]*Institut für Mikrobiologie, Universität Stuttgart, Allmandring 31, D-70550 Stuttgart, Germany*
E-mail: hans-peter.molitoris@biologie.uni-regensburg.de

Key words: deep sea, degradation, hydrostatic pressure, marine fungi, marine pollution, poly-β-hydroxy-butyric acid (PHB)

Abstract

Little is known about marine filamentous fungi and yeasts, almost nothing about their life and metabolism under deep sea conditions. Data on growth and metabolic activity give insight into the role of organisms in the marine habitat. Degradation studies on pollutants, such as polymeric thermoplasts, provide information about the self-cleaning capacity of a habitat. Therefore, recently isolated fungal strains from the deep sea and our newly developed methods and apparatus for investigation of fungi under simulated deep sea conditions were used to study fungal growth and degradation of a commercially produced thermoplastic polymer (poly-β-hydroxybutyric acid = PHB). Two deep sea isolates, a filamentous fungus (*Aspergillus ustus*) and one yeast (*Rhodosporidium sphaerocarpum*), and for comparison, two marine surface yeast isolates (*Candida guilliermondii, Debaryomyces hansenii*) and one terrestrial isolate of *Aspergillus ustus* were investigated. Growth (colony-forming units, dry weight), physiological parameters (oxygen saturation of the hydraulic fluid as oxygen reservoir, pH and consumption of total carbohydrate) and PHB degradation (clearing test: clearing of PHB-turbid agar medium; spectrophotometric test: PHB depolymerase activity) were followed after incubation in high-pressure autoclaves in artificial seawater medium at 27 °C and pressures of 0.1 MPa (= atmospheric pressure), 5 MPa, 10 MPa, 20 MPa, 30 MPa, 45 or 50 MPa and 100 MPa (∼ 10 000 m water depth) for a maximum of 21 days (yeasts) and 28 days (filamentous fungi), respectively. Irrespective of the marine or terrestrial origin of the isolates, growth decreased with increasing pressure with a limit between 30 MPa and 50 MPa for filamentous fungi and yeasts. Metabolic activity (consumption of medium components) started to decrease from 20 MPa, ceasing at growth-limiting pressures. Under atmospheric conditions, all strains degraded PHB in solid medium, in liquid medium degradation was less and decreased further and/or was delayed with increasing hydrostatic pressure; beyond 30 MPa, no PHB degradation could be observed. In summary, it could be shown that growth, metabolism and degradation of pollutants such as PHB by marine fungal isolates was impaired with increasing pressure, showing one aspect of the reduced self-cleaning capacity of the deep sea habitat.

Introduction

The deep sea constitutes one of the largest and vastly unexplored habitats on our planet. More than two thirds of the earths' surface is covered by oceans and more than 50% by oceans with a depth ranging from 3000 to 6000 m with an average depth of approximately 3200 m (Gage & Tyler, 1991). The deep sea conditions can be considered extreme due to high hydrostatic pressures, its low temperatures with the notable exception of hydrothermal vents, low nutrient concentrations and complete darkness.

Several substrates, such as wood, particulate organic matter (POM) and chitin (exoskeleton of marine crustacea), could participate in transportation and nutrition of marine fungi and have shown to be degraded by autochthonous fungi even in greater depths of the

* Dedicated to Prof. Dr Jan Kohlmeyer, Morehead City, USA, on the occasion of his 70th anniversary

oceans (Kohlmeyer, 1969; Kohlmeyer & Kohlmeyer, 1979; Kohlmeyer & Volkmann-Kohlmeyer, 1988).

There is strong evidence that fungi as prominent heterotrophic organisms in the oceans are transported passively from the surface to the deep sea by rapidly sinking water masses in the Arctic and Antarctic (Gage & Tyler, 1991) and/or by attachment to their sinking particulate substrate (Van Uden & Fell, 1968; Kohlmeyer & Kohlmeyer, 1979).

Fungi isolated from these substrates in deep sea habitats are the cosmopolitan yeasts *Debaryomyces hansenii* and *Rhodosporidium sphaerocarpum* (Fell et al., 1960; Fell, 1967; Meyers et al., 1967; Yamasato et al., 1974) and filamentous fungi mostly representing *Aspergillus* species (Sparrow, 1937; Roth et al., 1964). *In vitro* growth experiments with these organisms have revealed barotolerant growth under simulated deep sea conditions (low temperature, high hydrostatic pressure, salinity and oxygen concentration: Lorenz, 1993; Zaunstöck, 1994). Little, however, is known about the degradation of anthropogenic substrates in the deep sea. Aromatic hydrocarbons as components of natural fossil oil have been used as carbon and energy sources in laboratory experiments by microorganisms isolated from oil spills (Ahearn et al., 1971; Grüttner & Jensen, 1983). Conventional plastics produced chemosynthetically from these compounds, however, increasingly pose a great threat to the marine environment, because they are not degradable and are being dumped into the oceans each year in large quantities (several hundred thousand tons: Pruter, 1987). For their negative effects, it has been estimated that about one million marine animals are killed every year by choking on floating plastic items which were mistaken for food or by entanglement in plastic debris (Bean, 1987; Laist, 1987).

Biosynthetic thermoplasts such as the polyhydroxyalkanoates (PHA) are produced by bacteria from renewable resources and could replace conventional plastics because they are biodegradable. Microbial degradation of PHA has been investigated in many laboratories (Brandl et al., 1990; Brandl & Püchner, 1990; Matavulj & Molitoris, 1992; Neumeier, 1994, 1997; Molitoris et al., 1995) and the biochemistry and molecular biology of the extracellular PHA depolymerases has been summarized recently (Jendrossek et al., 1996). Poly-β-hydroxybutyric acid (PHB), the only PHA that is commercially available (BIOPOL®), sinks readily in aquatic environments (Brandl et al., 1990) and has proven to be biodegradable both *in situ* in the marine environment (Doi et al., 1992) and *in vitro* under marine conditions (Neumeier, 1994; Molitoris et al., 1995). Furthermore, PHA degradation could be shown under increased hydrostatic pressure in freshwater ecosystems (Brandl & Püchner, 1990).

However, hydrostatic pressure and salinity as key factors of deep sea environments have not yet been considered in degradation experiments of PHA. Therefore, as a starting point, the present study reports on the degradation of PHB by filamentous fungi and yeasts under simulated deep sea pressure, salinity and oxygen concentration. In addition, data on the viability of pure fungal cultures under these conditions could be obtained. Results should help to understand how and at which rate degradation of polluting biopolymers (PHA) may take place in the deep sea environment, therefore showing the endangering of this important natural habitat by plastic waste dumping.

Materials and methods

Fungal strains and media

Terrestrial isolate (T-): *Aspergillus ustus* (Bain.) Thom & Church. (T-221) (from greenhouse peat, Regensburg, 25 °C).

Marine isolates (M-): *A. ustus* (Bain.) Thom & Church (M-224) (deep sea isolate, Bay of Bengal, 650 m depth, 15 °C), *Candida guilliermondii* (Castellani) Langeron & Guerra (M-122) (surface sea water, North Sea, September, ca. 20 °C), *Debaryomyces hansenii* (Zopf) Lodder & Kreger-van Rij (M-113) (surface sea water, North Sea, July, ca. 20 °C), *Rhodosporidium sphaerocarpum* Newell & Fell (M-185) (deep sea isolate, Tahiti, 1494 m depth, near hydrothermal vent, 28 °C).

For all marine isolates (except *C. guilliermondii*) used in this project, it had been established that growth (*A. ustus* , M-224, also spore germination) could occur over a wide range of salinities and pressures with a temperature optimum between 15 and 27 °C (Böhm, 1991; Lorenz, 1993; Zaunstöck, 1994; Lorenz & Molitoris, 1997).

All strains are maintained in the culture collection of fungi at the University of Regensburg (strain number in parentheses).

For cultivation the GPYA liquid medium [1.0 g glucose (G), 0.5 g peptone (P) and 0.1 g yeast extract (Y)], in 1000 ml artificial seawater (A) according to Lorenz & Molitoris (1997) was used at a temperature of 27 °C. The pH was adjusted to 6.0 before autoclaving with addition of 0.1% PHB powder (w/v) as the degradable polymeric carbon source (as described by Neumeier, 1997). PHB was isolated from sodium-gluconate-grown cells of the bacterium *Ralstonia eutropha* H16 by sodium-hypochlorite treatment followed by acetone-ether (2:1, v/v) extraction.

High pressure equipment

The high pressure equipment of Lorenz & Molitoris (1997) was used for all experiments. All strains were incubated in a high pressure unit consisting of six interconnected stainless steel vessels of approximately 1000 ml internal volume at pressures of 5, 10, 20, 30, 45 (or 50) and 100 MPa (~10 000 m depth), respectively. Pressure was measured continuously in each vessel. Two similar vessels made of PVC were used for control experiments at atmospheric pressure (0.1 MPa). Pressure was generated with a membrane pump; a fluorocarbon liquid of high oxygen dissolving capacity (FC-77) served as hydraulic fluid and oxygen reservoir to provide a continuous supply of oxygen to the growing cultures.

For temperature control during incubation, water was pumped from a water bath through a thermal jacket surrounding each vessel. Internal vessel temperature was measured with a thermocouple introduced through the lower high pressure tubing of each vessel. Temperature was monitored with a digital thermometer and recorded on a chart recorder.

Cultivation method

Gas-permeable culture bags (10 × 3 cm) were made from high density polyethylene foil (Melitta Toppits Gefrier-Koch Beutel), autoclaved for 20 min at 121 °C, filled with 5 ml inoculated GPYA/PHB medium each and sealed with an impulse sealer. Three rows with six bags each were fixed on a stainless steel wire for even distribution of the bags in the pressure vessel. Maximal oxygen saturation of the hydraulic fluid was achieved by bubbling air through the hydraulic fluid for at least 3 h at 5 °C at atmospheric pressure (0.1 MPa). The liquid was then equilibrated with air under atmospheric conditions at the incubation temperature of 27 °C to avoid oversaturation with air and was then filled into the pressure vessels. A decrease in gas exchange through contact between the cultivation bags was prevented by individually positioning the bags on wire loops inside the cultivation vessel.

Inoculum

Stock cultures of yeasts were grown at 22 °C on GPYA agar plates and stored at 5 °C. The inoculum for the experiments was obtained from liquid cultures containing stationary growth phase yeast cells by transferring 1 ml cell suspension of this culture into 100 ml fresh 5×GPYA medium and incubating it at 27 °C on a rotary shaker until the logarithmic phase was reached (Lorenz, 1993). Then the cell suspension was diluted with sterile GPYA/PHB Medium to give approximately 10^6 cells ml^{-1}; aliquots of 5 ml were sealed in sterile plastic bags and stored on ice until used as inoculum in high pressure experiments.

For filamentous fungi, stock cultures were prepared by transferring single squares from the actively growing margin of an agar plate onto the middle of a fresh GPYA plate, incubating it at 22 °C and storing it at 5 °C. For pressure experiments, 6–8 squares from the margin of the stock culture were inoculated into fresh IM medium (10 g glucose, 10 g malt extract, 2 g peptone, 2 g yeast extract, 1 g asparagin monohyrate, 2 g potassium dihydrogen phosphate, 1 g magnesium sulphate and 1 mg thiamine-HCl in 1000 ml artificial seawater, pH adjusted to 5.5 before autoclaving) as described by Neumeier (1997) and incubated at 27 °C for at least one week until good growth was visible. A homogenate of this inoculum was then mixed with fresh incubation medium (GPYA with 0.1% PHB w/v; 1:10, v/v) to yield a suspension of actively growing hyphae; aliquots of 5 ml were sealed in plastic bags and stored on ice until used as inoculum in the experiment.

Sampling

Samples were taken at different time intervals depending on the strain investigated and incubated for a total of 21 days (yeasts) or 28 days (filamentous fungi), respectively. For sampling, pressure was released from all pressure vessels simultaneously and one or two bags according to the expected growth were harvested for each strain and treatment. Then, the original pressure was restored for further cultivation and the harvested bags were used for the determination of growth and degradation. Depressurization for sampling did not affect growth and other growth-related parameters significantly (Ogawa et al., 1990; Lorenz, 1993).

Growth

Growth (increase in cell number) was measured as number of colony-forming units (cfu) on GPYA agar plates, from which the total number of living cells per ml were calculated for the yeasts at atmospheric pressure (0.1 MPa) and near the growth-limiting pressure (30 MPa for the two marine surface strains, 50 MPa for the deep sea strain). For both *A. ustus* strains and *C. guilliermondii*, growth (increase in biomass) was also measured as dry weight which was determined by membrane filtration of 2–5 ml crude cell suspension before and after incubation. The filters with yeast cells were dried for 10 min under infrared light, filters with mycelium were dried at least for 4 h at 80 °C in an oven. Filters were stored over blue gel at room temperature, weighed when constant dry weight was reached and the resulting dry weight was given in mg ml^{-1}.

Viable cell numbers were determined for yeasts by counting colony forming units (cfu) on 5×GPYA agar plates. Three samples of 100 μl of incubated cell suspension (GPYA/PHB) were diluted to 10^{-6} (v/v) and duplicates of 100 μl of each dilution series were plated sterile on surface-dried 5×GPYA agar plates and incubated at 27 °C for 1 week. Colonies were counted thereafter and cell numbers of the original sample were calculated in cells ml^{-1}.

Oxygen saturation

The oxygen saturation of the hydraulic fluid was measured with an external oxygen electrode to determine the oxygen supply of the cultures in the pressure vessels during incubation. The oxygen measuring device consists of a thermostat, a glass vessel containing the oxygen electrode with two openings for entry and exit of the hydraulic fluid and the electrode itself. The electrode was equilibrated with N_2 gas (0% oxygen saturation) and pressurized air (100% oxygen saturation) before usage. After opening the pressure vessel for harvesting the culture bags, 100 ml of the hydraulic fluid was taken from the bottom of the respective vessel with a self-constructed glass syringe and pressed gently through the upper opening of the glass vessel until a continuous and gentle flow of hydraulic fluid along the electrode could be observed. The oxygen content of the hydraulic fluid was then measured in % saturation and recorded on a chart recorder.

Medium pH

Immediately after incubation, the bags were opened for analysis and the pH of the culture fluid was determined with a pH electrode.

Total carbohydrate content in the medium

The total carbohydrate content of the medium with glucose being the main carbohydrate was determined with the method of Dubois et al. (1956; modified by the authors). Measurements were taken in four parallels for each strain and pressure value whenever samples were taken. Then 0.5 ml phenol (5% w/v) was thoroughly mixed with 100 μl diluted culture filtrate (1:10 w/v, in artificial seawater) in glass tubes (volume 5 ml) and 2 ml of concentrated sulfuric acid (97%) was quickly poured onto the surface of the solution with a glass pipette, mixed again, and incubated for 30 min at room temperature. The extinction was measured at 492 nm in 1 cm glass cuvettes in a filter photometer against distilled water. Total carbohydrate concentration in the medium (in μg 100 μl^{-1}) was calculated from a standard curve using diluted glucose in artificial seawater.

PHB degradation

Culture filtrate (100 μl) was filled into each of five wells cut into minimal medium agar plates (MPY: Matavulj & Molitoris, 1992) which were opaque due to their content of 0.1% (w/v) PHB. Plates were sealed with parafilm and incubated at 35 °C for 48 h. The degradation activity of the culture filtrate was then quantified by measuring the diameter of clearing zones in mm around the wells.

In addition, PHB depolymerase activity was asserted by adding 300 μl of culture filtrate to 700 μl of 0.3% PHB suspension (w/v, in artificial seawater) in 10 mM Tris/HCl buffer (pH 7.4) in 1 cm plastic cuvettes. The two components were mixed in the cuvettes, and enzyme activity was measured by monitoring the extinction loss of the mixture at 30 °C and 650 nm in a filter photometer for 10 min (Neumeier, 1997). Resulting enzyme activities were given in units (1 U ml^{-1} = 100 $\Delta E\, t^{-1}$).

Type, manufacturer and locality of production for the instruments used is given in Lorenz & Molitoris (1997) or may be obtained from the corresponding author.

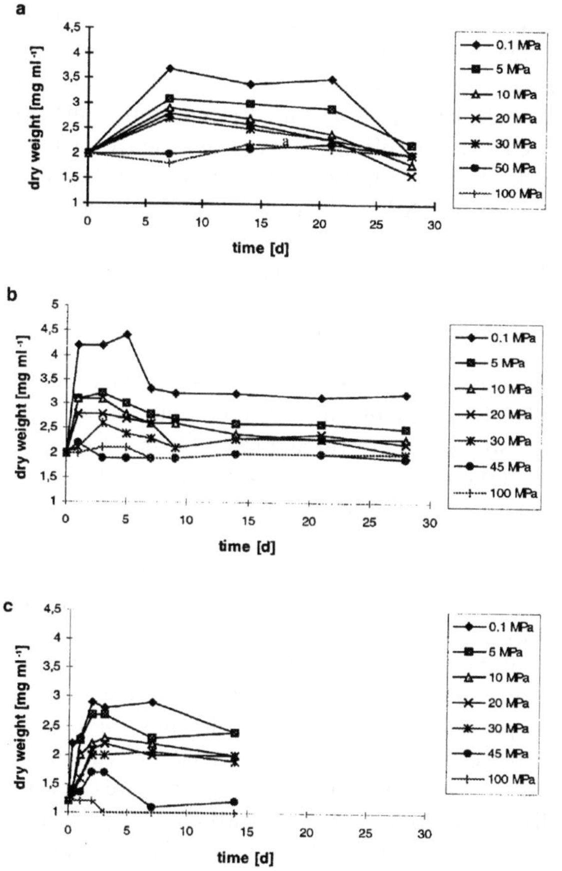

Figure 1. Dry weight of as function of incubation time and pressure of (a) *Aspergillus ustus* (T-221), (b) *Aspergillus ustus* (M-224) and (c) *Candida guilliermondii* (M-122).

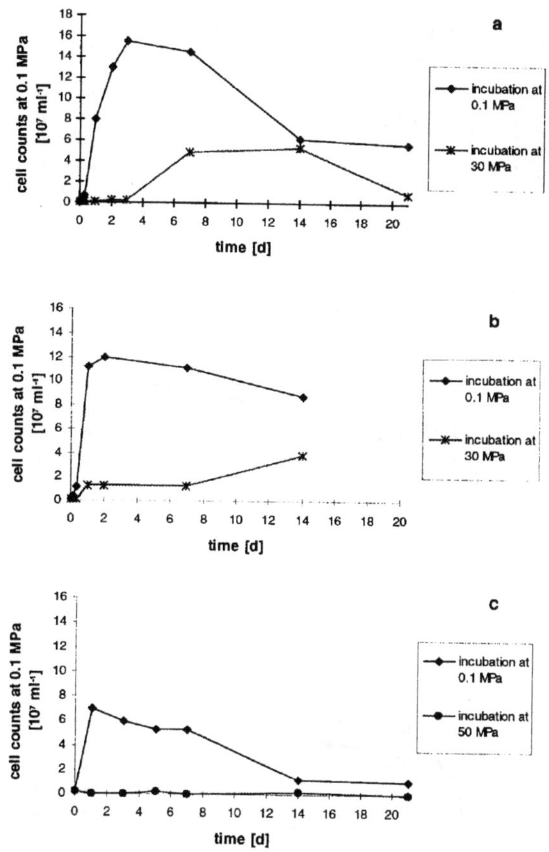

Figure 2. Viable cell counts as function of incubation time and pressure of (a) *Candida guilliermondii* (M-122), (b) *Debaryomyces hansenii* (M-113) and (c) *Rhodosporidium sphaerocarpum* (M-185).

Results

Dry weight (growth as increase in biomass)

Both the marine and the terrestrial isolate of *A. ustus* showed rather similar growth with an early maximum, followed by a later decrease, irrespective of pressure (Figure 1a, b). Growth values of the marine yeast isolate *C. guilliermondii* (Figure 1c) were between those of the two filamentous *A. ustus* strains. In all cases, best growth occurred at atmospheric pressure (0.1 MPa) and decreased with increasing pressure, with no growth at the highest pressures tested (45/50 and 100 MPa, respectively).

Viable cell counts (growth as increase in cell number)

All three marine yeast isolates (*C. guilliermondii*, *D. hansenii*, *R. sphaerocarpum*, Figure 2a–c) showed an early growth maximum at atmospheric pressure (0.1 MPa) within 1–3 days, followed by a later decrease in cell number. *D. hansenii* and *R. sphaerocarpum* grew fastest, *C. guilliermondii* reached the highest maximum, followed by *D. hansenii* and *R. sphaerocarpum*. The subsequent decrease in cell number was steepest in *R. sphaerocarpum*, least in *D. hansenii*. Higher pressure prolonged the lag phase (viable cell counts, Figure 2a–c). Growth-limiting pressures for the yeasts were beyond 30 MPa (*C. guilliermondii*, Figure 2a; *D. hansenii*, Figure 2b), whereas at 50 MPa, no growth occurred (*R. sphaerocarpum*, Figure 2c).

Oxygen saturation

The air-saturated hydraulic fluid (FC-77) supplied oxygen for the growing cultures through the semipermeable membrane of the culture bags, serving therefore as oxygen reservoir. In order to detect oxy-

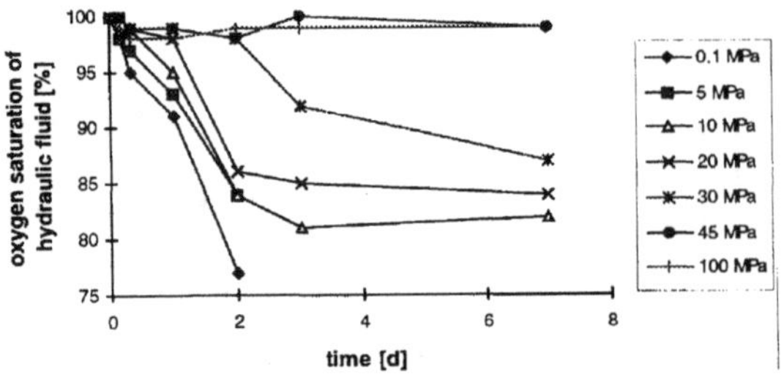

Figure 3. Oxygen content in the hydraulic fluid (FC-77) of *Candida guilliermondii* (M-122) as function of incubation time and pressure.

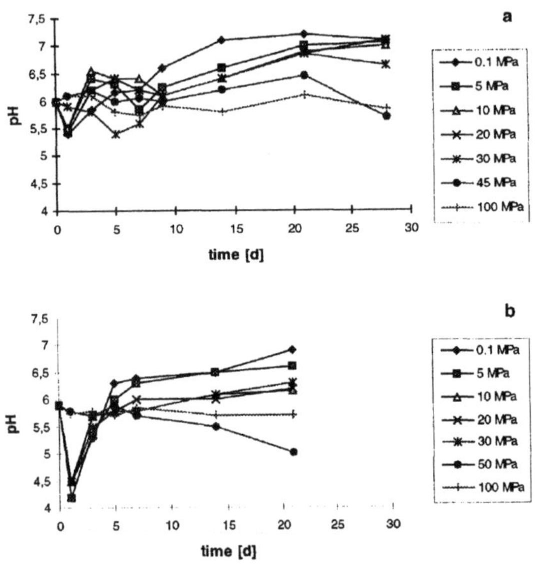

Figure 4. Medium pH for as function of incubation time and pressure of (a) *Aspergillus ustus* (M-224) and (b) *Rhodosporidium sphaerocarpum* (M-185).

gen exhaustion as a growth-limiting factor, oxygen saturation of the hydraulic fluid was determined after each harvest. Figure 3 (*C. guilliermondii*) shows a typical oxygen saturation curve at different hydrostatic pressures. Highest and fastest oxygen comsumption was noted at atmospheric pressure (0.1 MPa) corresponding with fastest growth. With increasing pressure, growth slowed down corresponding with less oxygen consumption. At the highest pressures (45 and 100 MPa) no growth occurred and oxygen consumption was zero.

pH

Changes in the pH of the culture medium may be indicative of metabolic activity. Therefore, pH was routinely determined after each harvest in the culture filtrate. Figure 4a gives the pressure-dependent pH curve for the filamentous marine isolate *A. ustus* (M-224), Figure 4b the pH curve for the marine yeast isolate *R. sphaerocarpum*. In both cases, growth was correlated with an initial steep drop in pH, followed by a later pH increase to values above the initial level. Strongest changes were observed at the lowest pressure (0.1 MPa), the changes becoming less with increasing pressure. At highest pressures (45/50 and 100 MPa), almost no pH changes occurred.

Total carbohydrate in the medium

The carbohydrates in the medium (almost exclusively consisting of glucose) constitute the main energy source for growth and metabolism. Decrease of the carbohydrate content, therefore, should be correlated with growth and was determined (as total carbohydrate) after each harvest.

Generally, in all cultures of filamentous fungi and yeasts, a fast decrease in the total carbohydrate content of the medium was observed at optimal growth conditions (pressures up to 20 MPa), reaching almost zero within 3–5 days (Figure 5a–e). With increasing pressure (and slower growth), this carbohydrate decrease was less pronounced and/or occurred later (Figure 5a–e). At growth-limiting pressures, specifically at 100 MPa, a distinct carbohydrate peak above starting level appeared for 1 or 2 days in the early growth phase (Figure 5a–e). Later, the carbohydrate content either dropped rather late and did not reach exhaustion (45/50 MPa, Figure 5b: marine isolate of *A. ustus*; Figure 5c: *C. guilliermondii*; Figure 5d: *D. hansenii*) or did not change at all (100 MPa).

PHB degradation and activity of PHB depolymerase

Degradation of PHB was tested for all isolates of filamentous fungi and yeast strains with the respective culture filtrates as clearing of PHB-turbid agar medium, and for the marine isolates of *A. ustus*, *C. guilliermondii* and *D. hansenii* also as PHB-depolymerase activity in a spectrophotometric test. The results for both *Aspergillus* strains (Figure 6a and b) and for both test methods (Figures 6b and 7) were similar as with increasing pressure, PHB degradation became less and occurred later with no degradation at the highest pressures tested (30 MPa and beyond). But whereas at atmospheric pressure (0.1 MPa) both isolates of *A. ustus* showed about the same maximum activity in the PHB clearing test between day 3 and 7 at the time of highest growth values (Figure 6a and b), the marine isolate of *A. ustus* (Figure 6b) exhibited an earlier

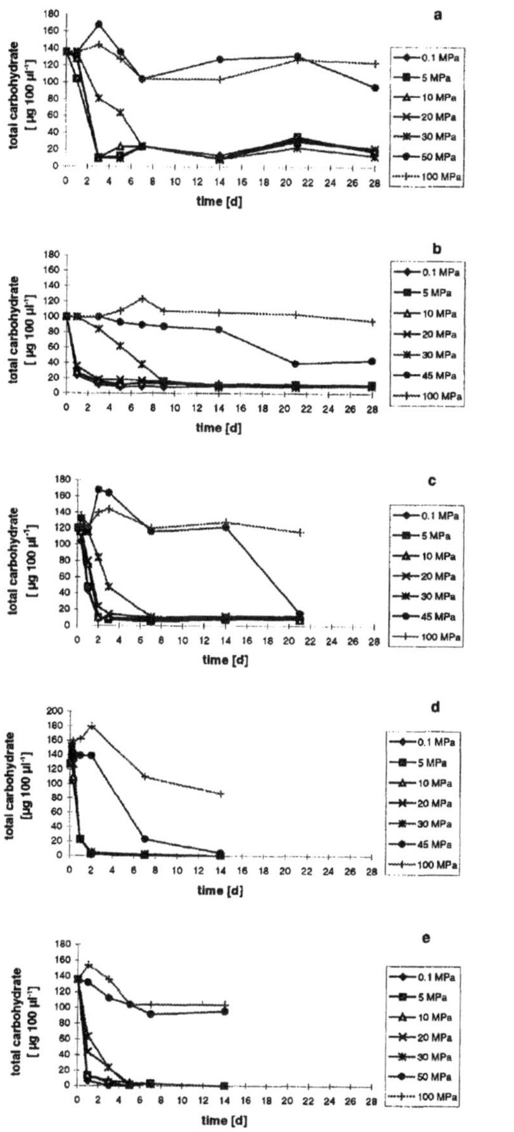

Figure 5. Carbohydrate content of the culture filtrate as function of incubation time and pressure of (a) *Aspergillus ustus* (T-221), (b) *Aspergillus ustus* (M-224), (c) *Candida guilliermondii* (M-122), (d) *Debaryomyces hansenii* (M-113) and (e) *Rhodosporidium sphaerocarpum* (M-185).

and almost twofold PHB degrading activity at higher pressures than the terrestrial isolate (Figure 6a).

When the two tests for PHB degrading activity of the marine isolate of *A. ustus*, the clearing test (Figure 6b) and the PHB-depolymerase activity (Figure 7) are compared, highest activity was reached on day 3–5 at lowest (atmospheric) pressure in both tests, followed by a decrease with increasing hydrostatic pressure; no PHB degradation was found at

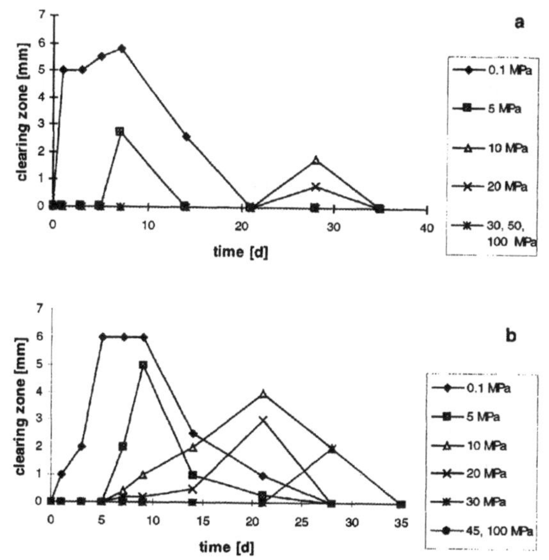

Figure 6. PHB degradation (clearing test) as function of incubation time and pressure by (a) *Aspergillus ustus* (T-221) and (b) *Aspergillus ustus* (M-224).

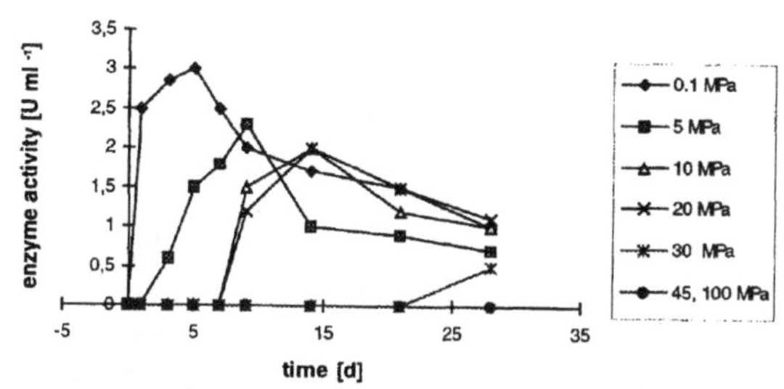

Figure 7. PHB depolymerase activity for *Aspergillus ustus* (M-224) as function of incubation time and pressure.

pressures of 30 MPa and beyond. In all cases of PHB degradation, the earliest (day 3–7) and highest activity occurred at lowest test pressure (atmospheric pressure, 0.1 MPa). In contrast to the results for the filamentous *Aspergillus* strains, the yeast strains did not show any PHB-degrading activity in both the clearing test and the spectrophotometric test.

Discussion

Growth and other growth-related parameters: In order to determine the influence of deep sea pressure on growth and growth-related parameters, a deep sea-simulating setup was used, where all conditions could be kept constant with hydrostatic pressure being the only variable (Lorenz & Molitoris, 1997).

In contrast to bacteria (Jannasch, 1987, Yayanos & Delong, 1987) 'true (obligate barophilic)' deep sea strains of fungi and yeasts are so far not available because sampling and retrieval is performed without keeping deep sea conditions (hydrostatic pressure, temperature, salinity and oxygen partial pressure) until isolation and purification. Methods and apparatus allowing experiments with fungi under simulated deep sea conditions in the laboratory became available only recently (Lorenz, 1993; Lorenz & Molitoris, 1997). Therefore, deep sea isolates (mostly facultative barophilic) of filamentous fungi and yeasts were studied for the effect of simulated deep sea pressure.

Generally, the temperature in the deep sea (below 200 m depth: Lorenz, 1993) is around 0 °C. However, in tropical or subtropical zones or at specific locations such as hydrothermal vents, the temperature may be considerably higher. In our experiments for growth and thermoplast degradation in liquid seawater medium, an incubation temperature of 27 °C was used for the following reasons: 1. For all the marine isolates used, the water temperature in the isolation area was 15 °C or higher; 2. For all the marine isolates used, *in vitro* experiments had shown that optimal growth (as well as enzyme production and thermoplast degradation) was highest at temperatures above 15 °C; 3. For some of the isolates used, it was shown that lowering the incubation temperature below 10 °C and increasing the hydrostatic pressure to 40 MPa did not lead to irreversible damage, but resulted only in retardation of growth and metabolic activities (growth of *R. sphaerocarpum*, M-185: Böhm, 1991; growth and ascospore germination of *A. ustus*, M-224: Zaunstöck, 1994; growth of *R. sphaerocarpum*, M-185 and *D. hansenii*, M-113: Lorenz, 1993; Lorenz & Molitoris, 1997). Optimal conditions for growth and metabolic activities are important, since changes in cultivation temperature otherwise could interfere with pressure effects (Marquis, 1976).

A growth-limiting effect of oxygen exhaustion was prevented by the use of FC-77 as hydraulic fluid and oxygen reservoir by its 16-time higher oxygen-dissolving capacity than water (Helmke, 1980). Use of semipermeable plastic foil for the cultivation bags allowed quick and efficient gas exchange between the hydraulic fluid and the cultivation medium. Even the lowest oxygen concentrations of around 24% saturation, as measured in our experiments in the late growth phase of *A. ustus* M-224 (Figure 1b), should not be growth-limiting according to Lorenz & Molitoris (1997).

Factors affecting growth other than pressure are exhaustion of nutrients at the end of the stationary phase and local aggregations of cells by sedimentation leading to possible oxygen deficiency inside the bags since the vessel contents were not constantly mixed.

Barotolerant growth was observed with a strain-specific growth limit depending on the hydrostatic pressure applied. Therefore, only the lowest and the highest pressures were identical for all strains, whereas the intermediate pressures were chosen depending on the growth-limiting pressure of each isolate. In contrast to bacteria where a correlation between isolation depth and pressure-related growth limit was found (Yayanos & Delong, 1987; Imada et al., 1989), this could not be confirmed for the filamentous strains and yeast isolates investigated in this study. This result is in agreement with the work of other authors who did not find a strict correlation between isolation depth and pressure-related growth in marine fungi (ZoBell & Johnson, 1949; Kriss & Zaichin, 1971; Yamasato et al., 1974). However, the low number of fungal strains investigated does not allow a general conclusion yet. Strain-specific and pressure-dependent growth limits, as found in our experiments for isolates of filamentous fungi and yeasts, have also been reported for yeasts by ZoBell & Johnson (1949), Yamasato et al. (1974) and Lorenz (1993).

Pressure-dependent growth limits as determined by dry weight and viable cell count measurements correlated well with other parameters, such as consumption of the medium carbohydrates and oxygen and with changes in the pH of the culture filtrate.

An unexpected result was the appearance of a short-lived but significant (standard deviation <5%) peak above the initial level in the carbohydrate content of the medium at growth-limiting pressures of 45 and 50 MPa and/or 100 MPa during the early growth phase for filamentous fungi and during the stationary phase for yeasts. The appearance of carbohydrates in the growth medium after application of higher than growth-limiting hydrostatic pressures could be the effect of cell breakage, followed by release of the intracellular material. This agrees with SEM observations on these and other isolates of filamentous fungi and yeasts showing an increasing number of dead cells and/or morphological aberrations during growth and spore germination at higher hydrostatic pressures (Lorenz, 1993; Zaunstöck, 1994; Lorenz & Molitoris, 1997). Shimada et al. (1993), however, working with *Saccharomyces cerevisiae*, conclude from the results of their SEM studies, membrane analyses and quantitative measurements of UV-absorbing material that a pressure-dependent release of internal carbohydrates generally is not likely, since there were no major differences in cell structure or medium composition at pressures ranging from 0.1 MPa to 100 MPa. Another possible explanation for the early carbohydrate peak is based on results of Iwahashi et al. (1997a,b), who observed the intracellular production and storage of trehalose, a non-reducing disaccharide, as a protective agent for membranes of stationary-phase yeast cells under stress conditions (e.g. high hydrostatic pressure). This carbohydrate could be released into the medium upon cell breakage caused by growth-limiting pressures.

PHB degradation

Growth is correlated with uptake and enzyme-mediated metabolism of substrate. This process has been investigated under simulated deep sea conditions through variation of hydrostatic pressure and salinity. As a first experimental approach to the effect of hydrostatic pressure on the breakdown of pollutants in the deep sea environment, the degradation of the thermoplastic polymer PHB (as copolymer in the commercially produced ® BIOPOL) was investigated.

Negative effects of salinity on the degradation of different substrates by fungi have been reported by several authors: Rohrmann (1993) observed a decrease in degradation of wood with increasing salinity. Decrease of growth, substrate degradation and enzyme activity with increasing salinity was also found by Molitoris et al. (1998) and Kurchenko et al. (1998) for a number of marine isolates. By use of several commercial dyes as model substrates for the self-cleaning capacity of the Dead Sea, an hypersaline marine environment, Rawal et al. (1998) found that with increasing salinity (and increasing temperature) less fungi degraded the dyes and less dyes were decolourized. Mergaert & Swings (1996), using polyhydroxyalkanoates (PHA) as another pollutant, were unable to isolate any marine fungus able to degrade these substances *in situ*. Neumeier (1994) and Molitoris et al. (1995), using pure fungal cultures, found that considerably less marine than terrestrial isolates of filamentous fungi and yeasts degraded PHB. The results of our study add to this general trend by showing less degradation of PHB by the marine and the terrestrial isolate of *A. ustus* (T-221, M-224) in artificial seawater as compared to the terrestrial isolate of *A. ustus* (T-221) grown in freshwater medium under otherwise identical conditions (0.1 MPa, 27 °C, GPY medium with PHB) (Neumeier, 1997). Even more convincing is the failure of the marine yeast isolates studied here to degrade PHB in artificial seawater medium at any pressure since on solid freshwater and seawater medium they all degraded PHB, at least in the clearing test (Neumeier, 1994).

Synthesis of PHB depolymerase, the enzyme responsible for hydrolytic cleavage of the PHB polymer is repressed by glucose (catabolite repression) as shown for bacteria by Jendrossek et al. (1996) and reported for the PHB depolymerase of the terrestrial isolate of *Aspergillus ustus* by Neumeier (1997). This is confirmed by our results where PHB degradation and PHB depolymerase activity were found only after almost complete exhaustion of the medium carbohydrates consisting almost exclusively of the PHB depolymerase-repressing glucose.

Our results showing inhibition of PHB degradation under marine conditions are confirmed and extended by including hydrostatic pressure as an additional factor affecting fungal live in the deep oceans. High hydrostatic pressure is known for its negative effects on substrate degradation and enzyme activity even in freshwater ecosystems such as deep lakes. Increased pressure, low temperatures and low oxygen content in deep lakes have been shown to inhibit degradation of PHA *in situ* considerably (Brandl & Püchner, 1990). The results of the present study – where for optimal growth and metabolic activity, a cultivation temperature of 27 °C was chosen – identify hydrostatic pressure as the main limiting factor for growth and PHB degradation, when other factors were kept constant (temperature, salinity and oxygen). Increasing hydrostatic pressure not only decreased PHB degradation by the filamentous *Aspergillus* strains but also retarded it. When the activity of PHB depolymerase of the marine isolate of *A. ustus* (M-224) was determined spectrophotometrically, it also decreased with increasing hydrostatic pressure. However, presently it cannot be decided whether this is due to lower enzyme production or enzyme activity or due to both factors. Furthermore, since all *in vitro* PHB degradation tests (clearing test, spectrophotometric test) were performed at optimal degradation temperatures (35° and 30 °C, respectively) one has to assume that under low deep sea temperatures PHB degradation is even slower.

Catabolite repression of PHB depolymerase synthesis by glucose in the medium cannot be the reason since despite glucose exhaustion (Figure 5b) less or later PHB degradation was observed with increasing incubation pressure (Figures 6b and 7). Ribosomes, as key organelles for any enzyme biosynthesis, are known to be compressed at high pressures (50 MPa and beyond) in the yeast *Rhodosporidium sphaerocarpum*, resulting in partial loss of their function (Gross, 1993). Furthermore, compression of the substrate-binding site of the enzyme and the resulting metabolic dislocation of the substrate (Jaenicke, 1991) could furthermore be a possible reason for the negative effect of hydrostatic pressure on enzyme activity. Denaturation of oligomeric enzymes, as shown by Erijman & Weber (1991), probably does not apply here because of the low molecular weight (ca. 50 kDA) of the PHB depolymerase and its monomeric structure (Neumeier, 1997).

Despite good growth, no PHB degradation was found even at atmospheric pressure (0.1 MPa) in liquid culture for the yeast strains investigated (*C. guilliermondii, D. hansenii, R. sphaerocarpum*). They all had shown on solid agar medium good PHB degradation under atmospheric pressure in the clearing test (Neumeier, 1994). Glucose, by catabolite repression of PHB depolymerase synthesis, cannot be the reason for the lack of PHB degradation, since medium glucose (as shown by carbohydrate determinations) was exhausted very early, at the end of the logarithmic growth phase. One possible explanation is that by using different culture conditions (solid as compared to liquid medium), this enzyme is not produced or at least not secreted. Adhesion of PHB depolymerase to its substrate, the PHB granula in liquid culture, which were filtered off during preparation of the culture filtrate, is another possible explanation (Neumeier, 1997).

The fact that PHB degradation at 30 MPa by the marine isolate of *A. ustus* M-224 was only observed when measured spectrophotometrically, but not when determined by a clearing test (clearing of PHB-turbid solid medium), may be attributed to a lower sensitivity of the clearing test. This has already been suggested by Neumeier (1997).

Conclusions

In conclusion, it could be shown by investigation of a marine and a terrestrial isolate of a filamentous fungus and of three marine yeasts isolates that hydrostatic pressure is a major factor for limitation of growth and metabolic activity in the deep sea environment. As a first approach, *in vitro* studies in liquid seawater medium under different hydrostatic pressures on the degradation of the thermoplastic polymer PHB serving as a model compound for pollution in the sea suggest, that the oceans generally, and the deep sea in particular, may be endangered because of the low self-cleaning capacity of this habitat under such extreme conditions.

Acknowledgements

The authors want to thank Inge Lauer, Regensburg, for invaluable technical help, Prof. A. Kurtz, Regensburg, for supplying the oxygen electrode and Melitta Haushaltsprodukte, Minden, for providing foil material for the cultivation bags.

References

Ahearn, D. G., S. P. Meyers & P. G. Standard, 1971. The role of yeasts in the decomposition of oils in marine environments. Devs. ind. Microbiol. 12: 126–134.

Bean, M. J., 1987. Legal strategies for producing persistent plastics in the marine environment. Mar. Poll. Bull. 18: 357–360.

Böhm, H., 1991. Untersuchungen an Tiefseepilzen. Diplomarbeit, Universität Regensburg: 95 pp.

Brandl, H., R. A. Gross, R. W. Lenz & R. C. Fuller, 1990. Plastics from bacteria and for bacteria: Poly(β-hydroxyalkanoates) as natural, biocompatible and biodegradable polyesters. Adv. Biochem. Engng. Biotechnol. 41: 77–93.

Brandl, H. & P. Püchner, 1990. The degradation of shampoo bottles in a lake. In Dawes, E. A (ed.), Novel Biodegradable Microbial Polymers. Kluwer Academic Publishers, Dordrecht, The Netherlands: 421–422.

Doi, Y., Y. Kanesawa & N. Tanahashi, 1992. Biodegradation of microbial polyesters in the marine environment. Polymer Degrad. Stabil. 36: 173–177.

Dubois, M., K. A. Gilles, J. K. Hamilton, P. A. Rebers & F. Smith, 1956. Colorimetric method for determination of sugars and related substances. Anal. Chem. 28: 350–356.

Erijman, L. & G. Weber, 1991. Oligomeric protein associations: Transition from stochastic to deterministic equilibrium. Biochemistry, N.Y. 30: 1595–1599.

Fell, J. W., 1967. Distribution of yeasts in the Indian Ocean. Bull. mar. Sci. 17: 454–470.

Fell, J. W., D. G. Ahearn, S. P. Meyers & F. J. Roth, Jr., 1960. Isolation of yeasts from Biscayne Bay, Florida and adjacent benthic areas. Limnol. Oceanogr. 5: 366–371.

Gage, J. D. & P. A. Tyler, 1991. Deep-Sea Biology: A Natural History of Organisms on the Deep Sea Floor: Cambridge University Press, Cambridge: 504 pp.

Gross, M., 1993. Die Wirkung hoher hydrostatischer Drücke auf biochemische Prozesse: Dissoziation von Assembly-Systemen versus Kompression aktiver Strukturregionen. Ph.D. Thesis, Universität Regensburg: 140 pp.

Grüttner, H. & K. Jensen, 1983. Effects of chronic oil pollution from refinery effluent on sediment microflora in a Danish coastal area. Mar. Poll. Bull. 14: 436–459.

Helmke, E., 1980. Wachstum von Actinomyceten mariner und terrestrischer Herkunft unter hydrostatischem Druck. Ph.D. Thesis, Technische Universität Braunschweig.

Imada, C., S. Kurasawa & Y. Okami, 1989. The effect of hydrostatic pressure on the growth of Actinomycetes. In Koyama, Y. (ed.), Trends in Actinomycetology in Japan. Society for Actinomycetes, Japan: 27–29.

Iwahashi, H., K. Obuchi, S. Fujii & Y. Komatsu, 1997a. Barotolerance is dependent on both trehalose and heat shock protein 104 but is essentially different from thermotolerance in *Saccharomyces cerevisiae*. Lett. Appl. Microbiol. 25: 23–47.

Iwahashi, H., K. Obuchi, S. Fujii & Y. Komatsu, 1997b. Effect of temperature on the role of Hsp 104 and trehalose in barotolerance of *Saccharomyces cerevisiae*. FEBS Lett. 416: 1–5.

Jaenicke, R., 1991. Protein stability and molecular adaptation to extreme conditions. Eur. J. Biochem. 202: 715–728.

Jannasch, H. W., 1987. Effects of hydrostatic pressure on growth of marine bacteria. In Jannasch, H. W., R. E. Marquis & A. M. Zimmermann (eds), Current Perspectives in High Pressure Biology. Academic Press, London: 1–15.

Jendrossek, D., A. Schirmer & H. G. Schlegel, 1996. Biodegradation of polyhydroxyalkanoic acids. Appl. Microbiol. Biotechnol. 46: 451–463.

Kohlmeyer, J., 1969. Deterioration of wood by marine fungi in the deep sea. In Materials Performance and the Deep Sea. Am. Soc. Test. Mater., Spec. Tech. Publ. 445: 20–29.

Kohlmeyer, J. & E. Kohlmeyer, 1979. Marine Mycology. The Higher Fungi. Academic Press, London: 690 pp.

Kohlmeyer, J. & E. Volkmann-Kohlmeyer, 1988. *Halographis (Opegraphales)*, a new endolithic lichenoid from corals and snails. Can. J. Bot. 66: 1138–1141.

Kriss, A. E. & E. I. Zaichkin, 1971. Distribution of barotolerant microorganisms in soil. Mikrobiologiya 40: 558–562.

Kurchenko, I., I. Lauer, A. S. Buchalo, S. Wasser, E. Nevo & H. P. Molitoris, 1998. Effect of salinities and temperatures on enzyme activities of fungi from the Dead Sea. Abstr. VI. Intern. Mycol. Congress, Jerusalem, Israel: p. 169

Laist, D., 1987. Overview of the biological effects of lost and discarded plastic debris in the marine environment. Mar. Poll. Bull. 18: 319–326.

Lorenz, R., 1993. Kultivierung mariner Pilze unter erhöhtem hydrostatischen Druck. Ph. D. Thesis, Regensburg: 267 pp.

Lorenz, R. & H. P. Molitoris, 1997. Cultivation of fungi under simulated deep sea conditions. Mycol. Res. 101: 1355–1365.

Marquis, R. E., 1976. High-pressure microbial physiology. Adv. Microb. Physiol. 14: 159–241.

Matavulj, M. & H. P. Molitoris, 1992. Fungal degradation of poly-β-hydroxyalkanoates (PHA) and a semiquantitative assay for screening their degradation by terrestrial fungi. FEMS Microbiol. Rev. 103: 323–332.

Mergaert, J. & J. Swings, 1996. Biodiversity of micro-organisms that degrade bacterial and synthetic polyesters. J. Ind. Microbiol. 17: 463–469.

Meyers, S. P., D. G. Ahearn & F. J. Roth, Jr., 1967. Mycological investigations of the Black Sea. . Bull. mar. Sci. 17: 576–596.

Molitoris, H. P., S. T. Moss, B. Zaunstöck, S. Neumeier, E. Straubinger & K. Rathberger, 1995. Decomposistion of thermoplastics in the sea. Abstr. IX. Intern. Congr. Marine Corrosion and Fouling. Portsmouth, England, 1995: 133.

Molitoris, H. P., I. Lauer, A. S. Buchalo, A. Oren, E. Nevo & S. Wasser, 1998. Growth of fungi from the Dead Sea at different salinities and temperatures. Abstr. VI. Intern. Mycol. Congress, Jerusalem, Israel: 169.

Neumeier, S., 1994. Abbau thermoplastischer Biopolymere auf Poly-β-Hydroxyalkanoat-Basis durch terrestrische und marine Pilze. Diplomarbeit, Universität Regensburg: 99 pp.

Neumeier, S., 1997. Pilzliche PHB Depolymerasen: Anzucht, Reinigung und Charakterisierung des Enzyms aus *Aspergillus ustus*. Ph. D. Thesis, Universität Regensburg: 148 pp.

Ogawa, H., K. Fukuhisa, Y. Kubo & H. Fumumoto, 1990. Pressure inactivation of yeasts, molds and pectinesterase in Satsuma mandarine juice: effects of juice concentration, pH and organic acids and comparison with heat sanitation. Agric. Biol. Chem. 54: 1219–1225.

Pruter, A., 1987. Sources, quantities and distribution of persistent plastics in the marine environment. Mar. Poll. Bull. 18: 305–310.

Rawal, B. S., I. Kurchenko, A. S. Buchalo, S. Wasser, A. Oren, E. Nevo & H. P. Molitoris, 1998. Effect of salinity on the dye degradation by fungi from the Dead Sea. Abstr. VI. Intern. Mycol. Congress, Jerusalem, Israel: 106.

Rohrmann, S., 1993. Holzabbau bei marinen Pilzen. Vergleichende Untersuchungen an ausgewählten marinen und terrestrischen Pilzen zum Abbau von Holz und zu daran beteiligten Enzymsystemen in Meerwasser- und Süßwassermedien. Ph. D. Thesis, Universität Regensburg: 180 pp.

Roth, F. J. Jr., P. A. Orpurt & D. G. Ahearn, 1964. Occurrence and distribution of fungi in a subtropical marine environment. Can. J. Bot. 42: 375–383.

Shimada, S., M. Andou, N. Naito, N. Yamada, M. Osumi & R. Hayashi, 1993. Effects of hydrostatic pressure on the ultrastructure and leakage of internal substances in the yeast *Saccharomyces cerevisiae*. Appl. Microbiol. Biotechnol. 40: 123–131.

Sparrow, F. K., Jr., 1937. The occurrence of saprophytic fungi in marine muds. Biol. Bull. mar. biol. lab., Woods Hole, 73: 242–248.

Van Uden, N. & J. W. Fell, 1968. Marine yeasts. In Droop, M. & E. J. F. Wood (eds), Advances in Microbiology of the Sea. Academic Press, London: 167–201.

Yamasato, K., S. Goto, K. Owada, D. Okuno, H. Araki & H. Iizuka, 1974. Yeasts from the Pacific Ocean. J. gen. appl. Microbiol. 20: 289–307.

Yayanos, A. A. & E. F. Delong, 1987. Deep-sea bacterial fitness to environmental temperatures and pressures. In Jannasch, H. W., R. E. Marquis & A. M Zimmermann (eds), Current Perspectives in High Pressure Biology. Academic Press, London: 17–32.

Zaunstöck, B., 1994. Keimung von Sporen mariner und terrestrischer Pilze in Abhängigkeit von hydrosta-tischem Druck, Temperatur, Medium und Wasserqualität. Diplomarbeit, Universität Regensburg: 105 pp.

ZoBell, C. E. & F. H. Johnson, 1949. The influence of hydrostatic pressure on the growth and viability of terrestrial and marine bacteria. J. Bacteriol. 57: 179–189.

Hydrobiologia **426**: 185–192, 2000.
G. Liebezeit, S. Dittmann & I. Kröncke (eds), Life at Interfaces and Under Extreme Conditions.

Effect of continuous nutrient enrichment on microalgae colonizing hard substrates

Helmut Hillebrand[1,*] & Ulrich Sommer
Institut für Meereskunde, Abteilung Meeresbotanik, Düsternbrooker Weg 20, D-24105 Kiel, Germany
[1]Present address: Erken Laboratory, Department of Limnology, EBC, University of Uppsala, Norr Malma 4200, S-76173 Norrtälje, Sweden
*E-mail: helmut.hillebrand@limno.uu.se (*author for correspondence)*

Key words: nutrient competition, periphyton, nitrogen, silicate, eutrophication, benthic microalgae, hard substrates

Abstract

In order to understand the effect of changing nutrient conditions on benthic microalgae on hard substrates, *in-situ* experiments with artificial substrates were conducted in Kiel Fjord, Western Baltic Sea. As an extension of previous investigations, we used artificial substrates without silicate and thus were able to supply nutrient media with different Si:N ratios to porous substrates, from where they trickled out continuously. The biofilm developing on these substrates showed a significant increase in biovolume due to N + P enrichment, while Si alone had only minor effects. The stoichiometric composition of the biomass indicated nitrogen limitation during most of the year. The C:N ratios were lowered by the N + P addition. The algae were dominated by diatoms in most cases, but rhodophytes and chlorophytes also became important. The nutrient treatment affected the taxonomic composition mostly at the species level. The significance of the results with regard to coastal eutrophication is discussed.

Introduction

Whereas benthic microalgae inhabiting sediments may receive nutrients from the water column and from the sediment porewater, the benthic microflora on hard substrates depends upon water column nutrients and the internal recycling of nutrients (Riber & Wetzel, 1987; Hillebrand & Sommer, 1997). The influence of nutrient supply on marine microalgae on hard substrates has rarely been investigated as the effect of nutrient competition on epipelic diatoms was thought to be minor compared to abiotic factors (Admiraal, 1984). It became clear, however, that nutrient limitation is possible in a variety of benthic habitats (Sundbäck & Snoeijs, 1991; Nilsson, 1995; Borchardt, 1996). It can be assumed that nutrients should be more important for the structure of benthic microflora on hard substrates lacking the sediment porewater as an additional nutrient pool.

It was shown in laboratory experiments with periphyton that diatoms are replaced by cyanobacteria and chlorophytes at low supply ratios of Si:N and Si:P (Sommer, 1996). In a recent field study, the effect of nutrients on biomass and species composition of epilithic microalgae was tested experimentally in Kiel Fjord, Western Baltic Sea (Hillebrand & Sommer, 1997). This study revealed a positive response of total algal biovolume to nitrogen enrichment, indicating previously N-limited conditions. The experimental supply of nitrogen and/or phosphorus changed the species composition on artificial substrates and influenced biomass stoichiometry and community diversity. However, changes mostly occurred at the species level. Even at highest N and P supply rates, the diatoms stayed dominant in all experiments except one, which may be due to the fact that the kieselgur substrates themselves released silicate (Hillebrand & Sommer, 1997).

Here we report results of *in-situ* experiments conducted with Si-free substrates and two different modes of silicate enrichment (with or without additional enrichment of N and P). We were interested in the question whether compositional changes on the level of higher taxa would occur if the Si:N ratio was altered.

Table 1. Names, dates and treatments of *in-situ* experiments. Nutrient concentrations in ambient seawater in the Kiel Fjord at the end of the experiments are given in μmol l^{-1}. Nutrient treatments are given combined for all experiments

Name of experiment	Duration	Ambient N (NO_2^- + NO_3^- + NH_4^+	Ambient P (PO_4^{3-})	Ambient Si Si $(OH)_4$	Nutrient treatment (for all 3 experiments)
Autumn 1997	2 Sep–14 Oct 1997	8.4	1.11	18.5	- C: control (n=2) without addition of nutrients
Winter 1997	14 Oct–9 Dec 1997	16.2	1.19	31.5	- N + P + Si: (n=6) 450 N (as $NaNO_3$) + 30 P (as KH_2PO_4) + Si (0, 10, 50, 100, 250, 500)
Spring 1998	17 Mar–28 Apr 1998	24.6	0.27	13.2	- Si: (n=5) (10, 50, 100, 250, 500). Si as $Na_2O_3Si * 5H_2O$

Methods

The experiments were conducted in Kiel Fjord, Western Baltic Sea, from September 1997 to May 1998 (Table 1). The experimental setup was similar to that used by Hillebrand & Sommer (1997). Artificial substrates were suspended from the pier of the institute at a water depth of ca. 1.5 m. The area is virtually non-tidal, but wind-driven fluctuations of the water-level occur. Nevertheless, the substrates were at least 0.5 m below the water surface at low water levels. The substrates consisted of lime tree wood blocks used for aeration of aquaria (15×15×30 mm, with a plastic connection for tubes on the top). These substrates were connected via silicone tubes to a darkened 10 l bottle placed upon the pier. These bottles were filled with 0.2 μm-filtered seawater (cellulose-acetate filters) from Kiel Fjord, to which nutrients were added (Table 1). The flow of these liquid media to the artificial substrates was controlled via a precise mechanic regulation device designed for intravenous infusion (Angiokard AK 5505), adjusting the flow rate to 1 (± 0.1) l d^{-1}. The supply bottle was refilled twice per week and the flow rate readjusted, if deviations >0.1 l d^{-1} were recorded. The nutrient supply rates were arranged in an increasing gradient of silicate concentrations, each with or without additional increase of N and P (Table 1). Additions of N and P were supplied in a 15:1 ratio, silicate concentrations in the media ranged from 0 to 500 μmol l^{-1}. Although these concentrations in the medium appear to be high, the maximum supply rates were low due to the low flow rate (N: 26 μmol cm^{-2} d^{-1}, Si: 29 μmol cm^{-2} d^{-1}, P: 1.7 μmol cm^{-2} d^{-1}).

At the end of the experiments, the artificial substrates were collected and the biomass was removed with a razor blade until no pigment colour could be detected on the substrate. The biomass was transferred to 100 ml of 0.2 μm-filtered seawater. The suspension was cautiously stirred, while subsamples were taken for microscopical analysis (fixation with Lugol's iodine), taxonomic identification and chemical analysis of biomass stoichiometry (processing of subsamples and determination of C, N and P as described in Hillebrand & Sommer, 1997). Very few wood particles were found in the counting samples, indicating that C:N:P measurements were not biased by wood scraped off the artificial substrate.

Cell abundance was determined with an inverted microscope (Leitz DMIRB) and standard Utermöhl counting chambers (Hydrobios), counting up to 1500 cells per sample at a magnificiation of 400×. Biovolume was calculated by measuring linear dimensions of 20 cells of each species and fitting nearest geometric models to the cell shape (Hillebrand et al., 1999).

Statistical analysis comprised two-factor ANOVA on gross variables (biovolume and cellular stoi-

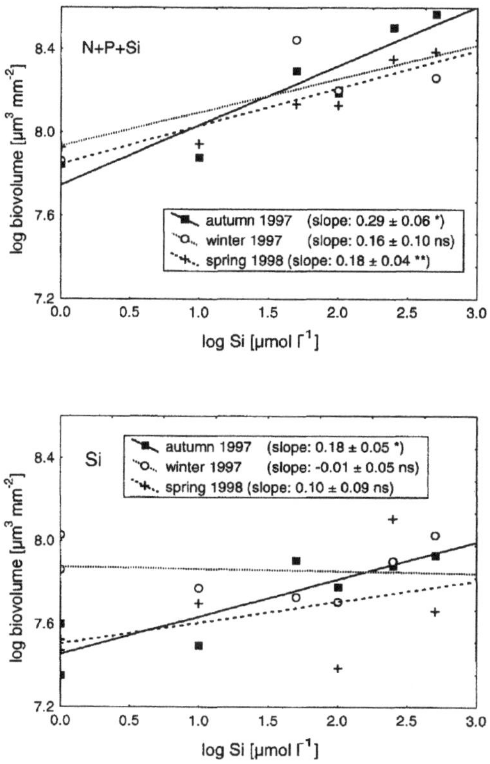

Figure 1. Total biovolume as a function of Si-concentrations, plotted for three experiments. Top: Enrichment of Si with additional supply of N+P. Bottom: Enrichment of Si alone. Regression slopes are given with standard error and significance level (* $p<0.05$; ** $p<0.01$; *** $p<0.001$; ns: not significant).

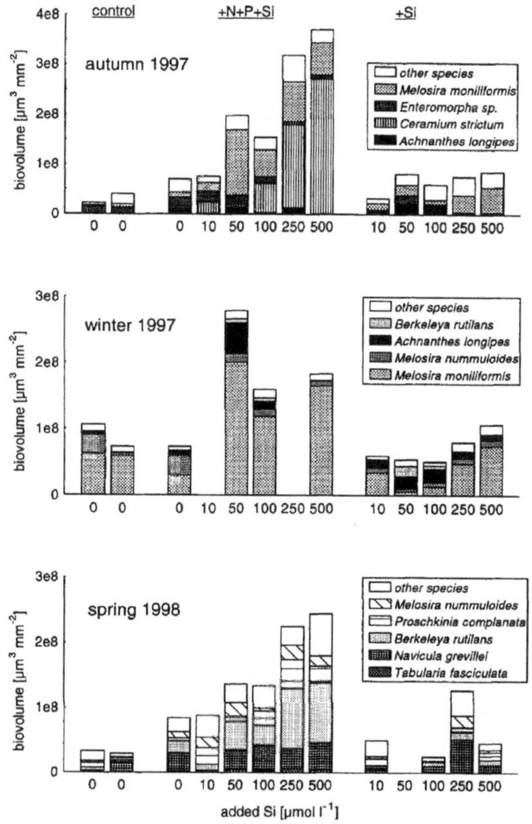

Figure 2. Species composition in three experiments conducted in the Kiel Fjord. Treatments are C (controls), N+P+Si (enrichment of N, P and Si), and Si (enrichment of Si alone). Si concentrations are given in μmol l^{-1}.

chiometry) and the factors nutrient treatment and season of experiment. Homoscedasticity was assured by Bartlett's χ^2-test. Combining different Si concentrations in one category is a conservative approach increasing the variation within categories. We applied linear regression models to evaluate the response of total biovolume to Si enrichment. Since single species could not be expected to response in a linear way to nutrient enrichment, we adopted Spearman rank correlation to test the response of dominant species to Si treatments, whereas the Mann–Whitney U-test was used to test for differences between treatments with or without N + P enrichment. For both non-parametric tests, the biovolume of each species was normalized to the control in order to compare experiments with different biomass accrual.

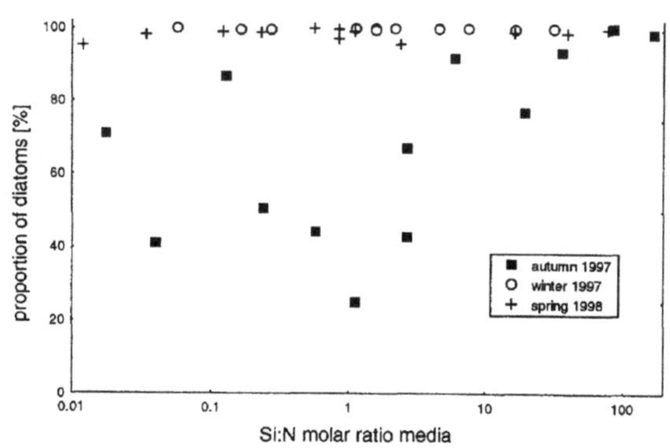

Figure 3. Proportion of diatoms in the biofilm as a function of the Si:N molar ratio in the medium.

Table 2. Results of a two-factor ANOVA on log-transformed total biovolume with treatment (control, N+P+Si or Si) and season of experiment (autumn, winter, spring) as factors. Log-transformation resulted in normal distribution ($\chi^2 = 1.95$, $p = 0.930$) and homogeneity of variances (Bartlett's $\chi^2 = 6.09$, $p = 0.637$)

Effect	df	MS effect	*F*-ratio	*p*-level
Treatment	2	0.907	18.218	<0.001
Season	2	0.109	2.192	0.131
Treatment × season	4	0.051	1.015	0.417
Error	27	0.050		

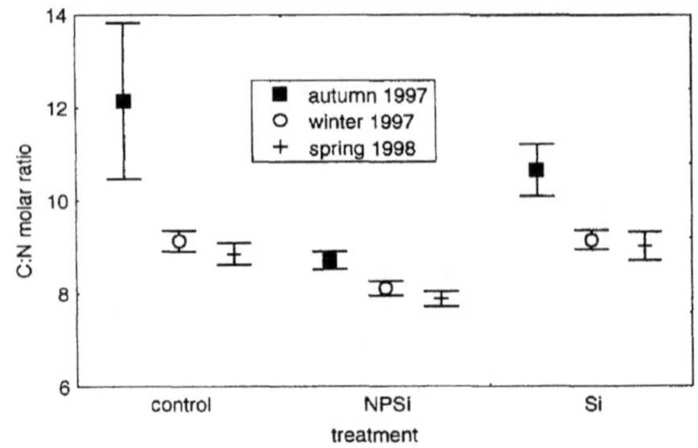

Figure 4. C:N ratios of algal biomass in control, N+P+Si and Si treatments of three experiments. Mean and standard errors are plotted.

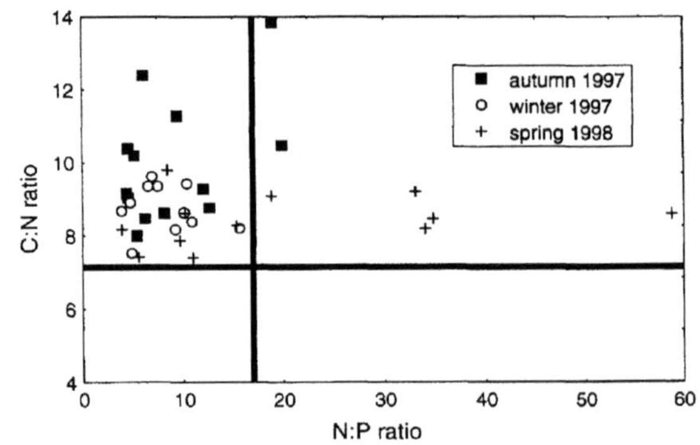

Figure 5. C:N and N:P ratios of algal biomass, with optimal ratios *sensu* Hillebrand & Sommer (1999) plotted as lines.

Results

Total biovolume of microalgae was affected by nutrient treatment, but not by the seasonal setting of the experiment (Figure 1, Table 2). The biomass increased significantly in the N + P-enriched treatments compared to controls and Si-enrichments (Tukey's HSD, $p < 0.01$). The ANOVA detected no significant difference between controls and Si-treatments (Figures 1 and 2). However, Si-enrichment resulted in a significant linear increase in total biovolume in autumn 1997 and in spring 1998; but in the latter experiment, only in combination with N + P enrichment (Figure 1). In winter 1997, the Si enrichment had no significant effects.

In general, diatoms were the most dominant group (Figure 3). They stayed dominant throughout all treatments in winter 1997, when ambient Si-concentrations were highest (Table 1). In spring 1998, the proportion of diatoms declined at the lowest Si:N supply ratios, but they still contributed more than 90% to total biovolume in every case. Only in autumn 1997 the proportion of diatoms decreased significantly with decreasing Si:N ratios (r: 0.588, $p = 0.035$, $n = 13$), although the proportion was still very variable.

The species composition was changed in response to the nutrient treatment (Figure 2, Table 3). The addition of N + P favoured *Berkeleya rutilans*, *Ceramium strictum*, *Melosira moniliformis*, *M. nummuloides*, *Navicula grevillei* and *Tabularia fasciculata*. The biomass increase of *Proschkinia complanata* in spring 1998 and of the chlorophyte *Enteromorpha* sp. due to N + P enrichment was slightly insignificant (Table 3). *Enteromorpha* dominated at low Si concentrations and was negatively correlated to Si enrichment. The *Enteromorpha* individuals were small (up to 5 cm) compared to the size reported in other studies conducted in the Baltic Sea (Lotze, 1998). The diatom *Melosira moniliformis* increased in high Si and medium Si + N + P treatments. At the highest Si + N + P conditions, the rhodophyte *Ceramium strictum* became dominant (Figure 2). The diatoms *Haslea crucigera* (in spring 1998) and *Pleurosigma elongatum* (in autumn 1997) showed significant positive correlations to Si-enrichment (Table 3). *Licmophora abbreviata* was most abundant in the control and decreased with enrichment. In winter 1997, the species composition showed no trends with nutrient enrichment.

In all experiments, the biomass stoichiometry changed with the nutrient treatment. C:N ratios decreased in the N + P-enriched treatments (Figure 4, Table 4), whereas the controls and Si treatments showed similar C:N ratios. The cellular C:N ratio was also influenced by the season of the experiments, since nutrient conditions changed during the year (Table 1). N:P ratios were generally lower than 17 in autumn and winter 1997 and varied broadly in spring 1998 (Figure 5).

Table 3. Response of dominant species to nutrient enrichment: Spearman-rank-correlation coefficients between species biovolume and Si-enrichment and Mann–Whitney *U*-test between samples enriched with N+P and those without. The table lists species name, rank correlation coefficient *r* for effect of Si enrichment, *z*-statistic for effect of N+P enrichment, with significance level (ns: not significant, * $p < 0.05$, ** $p < 0.01$). Moreover, the seasonal occurrence of the respective species is annotated together with the number of valid observations (*n*)

Species	Si *r* (Spearman)	N + P *z* (Mann–Whitney)	Annotation	*n*
Achnanthes longipes	0.14 ns	0.21 ns	only autumn	13
Berkeleya rutilans	0.27 ns	−2.42 *	all	36
Ceramium strictum	0.29 ns	−2.50 *	only autumn	13
Enteromorpha spec.	−0.56 *	−1.71 ns	only autumn	13
Haslea crucigera	0.75 **	−0.80 ns	only spring	12
Melosira moniliformis	0.49 **	−2.64 **	all	36
Melosira nummuloides	0.21 ns	−2.39 *	all	36
Navicula grevillei	0.24 ns	−2.88 **	only spring	12
Pleurosigma elongatum	0.71 **	−0.71 ns	only autumn	13
Proschkinia complanata	0.50 ns	−1.92 ns	only spring	12
Tabularia fasciculata	0.41 *	−2.39 *	all	36

Table 4. Results of a two-factor ANOVA on C:N ratios with treatment (control, N + P + Si or Si) and season of experiment (autumn, winter, spring) as factors. Homogeneity of variances was achieved by squareroot transformation of the dependent variable (Bartlett's $\chi^2 = 12.58$, $p = 0.127$), whereas normal distribution was not achieved ($\chi^2 = 13.80$, $p = 0.008$)

Effect	df	MS effect	*F*-ratio	*p*-level
Treatment	2	0.060	20.667	<0.001
Season	2	0.057	19.637	<0.001
Treatment × season	4	0.006	2.146	0.101
Error	28	0.003		

Discussion

Evaluation of the experimental setup

The method of direct and continuous supply of liquid media to benthic microalgae was already discussed earlier (Hillebrand & Sommer, 1997). The major change to previous experiments concerned the type of the substrate. Searching for a substrate with a surface not too smooth to hinder the attachment of microalgae (Snoeijs, 1991), we decided to use wood substrates, which additionally allowed the connection of tubes. The microbenthic community on the artificial substrates was generally similar to the community found on kieselgur substrates (Hillebrand & Sommer, 1997). Dominant species were largely the same, with special emphasis on the diatoms *Berkelya rutilans* and *Proschkinia complanata* in spring and *Melosira moniliformis* in autumn, as well as the rhodophyte *Ceramium strictum* in autumn. However, *Enteromorpha* sp. dominated at low Si-enrichment, but occurred only rarely on kieselgur substrates (Hillebrand & Sommer, 1997). Wood substrates were thus equally suitable for the attachment of microalgae as the kieselgur substrates used in previous experiments. Differences in colonization on rock and wood substrates were reported for stream peripyhton, but it was also indicated that these initial differences may vanish with colonization time (Sabater et al., 1998).

As reported earlier, suspending the substrates effectively prevented the appearance of slow moving grazers like gastropods or large isopods (Hillebrand & Sommer, 1997). Few micrograzers like amphipods, nematodes and ciliates were recorded, but they never contributed more than 1% of total organism abundance. Bacteria were not analyzed in our experiments, but it can be assumed that bacteria may profit from the enhanced algal production (Hepinstall & Fuller, 1994). However, the microbial response to these kinds of enrichment experiments has rarely been assessed (but see Nilsson et al., 1991) and it remains a future task to unravel the interaction between periphyton and benthic bacteria under different degrees of eutrophication.

Effect of nutrient treatment

Biomass stoichiometry is a useful tool for detecting nutrient limitation and has been widely applied to phytoplankton (Hecky et al., 1993). Recently, this concept has been transferred to benthic microalgae (Kahlert, 1998; Hillebrand & Sommer, 1999). Optimal C:N:P ratios in periphyton were proposed as indicators of saturated internal nutrient pools and thus optimal growth conditions. Hillebrand & Sommer (1999) derived optimal ratios of C:N:P of 119:17:1 from laboratory experiments, while Kahlert (1998) surveyed *in situ* freshwater data and found an optimal ratio of 158:18:1. Indication of N-limitation was proposed by Hillebrand & Sommer (1999) with a C:N >9 and a N:P <13.

The high C:N ratios and low N:P ratios in the control and the Si-treatments indicated nitrogen limitation in the autumn experiment. No N-limitation was detected during winter (C:N decreased) and spring, where high N:P ratios with high C:P ratios indicated P-limitation in some treatments. In early spring, a shift from N- to P-limitation in Kiel Fjord was observed before (Hillebrand & Sommer, 1997; Deegen, 1997) and was indicated also by ambient nutrient concentrations (Table 1). A shift from spring P-limitation to summer N-limitation was also reported from other sites, especially those with a freshwater inflow (D'Elia et al., 1986; Malone et al., 1996). Furthermore, the decrease in C:N:P ratios following the experimental addition of N indicated the effectiveness of the nutrient treatment and created a more nutrient replete situation for the periphyton.

Nutrient limitation is additionally indicated by the biomass increase following the N + P enrichment. Additional Si increased the total biovolume as well, but this increase was significant only in autumn 1997 and in spring 1998, for the latter time only in combination with N + P enrichment. Generally, Si enrichment was more effective (higher slopes, Figure 1) when N + P were enriched simultaneously. This indicates that Si limitation was less important than N limitation. Si-enrichment resulted in an biovolume increase only in autumn 1997 (Figure 1), when water column concentrations were >15 μmol l^{-1} Si (Table 1). This concentration was obviously not sufficient to meet the demands of the developing biofilm, so there must be high diffusive barriers preventing the effective uptake of water column Si by the biofilm. Freshwater periphyton was shown to have limited access to water column nutrients (Bothwell, 1985, 1989; Riber & Wetzel, 1987). The degree of limitation was shown to depend on low diffusion and low flow within the microbenthic assemblages, which is dependent on the biomass or the height of the mat (Wetzel, 1996). By supplying nutrients from the bottom, we presumably overcame this nutrient barrier, but also changed the vertical gradient of nutrient supply. Nevertheless, mainly erect growing species profited from the nutrient supply, indicating the importance of other factors related to vertical position in the periphyton (e.g. light).

Nutrient competition

From the nutrient-limited scenario shown in this and a previous study (Hillebrand & Sommer, 1997), it can be assumed that nutrient competition may play an important role for the structure of periphyton. However, the factors influencing the taxonomic composition of benthic microflora are not as well understood as those influencing the pelagic microalgae. For sediment-inhabiting microflora, Admiraal (1984) emphasized the role of physical and chemical conditions rather than nutrient competition, whereas the importance of nutrient supply was stressed later on (Sundbäck & Snoeijs, 1991; Flothmann & Werner, 1992; Nilsson, 1995). Whereas several studies on nutrient competition in freshwater periphyton have been published (see reviews in Borchardt, 1996; McCormick, 1996), only few studies have compared the combined effects of Si- and N + P- enrichment on benthic microalgae. In fact, few studies have regarded Si in any way, which is surprising since diatoms are often very dominant in periphyton (Borchardt, 1996). In experiments with combined control of Si and N + P, the community composition was mainly altered by enrichment of N + P (Carrick et al., 1988; Nilsson, 1995). The enrichment of Si alone did not increase the biomass (Carrick & Lowe, 1988) and enrichment of N + P + Si did not change the dominance of diatoms (Nilsson, 1995). A dominance shift to other taxa, e.g. cyanobacteria or chlorophytes, took place only at very low Si:N supply ratios (Nilsson, 1995; Sommer, 1996). These results are in agreement with our experiments, which showed a greater influence of N + P enrichment as well. The effect of Si on higher taxonomic levels was comparably low since only in one of our experiments a distinct dominance shift could be observed (Figure 3).

Also, the increase of biovolume with Si addition was more pronounced in combination with N + P enrichment (Figure 1). The variability of diatom dominance (Figure 3) in autumn 1997 was mostly due to the unexpected dominance of *Ceramium* (see below).

On the species level, several trends could be affirmed, which became already visible in previous experiments. *Berkeleya rutilans, Ceramium strictum, Melosira moniliformis, Navicula grevillei* and *Tabularaia fasciculata* responded positively to N + P enrichment in our experiments (Table 3) and in similar studies (Sommer, 1996; Hillebrand & Sommer, 1997). Although species specific information about nutrient demands of benthic microalgae is widely lacking, the consistency of the taxonomic response to nutrient treatments between studies makes clear that physiological traits determine the competitive outcome. For *Berkeleya rutilans*, a high N-demand can be proposed from analysis of the gelatinous tubes of this species (Daniel et al., 1987). The chlorophyte *Enteromorpha* sp. decreased with increasing Si-concentrations, replaced by species with high Si-demand. These were mainly diatoms, especially *Haslea crucigera, M. moniliformis* and *Pleurosigma elongatum*.

Quite unexpectedly, however, the red algae *Ceramium strictum* became increasingly dominant with N + P enrichment and increasing Si addition. *Ceramium strictum* was also dominant in late summer 1996 at high N concentrations on Si-containing substrates (Hillebrand & Sommer, 1997). The reliance on high N-concentrations can be explained by high N-demands of *Ceramium* species (Pedersen & Borum, 1997). For the positive response to Si-enrichment, two possible explanations can be discussed: a direct explanation assumes Si-utilization by this rhodophyte, which to our knowledge has not been reported up to now. There are in fact hints that Si is utilized by filamentous algae, but these reports do not comprise rhodophytes (Parker, 1969). An indirect explanation is based on the succession often found during the colonization of free substrates by periphyton (e.g. Hudon & Bourget, 1981; Hoagland et al., 1982). This would mean that microorganisms (in this case diatoms profiting from the additional Si) alter the surface of the substrate and thus facilitate the adherence of *Ceramium*, by this increasing the abundance of attached individuals, as well as prolonging the growth phase of the attached individuals until the experiment was harvested.

Eutrophication and silicate

Eutrophication represents an imbalanced nutrient supply, since only N and P are anthropogenically enriched, while Si and several micronutrients are not. This imbalance is intensified by an alteration of biogeochemical cycles due to enhanced diatom production, transferring more dissolved Si (DSi) to particulate Si (PSi) (Conley et al., 1993). Since the turnover time of PSi is quite long (Conley et al., 1993), the pool of DSi is gradually depleted. Together with increased loadings of N and P, this leads to a decrease of the DSi:DIN ratio, as has been found in the Baltic Sea for the time period 1970–1990 (Rahm et al., 1996). For phytoplankton, the consequences of this development for species composition and algal blooms became increasingly obvious (Riegman et al., 1992; Bodeanu, 1993; Sommer, 1994). For benthic microalgae, however, predictions are less clearcut. A decrease of DSi in the water column may affect benthic diatoms lacking the sediment-associated pool of PSi. A decrease of diatoms at N + P enriched conditions has been shown experimentally on a local scale (Carrick et al., 1988; Nilsson, 1995; Sommer, 1996). However, the decrease of Si:N supply ratios may be less influential for diatoms inhabiting sediments which represents an undepletable pool of PSi. This pool may prevent Si-limitation (Sundbäck & Snoeijs, 1991), if the temperature-dependent dissolution rates meet the demand of the biofilm (Sigmon & Cahoon, 1997) or if bacterial mediated PSi-dissolution is fast enough (Bidle & Azam, 1999).

Conclusions

Total biovolume of microalgae colonizing hard substrates was more strongly influenced by N + P enrichment compared to Si enrichment. Only in one experiment the dominance of diatoms decreased with decreasing Si:N supply ratios. N-limitation was indicated by the enrichment experiments and by biomass stoichiometry of the benthic microalgae. However, the supply of nutrients had strong effects on the species composition of the periphyton.

Acknowledgements

The technical assistance of Petra Deegen, Cordula Stielau and Thomas Hansen (especially for CN-analysis) is gratefully acknowledged. The work was funded by Deutsche Forschungsgemeinschaft (DFG grant no. SO-145/15-1). The manuscript profited from comments by G. C. Cadée and an anonymous reviewer.

References

Admiraal, W., 1984. The ecology of estuarine sediment-inhabiting diatoms. Progr. Phycol. Res. 3: 269–322.

Bidle, K. D. & F. Azam, 1999. Accelerated dissolution of diatom silica by marine bacterial assemblages. Nature 397: 508–512.

Bodeanu, N., 1993. Microalgal blooms in the Romanian area of the Black Sea and contemporary eutrophication conditions. In Smayda, T. J. & Y. Shimizu (eds), Toxic Phytoplankton Blooms in the Sea. Elsevier Science Publ., Amsterdam: 203–209.

Borchardt, M. A., 1996. Nutrients. In Stevenson, R. J., M. L. Bothwell & R. L. Lowe (eds), Algal Ecology – Freshwater Benthic Ecosystems. Academic Press, London: 183–227.

Bothwell, M. L., 1985. Phosphorus limitation of lotic periphyton groth rates: an intersite comparison using continuous flow throughs (Thompson River, British Columbia). Limnol. Oceanogr. 30: 527–542.

Bothwell, M. L., 1989. Phosphorus-limited growth dynamics of lotic periphyton diatom communities: Areal biomass and cellular growth rate response. Can. J. Fish. Aquat. Sci. 46: 1293–1301.

Carrick, H. J., R. L. Lowe & J. T. Rotenberry, 1988. Guilds of benthic algae along nutrient gradients: relationships to algal community diversity. J. n. am. Benthol. Soc. 7: 117–128.

Carrick, H. J. & R. L. Lowe, 1988. Response of Lake Michigan benthic algae to *in situ* enrichment with Si, N and P. Can. J. Fish. aquat. Sci. 45: 271–279.

Conley, D. J., C. L. Schelske & E. F. Stoermer, 1993. Modification of the biogeochemical cycle of silica with eutrophication. Mar. Ecol. Prog. Ser. 101: 179–192.

D'Elia, C. F., J. G. Sanders & W. R. Boynton, 1986. Nutrient enrichment studies in a coastal plain estuary: phytoplankton growth in large scale, continuous cultures. Can. J. Fish. aquat. Sci. 43: 397–406.

Daniel, G. F., A. H. L. Chamberlain & E. B. G. Jones, 1987. Cytochemical and electron microscopical observations on the adhesive materials of marine fouling diatoms. Br. Phycol. J. 22: 101–118.

Deegen, P., 1997. Nährstoffansprüche von Aufwuchsalgen der Kieler Förde. Msc thesis. Christian-Albrechts-Universität zu Kiel: 90 pp.

Flothmann, S. & I. Werner, 1992. Experimental eutrophication on an intertidal sandflat: effects on microphytobenthos, meio- and macrofauna. In Colombo, G., I. Ferrari, V. U. Ceccherelli & R. Rossi (eds), Marine Eutrophication and Population dynamics. Proceed. 25th Europ. Mar. Biol. Symp. Olsen & Olsen, Fredenborg: 93–100.

Hepinstall, J. A. & R. L. Fuller, 1994. Periphyton reactions to different light and nutrient levels and the response of bacteria to these manipulations. Arch. Hydrobiol. 131: 161–173.

Hecky, R. E., P. Campbell & L. L. Hendzel, 1993. The stoichiometry of carbon, nitrogen and phosphorus in particulate matter of lakes and oceans. Limnol. Oceanogr. 38: 709–724.

Hillebrand, H. & U. Sommer, 1997. Response of epilithic microphytobenthos of the western Baltic Sea to *in situ* experiments with nutrient enrichment. Mar. Ecol. Prog. Ser. 160: 35–46.

Hillebrand, H. & U. Sommer, 1999. The nutrient stoichiometry of benthic microagal growth: redfield proportions are optimal. Limnol. Oceanogr. 44: 440–446.

Hillebrand, H., C. D. Dürselen, D. Kirschtel, U. Pollingher & T. Zohary, 1999. Biovolume calculation for pelagic and benthic microalgae. J. Phycol. 35: 403–424.

Hoagland, K. D., S. C. Roemer & J. R. Rosowski, 1982. Colonization and community structure of twoperiphyton assemblages, with emphasis on the diatoms (Bacillariophyceae). Am. J. Bot. 69: 188–213.

Hudon, C. & E. Bourget, 1981. Initial colonization of artificial substrate: community development and structure studied by Scanning Electron Microscopy. Can. J. Fish. aquat. Sci. 38: 1371–1384.

Kahlert, M., 1998. C:N:P ratios of freshwater benthic algae. Arch. Hydrobiol. Spec. Issue. Adv. Limnol. 51: 105–118.

Lotze, H. K., 1998. Population dynamics and species interactions in macroalgal blooms: abiotic *versus* biotic control at different lifecycle stages. Ph. D. thesis Christian-Albrechts-Universität Kiel: 134 pp.

Malone, T. C., D. J. Conley, T. R. Fisher, P. M. Glibert, L. W. Harding & K. G. Sellner, 1996. Scales of nutrient-limited phytoplankton productivity in Chesapeake Bay. Estuaries 19: 371–385.

McCormick, P. V., 1996. Resource competition and species coexistence in freshwater algal assemblages. In Stevenson, R. J., M. L. Bothwell & R. L. Lowe (eds), Algal Ecology – Freshwater Benthic Ecosystems. Academic Press, London: 229–252.

Nilsson, C., 1995. Microbenthic communities with emphasis on algal-nutrient relations. Ph. D. thesis, Göteborg University: 109 pp.

Nilsson, P., B. Jönsson, I. Lindström Swanberg & K. Sundbäck, 1991. Response of a marine shallow-water sediment system to an increased load of inorganic nutrients. Mar. Ecol. Prog. Ser. 71: 275–290.

Parker, B. C., 1969. Occurrence of silica in brown and green algae. Can. J. Bot. 47: 537–540.

Pedersen, M. F. & J. Borum, 1997. Nutrient control of estuarine macroalgae: growth strategy and the balance between nitrogen requirements and uptake. Mar. Ecol. Prog. Ser. 161: 155–163.

Rahm, L., D. J. Conley, P. Sanden, F. Wulff & P. Stålnacke, 1996. Time series analysis of nutrient inputs to the Baltic Sea and changing Dsi: DIN ratios. Mar. Ecol. Prog. Ser. 130: 221–228.

Riber, H. H. & R. G. Wetzel, 1987. Boundary-layer and internal diffusion effects on phosphorus fluxes in lake periphyton. Limnol. Oceanogr. 32: 1181–1194.

Riegman, R., A. A. M. Noordeloos & G. C. Cadée, G. C., 1992. *Phaeocystis* blooms and eutrophication of the continental coastal zones of the North Sea. Mar. Biol. 112: 479–484.

Sabater, S., S. V. Gregory & J. R. Sedell, 1998. Community dynamics and metabolism of benthic algae colonizing wood and rock substrata in a forest stream. J. Phycol. 34: 561–567.

Sigmon, D. E. & L. B. Cahoon, 1997. Comparative effects of benthic microalgae and phytoplankton on dissolved silica fluxes. Aquat. Microb. Ecol. 13: 275–284.

Snoeijs, P., 1991. Monitoring pollution effects by diatom community composition. A comparison of sampling methods. Arch. Hydrobiol. 121: 497–510.

Sommer, U., 1994. Are marine diatoms favoured by high Si:N ratios? Mar. Ecol. Prog. Ser. 115: 309–315.

Sommer, U., 1996. Nutrient competition experiments with periphyton from the Baltic Sea. Mar. Ecol. Prog. Ser. 140: 161–167.

Sundbäck, K. & P. Snoeijs, 1991. Effects of nutrient enrichment on microalgal community composition in a coastal shallow-water sediment system: an experimental study. Bot. Mar. 34: 341–358.

Wetzel, R. G., 1996. Benthic algae and nutrient cycling in lentic freshwater ecosystems. In Stevenson, R. J., M. L. Bothwell & R. L. Lowe (eds), Algal Ecology – Freshwater Benthic Ecosystems. Academic Press, London: 641–667.

Hydrobiologia **426:** 193–201, 2000.
G. Liebezeit, S. Dittmann & I. Kröncke (eds), *Life at Interfaces and Under Extreme Conditions.*

Relevance of crustacean carapace wettability for fouling

Klaus Becker[1,*], Twee Hormchong[2] & Martin Wahl[3]
[1]*Biolab Research Institute, Kieler Strasse 51, D-24594 Hohenwestedt, Germany*
*E-mail: kb.bioklaus@t-online.de (*author for correspondence)*
[2]*Siam University, 235 Phetkasem Rd., Bangkok 10163, Thailand*
[3]*Zool. Institut, Univ. Kiel, Olshausenstrasse 40, D-24098 Kiel, Germany*
E-mail: mwahl@unam.na

Key words: crustaceans, carapace, biofouling, surface wettability

Abstract

Carapace wettability and density of fouling organisms (bacteria, diatoms, protozoa, fungi, macro-organisms) were investigated for 45 crustacean species (Hoplocarida, Decapoda) from 15 families in the Gulf of Thailand. The results show that crustaceans can create and maintain characteristic carapace wettabilities. About 21 species (47%) possess highly wettable carapaces with contact angles below 20°. Contact angles between 20° and 40° were recorded for four species (2%), angles between 40° and 60° for eight species (4%) and from 60° to 70° for 11 (24%) species. One species, *Alpheus euphrosyne* (Alpheidae, Decapoda), exhibited an extremely low surface wettability (contact angle: 91°). Densities of colonisers and contact angles did not correlate. Very low wettability by water ($\theta > 90°$) may only contribute little to fouling reduction in *A. euphrosyne* which showed the most hydrophobic carapace surface and was colonised by the lowest numbers of bacteria among all species and no other colonisers at all. We conclude that surface wettability is of little relevance for antifouling defence in crustaceans.

Introduction

Many artificial and natural surfaces become rapidly colonised upon exposure to the sea. Sessile marine organisms secrete a variety of adhesive materials to attach to surfaces (Hascall, 1973: Chamberlain, 1976; Corpe, 1980; Sutherland, 1980; Lindner, 1984; Webster et al., 1985; Cooksey & Cooksey, 1986; Waite, 1987, 1990; Jensen & Morse, 1988; Young et al., 1988; Decho, 1990; Abu et al., 1991; Fletcher et al., 1991; Neu & Marshall, 1991; Read et al., 1991; Hoagland et al., 1993). Typically, adhesives produced by bacteria and diatoms consist of polysaccharides, but proteins and lipids may also be present. Glycoproteins and sulfur containing protein-polysaccharide-complexes have been described in protozoan attachment. Fungi secrete similar types of adhesives. Glues of macro-organisms frequently consist of mucopolysaccharides, glycoproteins, silk and/or quinone-tanned proteins. These substances have to provide strong attachment to a variety of surfaces to avoid detachment by water currents or predators.

As a prerequisite for the establishment of adhesive bonds, a glue must be able to wet a surface (Wistuba, 1980). Adhesion is mediated by intermolecular forces which are acting across the glue/substratum interface. When a droplet of a liquid is placed on a surface, a characteristic equilibrium contact angle (θ) can be observed between the liquid and the surface. This contact angle reflects the energetic state of the glue/surface/medium system. An angle of $\theta = 0°$ indicates complete wetting. Larger contact angles ($0° < \theta < 180°$) indicate progressively poorer (incomplete) surface wetting. The smaller θ, the more energy is required to separate the glue from the surface. Furthermore, a non wetting adhesive leads to the formation of small bubbles at the interface during the hardening of the glue. The resulting smaller contact area and higher stress within the molecular structure of the glue increase the risk of fracture (Wistuba, 1980).

Several studies reported that surface wettability influences substratum preferences and/or attachment strength of fouling organisms on artificial substrata. However, there is considerable uncertainty concerning the efficiency of substratum wettability as a broad spectrum antifouling mechanism because some organisms are more likely to settle on hydrophobic while others prefer hydrophilic materials (Fletcher & Loeb, 1979; Absolom et al., 1983; Brewer, 1984; Fletcher & Baier, 1984; Crisp et al., 1985; Rittschof & Costlow, 1989; Roberts et al., 1991). One approach to assess the significance of wettability in the marine environment is to study the role of surface wettability of long-lived marine organisms as a defence mechanism against colonisation. Crustaceans were selected as study organisms because many species appear remarkably clean despite a strong colonisation pressure and a carapace longevity of several months (Shields, 1992; Becker, 1996; Carman & Dobbs, 1997). Wolff (1959) reported that crustaceans living under a strong current regime are less infested by other organisms than usual. There are several small crustacean species with highly hydrophobic surfaces, e.g. copepods of the genera Halectinosoma and Ectinosoma (Ectisomatidea, Harpactoidea), (G. Sach, pers. comm.).

The hypothesis was that if a given wettability could efficiently deter most potential foulers, then it could be expected to have evolved as an antifouling adaptation in some species. A body surface which provides weak attachment to colonisers could produce this effect. Antifouling mechanisms should be beneficial if they reduce otherwise adverse effects by fouling organisms, e.g. increased energy expense for locomotion, damage of the body surface with increasing risk of infections, lower egg production, disruption of the moulting cycle (Glynn, 1970; Turner et al., 1979; Nagasawa, 1987; Xu & Burns, 1991; Weissmann et al., 1993).

The following questions were addressed in the present study:

1. Are crustaceans able to create characteristic surface wettabilities despite adsorption and colonisation processes (Baier, 1970) which should lead to convergent carapace wettabilities in different species?

2. Do crustaceans possess surface wettabilities which reduce or impede colonisation (e.g. by interfering with the adhesion of colonisers glues)?

3. What relevance does surface wettability have as a defence mechanism compared to other mechanisms?

Materials and methods

Crustaceans were collected between March 1991 and April 1993 along the eastern coast of the Gulf of Thailand (Becker, 1996). They were collected from fishing ports, gill nets, by SCUBA-diving and by hand along the shoreline. All crabs were in intermoult stage with hard carapaces. The specimens designated for enumeration of colonisers other than fungi were preserved in 4% formalin. Crustaceans which were used for wettability measurements were also taken alive to the laboratory and killed by deep freezing (−15 °C) to avoid alterations of the molecular surface structure as far as possible. Carapace pieces were cut off, treated ultrasonically (Iuchi 20, 40 W, 30 min), rinsed with distilled water and dried in a dry oven at 30 °C, which was slightly above the temperature of the sea water (25 °C–28 °C). Carapace wettability was estimated by contact angles measurements with bidistilled water as reference liquid through a stereomicroscope equipped with a goniometer eyepiece. Droplets (1 μl) of doubly distilled water were placed on the carapaces and the left and right angle of one droplet were measured. Measurements were considered for data evaluation if the difference between the right and left angle did not exceed 5°. At least 10 measurements were made from each of 3–5 different specimens per species.

The number of fouling organisms on the carapaces were estimated as described earlier (Becker, 1996). Bacteria, diatoms and protozoa were collected quantitatively by embedding in Parlodion (Sechler & Gundersen, 1971), stained with acridine orange (bacteria) or Alcian blue and Ziehl Neelsen (diatoms, protozoa) and counted by epifluorescence and light microscopy. Counts were made on 3–5 randomly selected specimens of each species. In order to estimate fungal densities, live specimens were transported to the laboratory. Fungal densities were estimated by isolating them on a selective medium (DIFCO 2216 marine agar with 0.5 gl^{-1} Gentamicin) according to Marszalek et al. (1979). Unfilmed carapace pieces (1–3 cm^2) were rinsed with sterile seawater (filtered through 0.2 millepore filters and sterilized in a Yamato SD-41 autoclave) to remove unattached spores. The pieces were scraped over the medium and finally placed on the agar. Agar plates were incubated at 37 °C and the number of fungal colonies was recorded daily with a digital colony counter (EUDS DE-3, Kayaguchi) until the colony number levelled out.

Macro-organisms were counted directly on the carapaces using a stereomicroscope (VMZ Japan).

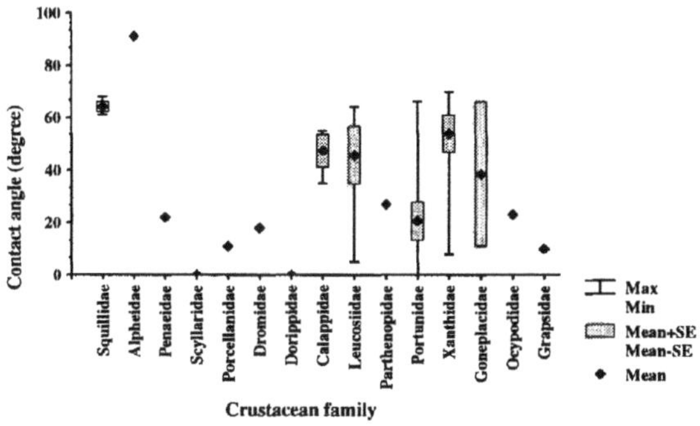

Figure 1. Carapace wettability of investigated crustacean families.

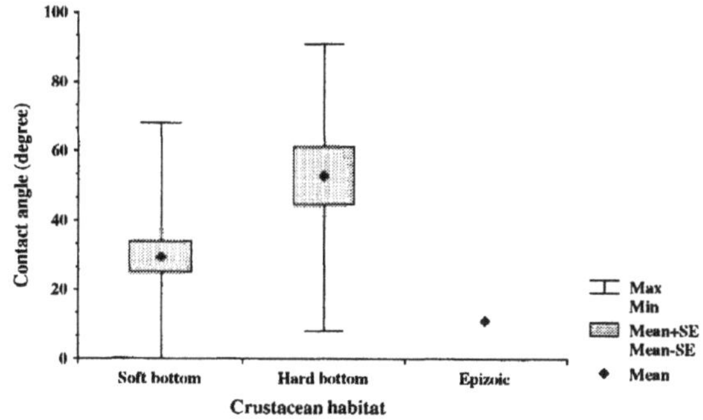

Figure 2. Comparison of carapace wettability between soft and hard bottom inhabiting species.

Coverage (%) of the carapaces by macro-organisms was determined by the dot method adopted from random sampling systems (Nair et al., 1984). Statistical analyses (Spearman rank test, Mann–Whitney-*U*-test, linear regression analyses) were carried out using Statistica software package. An index called 'colonisation degree' was used to get an estimate of the colonisation by all categories of colonisers (bacteria, diatoms, protozoa, fungi, macro-organisms). The 'colonisation degree' was calculated as follows: the highest mean density of each fouling group found on any species was set as 100%. Densities of the same fouling group on other crab species were calculated proportionally. The mean %-coverage by the five fouling groups yielded the 'colonisation degree' on a given host species.

Results

Contact angle measurements

A total of 45 crustacean species from 15 families was investigated. Mean contact angles ranged from 0° to 91° (Table 1). Although a few species showed a wide range of contact angles (e.g. *Harpiosquilla harpax, Leucosia craniolaris, Acrania novemspinosa, Ixa cylindrus*) most species exhibited narrowly defined, specific carapace wettability. Some variability was presumably caused by surface structures, roughness, adsorbed molecules and colonisers. A narrow species-specific carapace wettability range permits the assumption that those conspecifics that had served for fouling assessment fall within the same range of carapace wettability.

Twenty-one species possess highly wettable surfaces (θ <20°). Less wettable surfaces with mean contact angles between 20° and 60° were found on 12 species. Contact angles between 60° and 70° were recorded on 11 of the species. Very low wettability was found only on *Alpheus euphrosyne* (θ = 91°). Consistent differences of contact angles between different families were not detected (Spearman rank: $p = 0.785$, Figure 1). Hard bottom species tended to exhibit slightly higher contact angles than soft bottom species (Figure 2), but this was not significant (Spearman rank: $p = 0.137$, *U*-test: $p = 0.551$).

Fouling assessment

Colonisation data have been already published elsewhere (Becker, 1996) and will only be summarized here. One hundred percent of the species investigated were colonised by bacteria. Bacterial densities ranged from $7 * 10^2$ to $2.87 * 10^5$ mm^{-2}. More than 10^4 cells mm^{-2} were recorded on nine species. Lowest densities were found on *Alpheus euphrosyne*. Diatoms were found on 18 species (40%) ranging from 20 per cm^2 to $7.4 * 10^3$ cm^{-2}. High diatom densities ($4.1–7.4 * 10^3$ cm^{-2}) occurred on *Etisus* c.f. *laevimanus, Thalamita crenata*, and *Metapograpsus quadridentatus*. Sessile protozoa were detected on eight species (18%). The number of protozoa ranged from 10 to 500 per cm^{-2} on fouled crustaceans. Hyphens of fungi were not detected. Attached spores of fungi were discovered on 17 species (38%). The densities remained below 1 per cm^2 carapace surface on all species but Conchoecetes artificiosus (2.2 cm^{-2}). Macroorganisms were found on eight species. *Hexapus anfractatus* was regularly (23% of all individuals) colonised by pedunculate barnacles. An unidentified actinian species occurred on 18% of the *Dorippe facchino* specimens. Xanthid crabs were regularly colonised by bryozoa (*Akatopora* c.f. *tincta*, Hastings; Key, pers. com.), covering 2–11% of the carapaces. Very few individuals of *Ater-*

Table 1. Overview about investigated crustacean species, number of collected specimens (*N*), carapace size, carapace contact angle measurements (mean, standard error (S.E.), range) and natural substratum of the crustaceans

Crustacean species	*N*	Min.–Max. carapace width [cm]	Mean	S.E. (Contact angle [degree])	Range	Substratum
Squillidae (Hoplocarida)						
Oratosquilla interrupta (Kemp)	58	1.0–1.4	61	7	29–81	Sand; in burrows
Oratosquilla nepa (Latreille)	75	2.0–2.5	68	3	46–77	Sand; in burrows
Harpiosquilla harpax (de Haan)	50	1.4–1.7	63	12	20–93	Sand; in burrows
Alpheidae (Decapoda)						
Alpheus euphrosyne, de Man	48	2.1–2.4	91	3	81–96	underneath rocks
Penaeidae (Decapoda)						
Trachypenaeus fulvus, Dall	23	1.8–2.3	22	7	3–59	nectobenthic, sand
Scyllaridae (Decapoda)						
Scyllarus c.f. *sordidus* (Stimpson)	75	1.2–1.5	0	0	0–36	unspecific
Thenus orientalis (Lund)	29	3.0–4.0	0	0	0	muddy sand; partly burying
Porcellanidae (Decapoda)						
Porcellanella picta (Stimpson)	26	0.5–0.7	11	3	5–17	exposed, on Anthozoa
Dromidae (Decapoda)						
Conchoecetes artificiosus (Fabricius)	13	2.2–3.3	18	6	0–58	sand, mud; partly burying
Dorippidae (Decapoda)						
Dorippe facchino (Herbst)	63	1.0–1.5	0	0	0–8	sand, mud; partly burying
Dorippe frascone (Herbst)	19	2.5–3.0	0	0	0	sand, mud; partly burying
Calappidae (Decapoda)						
Calappa philargius, L.	27	6.5–8.2	35	3	18–43	sand, broken shells; partly burying
Matuta lunaris (Forskål)	22	3.0–4.1	52	9	5–76	sand; partly burying
Matuta banksii, Leach	17	3.2–3.8	55	3	39–68	sand; partly burying
Leucosiidae (Decapoda)						
Leucosia craniolaris (L.)	41	1.2–2.0	59	6	7–91	sand; partly burying
Acrania novemspinosa, Adams & White	75	2.6–2.9	64	15	5–109	sandy mud; partly burying
Acrania erinaceus (Fabricius)	18	1.6–2.1	60	3	49–66	sand; partly burying
Ixa cylindrus (Fabricius)	200	5.0–6.1	40	6	5–71	mud; partly burying
Iphiuculus spongiosus, Adams & White	19	2.9–3.7	5	1	0–8	sand, mud, algae; partly burying
Parthenopidae (Decapoda)						
Cryptopodia fornicata (Fabricius)	39	5.3–6.5	27	6	7–58	sand, broken shells; partly burying
Portunidae (Decapoda)						
Scylla serrata (Forskål)	154	6.5–12.3	63	3	50–78	mud; partly burying
Portunus pelagicus (L.)	207	7.5–15.6	0	0	0–5	sand, mud; partly burying
Portunus pulchicristatus, Gordon	123	2.8–1.6	0	0	0	sandy mud; partly burying
Portunus tweediei (Shen)	213	2.0–1.4	3	1	0–8	sand, mud; partly burying
Portunus gracilmanus (Stimpson)	31	3.2–2.2	5	2	0–8	sand, mud; partly burying
Portunus hastatoides (Fabricius)	52	3.5–1.9	3	1	0–7	sand, mud; partly burying
Charybdis feriatus (L.)	14	4.2–8.7	44	7	0–78	hard bottom
Charybdis truncata, Fabricius	19	3.4–4.0	3	1	0–8	sandy mud; partly burying
Charybdis c.f. *affinis*, Dana	23	2.9–4.0	4	1	0–7	sandy mud; partly burying
Charybdis anisodon (de Haan)	204	2.8–4.5	66	5	47–80	sandy mud; partly burying
Thalamita crenata (Latreille)	23	2.6–3.5	58	11	7–83	between rocks

Continued on p. 197

Table 1. Continued

Crustacean species	*N*	Min.–Max. carapace width [cm]	Mean S.E.	Contact angle [degree]	Range	Substratum
Thalamita sima, Milne–Edwards	18	2.9–4.2	6	2	0–9	between rocks
Podophthalmus vigil (Fabricius)	53	4.2–8.3	14	8	5–25	sand, sandy mud; partly burying
Xanthidae (Decapoda)						
Atergatis intergerrimus (Lamarck)	6	8.0–12.5	64	5	53–74	hard bottom, coral reefs, gravel
Atergatis floridus (L.)	3	5.1–7.3	66	3	49–79	coral gravel
Lophozymus pictor (Fabricius)	17	5.2–8.3	68	3	60–80	hard bottom, coral reefs
Halimede ochtodes (Herbst)	7	4.1–5.7	47	2	42–62	sand, mud; partly burying
Galene bispinosa (Herbst)	18	3.8–5.6	59	7	36–61	sand, mud; partly burying
Etisus c.f. *laevimanus*, Randall	108	3.7–5.8	49	5	36–61	intertidal; between rocks
Sphaerozius nitidus, Stimpson	32	1.0–1.5	8	3	3–9	between rocks
Sphaerozius sp.	9	5.6–7.4	70	3	61–83	between rocks
Goneplacidae (Decapoda)						
Hexapus anfractatus, Rathbun	229	1.1–2.5	11	7	3–38	mud; in holothurian burrows
Eucrate alocki, Serène	37	2.0–4.1	66	5	51–81	sand, broken shells, partly burying
Ocypodidae (Decapoda)						
Dotilla wichmani, de Haan	33	0.4–0.6	23	10	0–60	intertidal, muddy sand; in burrows
Grapsidae (Decapoda)						
Metapograpsus quadridentatus, Stimpson	113	1.9–3.2	10	4	3–27	intertidal; between rocks

gatis intergerrimus, *A. floridus*, *Lophozymus pictor* and *Sphaerozius* sp. were without bryozoa. One individual of *L. pictor* (Xanthidae) was also colonised by *Spirorbis* sp. (Serpulidae, Polychaeta). *M. quadridentatus* was frequently colonised by barnacles (*Balanus variegatus*) accounting for 4% average cover of carapace and macroalgae (5% cover). *B. variegatus* was also detected on one specimen of *Ixa cylindrus*.

Wettability and carapace colonisation

Statistical analyses (linear regression models, Spearman-rank-test) failed to show any correlation ($p > 0.05$) between contact angles and any of the coloniser taxa category (bacteria through macrofauna) or the colonisation degree (Table 2, Figures 3 and 4). Transformation of data (log, ln, square root) also failed to yield a significant correlation between contact angles and density of foulers ($p > 0.05$). These analyses strongly suggest that wettability has very little influence on the colonisation by fouling organisms. This conclusion will be graphically supported by the following distance weighted least square line fits (Figures 3 and 4): bacterial densities remained almost constant between $\theta = 0°$ and 68° but decreased towards higher contact angles. However, a contact angle of $\theta > 70°$ was only detected on a single species. On carapaces with contact angles between 40° and 60°, more diatoms and protozoa settled than on other carapaces. A maximum of fungi was recorded within 10–30°. Highest cover by macro-organisms was found on carapaces with contact angles from 66° to 68°. However, these maxima for diatoms, protozoa, fungi and macro-organisms remained small. The overall colonisation degree (Figure 4b) resembles the plot which was obtained for bacteria, partly due to the lack of other colonisers on many species.

A comparison of the colonisation on crustaceans with that on artificial substrata exposed in the same area suggest that wettability contributes little (if at all) to antifouling defence (Figure 5). Two crustacean species (*Alpheus euphrosyne*, *Cryptopodia fornicata*) showed wettabilities similar to those of two artificial

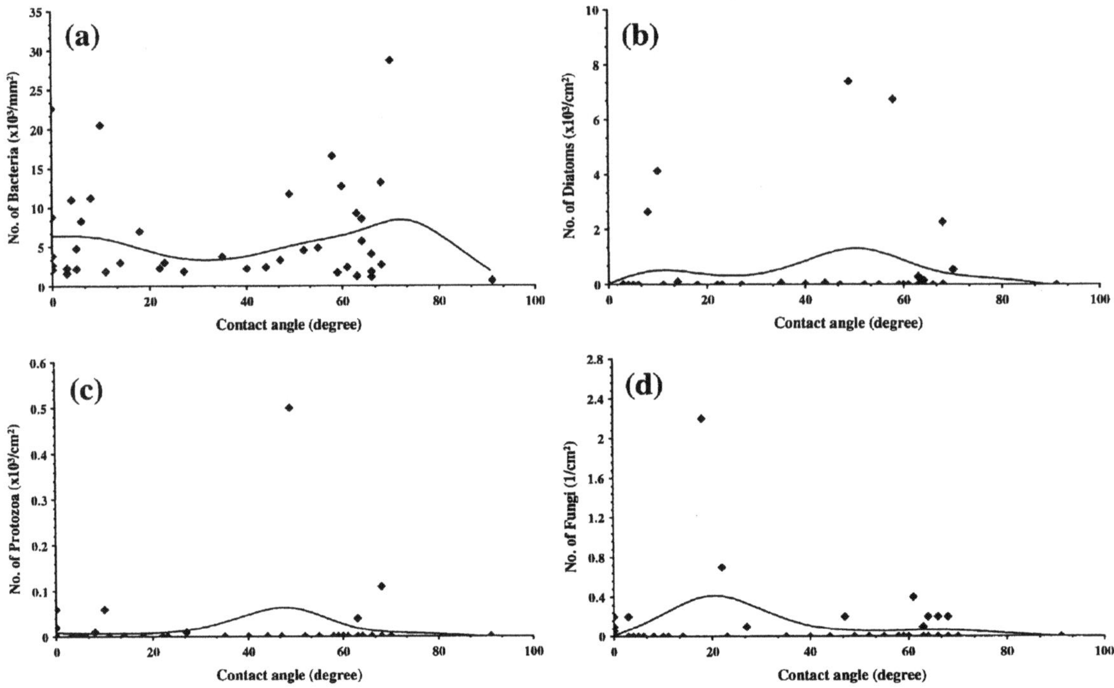

Figure 3. Distance weighted least square line fit of carapace wettability and densities of bacteria (3a), diatoms (3b), protozoa (3c) and fungi (3d).

Table 2. Results of Spearman Rank test and regression analysis for correlation between carapace wettability and density of epibionts

Group of epibionts	Spearman correlation	*p*-level	Linear regression (r)	*p*-level
Bacteria	−0.0859	0.5747	−0.0618	0.6903
Diatoms	0.1949	0.1995	0.0992	0.5224
Protozoa	−0.0616	0.6876	0.0846	0.5850
Fungi	−0.0235	0.8784	−0.0830	0.5921
Macro-organisms	0.1445	0.3438	0.1223	0.4292
Colonisation degree	−0.0936	0.5410	0.0158	0.9191

materials (glass; ETFE, a fluorpolymer) exposed in parallel experiments (see Becker et al., 1997). Both artificial materials became (*U*-test: $p < 0.05$) much more densely colonised by all groups of foulers within 5–8 days of exposure than carapaces of corresponding wettability.

Discussion

Surprisingly, some crustaceans can maintain a characteristic carapace surface wettability despite adsorption (molecular fouling) and colonisation processes. In general, upon exposure to natural waters, wettability quickly decreases on wettable surfaces and increases on non-wettable surfaces. Thus, the contact angles on various surfaces should converge towards approximately 75°–85° (see Baier, 1970). This implies that crustaceans possess mechanisms to restrict alterations of surface characteristics by adsorption and colonisation processes. It still has to be determined which processes are involved in maintaining the observed species specific wettability. Mechanisms similar to those involved in repair mechanisms or release of 'coating-substances' through pore channels within the cuticle (Green & Neff, 1972; Stevenson, 1985) may be involved. The composition of the epicuticle which is the top layer of the integument will determine the wettability of crustaceans. Epicuticles contain both hydrophobic and hydrophilic compounds. They consist of quinone-tanned proteins, lipids, glycoproteins and calcium salts (Denell, 1960; Stevenson, 1985). Recently, Compere & Goffinet (1995) detected that

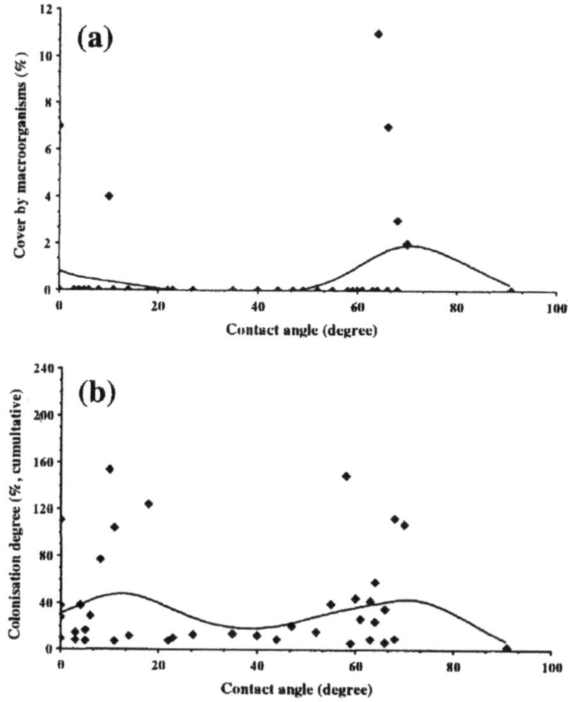

Figure 4. Distance weighted least square line fit of carapace wettability and densities of macroorganisms (4a) and the colonisation degree (4b).

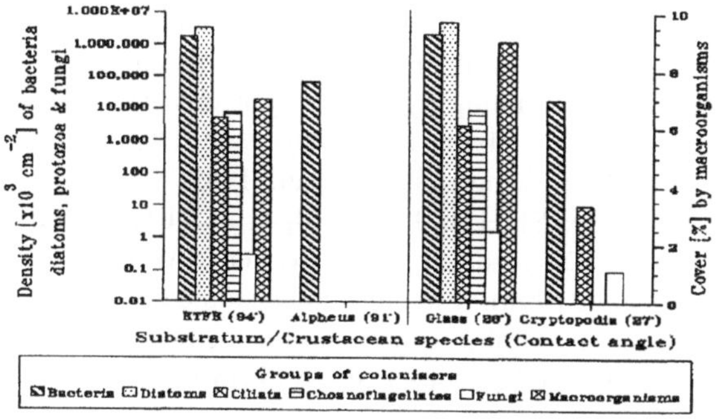

Figure 5. Colonisation of crustacean carapaces and artificial substrata with similar wettability after 8 days of exposure (ETFE: Ethylenefluorethylene).

the surface coat in *Carcinus maenas* contains polyanionic sites and acid mucopolysaccharides. The latter substances and calcium are likely to increase wettability while lipids and some hydrophobic proteins will decrease wettability. Surface wettability in recently moulted crustaceans may differ from surface wettability of intermoult specimens. The present study considered only crustaceans with fully calcified carapaces. If carapace wettability would be a major defense mechanism, one may expect that it is established at an early stage of the intermoult period to provide protection.

The present study strongly suggests that surface wettability is not an efficient antifouling adaptation in crustaceans. While the absence of colonisers may be caused by defence mechanisms other than carapace wettability, the presence of fouling organisms on a given surface proves that its wettability is not by itself a sufficiently strong antifouling mechanisms. Thus, according to the present results, wettabilities with water contact angles of $\theta < 70°$ may be rejected as possible broad-spectrum defense mechanism. Very low wettability by water ($\theta > 90°$) may contribute to fouling reduction to some extent (see *Alpheus euphrosyne*). The wettability of *Alpheus euphrosyne* is close to a range (approx.: $95° < \theta < 104°$) reported to be less densely fouled by bacteria than surfaces with other wettabilities (Dexter, 1979). Indeed, *A. euphrosyne* was the least fouled species with the lowest number of bacteria, while other fouling groups were lacking. However, some other crustacean species (e.g. *Harpiosquilla harpax*, *Leucosia craniolaris*, *Eucrate alocki*) showed more wettable carapaces than *A. euphrosyne* but were similarly little fouled.

So far, only Vrolijk et al. (1990) suggested that two gorgonian corals possess minimal adhesive surfaces which prevent fouling in conjunction with other mechanisms. Despite some promising results in antifouling research (Dexter, 1979; Bultmann et al., 1984; Fletcher & Baier, 1984; Rittschof & Costlow, 1989; Roberts et al., 1991; Lindner, 1992), there is some doubt about the efficiency of surface wettability as an antifouling mechanism under natural conditions (Becker, 1993; Becker et al., 1997). Wettability effects may be easily dwarfed by other abiotic factors and biological interactions under natural conditions. A substrate will eventually become colonised even if it initially offers unfavorable conditions in terms of adhesion. Many colonisers can adapt their attachment mechanisms to the surface properties (Lindner, 1984; Paul & Jeffrey, 1985; Webster et al., 1985; Van Loosdrecht et al., 1987).

If wettability is inefficient or its effects inconsistent in colonisation processes under natural conditions, it is unlikely that it has evolved as an antifouling mechanism in the first place. Low densities of colonisers suggest that the investigated crustaceans possess other antifouling mechanisms to restrict colonisation. Several studies indicated mechanisms by which crustaceans may fend off fouling species. These may be: moulting, grooming, behaviour (hiding, burying), immune response, chemical defence by bioactive compounds and cleaning by other organisms (Barnes & Bagenal, 1951; Glynn, 1970; Stevenson, 1985; White et al., 1985; Weng, 1987; Bauer, 1989; Gil-Turnes et al., 1989; Shields, 1992; Gili et al., 1993; Svarvarsson & Davidsdottir, 1994; Becker & Wahl, 1996; Wahl et al., 1998). It remains to be studied how some crustaceans are able to maintain a distinct carapace wettability and its ecological significance.

References

Absolom, D. R., F. V. Lamberti, Z. Policova, W. Zingg, C. J. Van Oss & A. W. Neumann, 1983. Surface thermodynamics of bacterial adhesion. Apl. envir. Microbiol. 10: 90–97.

Abu, G. O., R. M. Weiner, J. Rice & R. R. ColwellL, 1991. Properties of an extracellular adhesive polymer from the marine bacterium *Schewanella colwelliana*. Biofouling 3: 69–84.

Baier, R. E., 1970. Surface properties influencing biological adhesion. In Manly, R. S. (ed.), Adhesion in Biological Systems. Academic Press, New York, London: 15–48.

Barnes, H. & T. B. Bagenal, 1951. Observations on *Nephrops norvegica* (L.) and on an epizooic population of *Balanus crenatus* Brug. J. mar. biol. Ass. U.K. 30: 369–380.

Bauer, R. T., 1989. Decapod crustacean grooming: functional morphology, adaptive value and phylogenetic significance. In Felgenhauer, B. E., L. Walting & A. B. Thistle (eds), Functional Morphology of Feeding and Grooming in Crustacea, Crustacean Issues 6, F. R. Schram (ed.), A. A. Balkema, Rotterdam, Brookfield: 49–74.

Becker, K., 1993. Attachment strength and colonisation pattern of two macrofouling species on substrata with different surface tension (in situ studies). Mar. Biol. 117: 301–309.

Becker, K., 1996. Epibionts on carapaces of some malacostracan crustaceans from the Gulf of Thailand. J. crust. Biol. 16: 92–104.

Becker, K. & M. Wahl, 1996. Behavioural patterns as natural antifouling mechanisms of tropical marine crabs. J. exp. mar. Biol. Ecol. 203: 245–258.

Becker, K., S. Siriratanachai & T. Hormchong, 1997. Influence of initial substratum surface tension on marine micro- and macrofouling in the Gulf of Thailand. Helgoländer. wiss. Meeresunters. 51: 445–461.

Brewer, R. H., 1984. The influence of the orientation, roughness and wettability of solid surfaces on the behaviour and attachment of planulae of *Cyanea* (Cnidaria: Scyphozoa). Biol. Bull. 166: 11–21.

Bultman, J. D., J. R. Griffith & D. E. Field, 1984. Fluoropolymer coatings for the marine environment. In Costlow, J. D. & R. C. Tipper (eds), Marine Corrosion and Biodeterioration – An Interdisciplinary Study. E. & F. N. Spon. Ltd., London: 237–243.

Carman, K. R. & F. C. Dobbs, 1997. Epibiotic microorganisms on copepods and other marine crustaceans. Microscopy Res. Techn. 37: 116–135.

Chamberlain, A. H. L., 1976. Algal settlement and secretion of adhesive materials. In Sharpley, J. M. & A. M. Kaplan (eds), Proc. 3rd Intern. Biodegrad. Symp., Appl. Sci., London: 417–432.

Compere, P. & G. Goffinet, 1995. Cytochemical demonstration of acid mucopolysaccharides in the epicuticular surface coat of the crab *Carcinus maenas* (L.) (Crustacea, Decapoda). Belg. J. Zool. 125: 95–100.

Cooksey, K. E. & B. Cooksey, 1986. Adhesion of fouling diatoms to surfaces: some biochemistry. In Evans, L. V. & K. D. Hoagland (eds), Algal Biofouling. Elsevier, Amsterdam: 41–53.

Corpe, W. A., 1980. Microbial surface components involved in adsorption onto surfaces. In Bitton, G. & K. C. Marshall (eds), Adsorption of Micro-organisms to Surfaces. Wiley Interscience Publ., New York: 105–143.

Crisp, D. J., G. Walker, G. A. Young & A. B. Yule, 1985. Adhesion and substrate choice in mussels and barnacles. J. Coll. Interf. Sci. 104: 40–50.

Decho, A. W., 1990. Microbial exopolymer secretions in ocean environments: their role(s) in the food webs and marine processes. Oceanogr. mar. Biol. Ann. Rev. 28: 73–153.

Denell, R., 1960. Integument and exoskeleton. In Waterman, T. H. (ed.), The Physiology of Crustacea, Vol. I. Academic Press, New York-London: 449–473.

Dexter, S. C., 1979. Influence of substratum critical surface tension on bacterial adhesion – In situ studies. J. Coll. Interf. Sci. 70: 346–354.

Fletcher, M. & G. I. Loeb, 1979. Influence of substratum characteristics on the attachment of a marine *Pseudomonad* to solid surfaces. Apl. envir. Microbiol. 37: 67–72.

Fletcher, M., J. M. Lessmann & G. I. Loeb, 1991. Bacterial surface adhesives and biofilm matrix polymers of marine and freshwater bacteria. Biofouling 4: 120–140.

Fletcher, R. L. & R. E. Baier, 1984. Influence of surface energy on the development of the green alga *Enteromorpha*. Mar. Biol. Lett. 5: 251–254.

Gil-Turnes, M. S., M. E. Hay & W. Fenical, 1989. Symbiotic marine bacteria defend crustacean embryos from a pathogenic fungus. Science 240: 116–118.

Gili, J. M., P. Abello & R. Villanueva, 1993. Epibionts and intermoult duration in the crab *Bathynectes piperitus*. Mar. Ecol. Progr. Ser. 98: 107–113.

Glynn, P. W., 1970. Growth of algal epiphytes on a tropical marine isopod. J. exp. mar. Biol. Ecol. 5: 88–93.

Green, P. J. & M. R. Neff, 1972. A survey of the fine structure of the integument of the fiddler crab. Tissue Cell 4: 137–171.

Hascall, G. K., 1973. The stalk of the suctorian *Tokophyra infusionum*: histochemistry, biochemistry and physiology. J. Protozool. 20: 701–704.

Hoagland, K. D., J. D. Rosowski, M. R. Gretz & S. C. Roemer, 1993. Diatom extracellular polymeric substances: function, fine structure, chemistry and physiology. J. Phycol. 29: 537–566.

Jensen, A. R. & D. E. Morse, 1988. The bioadhesive of *Phragmotopoma californica* tubes: a silk cement containing L-Dopa. J. comp. Physiol. B. 158: 317–324.

Lindner, E., 1984. The attachment of macrofouling invertebrates. In Costlow, J. D. & R. C. Tipper (eds), Marine Corrosion and Biodeterioration – An Interdisciplinary Study. E. & F. N. Spon. Ltd., London: 184–201.

Lindner, E., 1992. A low surface energy approach in the control of marine biofouling. Biofouling 6: 193–205.

Marszalek, D. S., S. M. Gerchakov & L. R. Udey, 1979. Influence of substrate composition on marine microfouling. Apl. envir. Microbiol. 38: 987–995.

Nagasawa, S., 1987. Exosceletal scars by bacterial attachment to copepods (Short communication). J. Plankton. Res. 9: 749–753.

Nair, N. B., K. Dharmaraj, P. K. Abdul Azis, M. Arunchalam & K. Krishna Kumar, 1984. Ecology of biofouling on *Crassostrea madrasensis* (Preston) (Mollusca: Bivalvia) in a tropical backwater. Proc. Indian Acad. Sci. (Anim. Sci.) 93: 419–430.

Neu, T. R. & K. C. Marshall, 1991. Microbial 'footprint': a new approach to adhesive polymer. Biofouling 3: 101–112.

Paul, J. H. & W. H. Jeffrey, 1985. Evidence for seperate adhesion mechanisms for hydrophilic and hydrophobic surfaces in *Vibrio proteolytica*. Apl. envir. Microbiol. 50: 431–437.

Read, S., S. T. Moss & E. B. G. Jones, 1991. Attachment studies of aquatic hyphomycetes. Phil. Trans. r. Soc., Lond. B 344: 449–457.

Rittschof, D. & J. D. Costlow, 1989. Bryozoan and barnacle settlement in relation to initial surface wettability: a comparision of laboratory and field studies. In Ros, E. D. (ed.), Topics in Marine Biology. Scient. Mar. 53: 411–416.

Roberts, D., D. Rittschof, E. Holm & A. R. Schmidt, 1991. Factors influencing initial larval settlement: temporal, spatial and surface molecular components. J. exp. mar. Biol. Ecol. 150: 203–211.

Sechler, G. E. & K. Gundersen, 1971. New technique for microscopic examination of the fouling community of submerged opaque surfaces. Appl. Microbiol. 20: 140–143.

Shields, J. D., 1992. Parasites and symbionts of the crab *Portunus pelagicus* from Moreton Bay, Eastern Australia. J. crust. Biol. 12: 94–100.

Stevenson, J. R., 1985. Dynamics of the integument. In Bliss, D. E. & L. H. Mantel (eds), The Biology of Crustacea. Academic Press, London: 2–42.

Sutherland, I. W., 1980. Polysaccharides in the adhesion of marine and freshwater bacteria. In Berkeley, R. C. W., J. M. Lynch, J. Melling, P. R. Rutter & B. Vincent (eds), Microbial Adhesion to Surfaces. Ellis Horwood, Chichester: 330–338.

Svarvarson, J. & B. Davidsdottir, 1994. Foraminiferan (Protozoa) epizoites on arctic isopods (Crustacea) as indicators of isopod behaviour. Mar. Biol. 118: 239–246.

Turner, J. T., M. T. Poster & S. B. Collard, 1979: Infestation of the estuarine copepod *Acartia tonsa* with the ciliate *Epistylis*. Trans. am. microsc. Soc. 98: 136–138.

Van Loosdrecht, M. C. M., J. Lyklema, W. Norde, G. Schraa & A. Zehnder, 1987. Electrophoretic mobility and hydrophobicity as a measure to predict the initial steps of bacterial adhesion. Apl. envir. Microbiol. 53: 1898–1901.

Vrolijk, N. H., N. M. Targett, R. E. Baier & A. E. Meyer, 1990. Surface characteristics of two gorgonian coral species: Implications for a natural antifouling defence. Biofouling 2: 39–54.

Wahl, M., K. Kroeger & M. Lenz, 1998. Non-toxic protection against epibiosis. Biofouling 12: 205–226.

Waite, J. H., 1987. Nature's underwater adhesive specialist. Int. J. Adhesion Adhesives 7: 9–15.

Waite, J. H., 1990. The phylogeny and chemical diversity of quinone-tanned glues and varnishes. Comp. Biochem. Physiol. 97B: 19–29.

Webster, D. R., K. E. Cooksey & R. W. Rubin, 1985. An investigation of the involvement of cytoskeletal structures and secretion in gliding motility of the marine diatom, *Amphora coffaeformis*. Cell Motility 5: 103–122.

Weissmann, P., D. J. Lonsdale & J. Yen, 1993. The effect of peritrich ciliates on the production of *Acartia hudsonica* in Long Island Sound. Limnol. Oceanogr. 38: 613–622.

Weng, T. H., 1987. The parasitic barnacle, *Sacculina granifera* Boschma, affecting the commercial sand crab, *Portunus pelagicus* (L.), in populations from two different environments in Queensland. J. Fish Diseases 10: 221–227.

White, K. N., N. A. Ratcliffe & M. Rossa, 1985. The antibacterial activity of haematocyte clumps in the gills of the shore crab, *Carcinus maenas*. J. mar. biol. Ass. U. K. 65: 857–870.

Wistuba E., 1980. Kleben und Klebstoffe. Chemie in unserer Zeit 14: 124–133.

Wolff, T., 1959. Epifauna on certain decapod crustacea. Proc. XVth Congr. Zool. London: 1060–1061.

Young, G. A., A. B. Yule & G. Walker, 1988. Adhesion in the anemones *Actinia equina* L. and *Metridium senile* (L.). Biofouling 1: 137–146.

Xu, Z. & C. W. Burns, 1991. Effect of the epizoic ciliate, *Epistylis daphniae*, on growth, reproduction and mortality of *Boeckella triarticulata* (Thomson) (Copepoda: Calanoidea). Hydrobiologia 209: 183–189.

Hydrobiologia **426:** 203–210, 2000.
G. Liebezeit, S. Dittmann & I. Kröncke (eds), Life at Interfaces and Under Extreme Conditions.

Decline of the intertidal blue mussel (*Mytilus edulis*) stock at the coast of Lower Saxony (Wadden Sea) and influence of mussel fishery on the development of young mussel beds

M. Herlyn[1,*] & G. Millat[2]
[1]*Forschungszentrum Terramare, Schleusenstr. 1, D-26382 Wilhelmshaven, Germany*
[2]*Nationalparkverwaltung Niedersächsisches Wattenmeer, Virchowstr. 1, D-26382 Wilhelmshaven, Germany (*author for correspondence)*

Key words: mussel bed, mussel fishery, *Mytilus edulis*, seed mussels, stock development, Wadden Sea of Lower Saxony

Abstract

Since the middle of the 1980s, the intertidal blue mussel stock has decreased on the coast of Lower Saxony. In spring 1996, the remaining mussel beds covered an area of 1.7 km^2 with a total fresh weight of about 1000 t. This is equivalent to a loss of nearly 95% of the mussel bed area and 98% of the biomass compared with 1989–91. In autumn 1995, a comparison between fished and non-fished young intertidal mussel beds at sites of former mature beds was carried out in the Lower Saxonian Wadden Sea: most of the investigated fished young mussel beds and bed areas disappeared completely or were next to extinction after seed mussel fishery happened. In contrast to this development, the non-fished beds and bed areas remained.

Introduction

Since the middle of the 1980s, the blue mussel stock of the Lower Saxonian tidal flats has declined (Obert & Michaelis, 1991; Michaelis et al., 1995; Herlyn, 1996; Zens et al., 1997). In spring 1994, only 13 km^2 were left from the total bed area of 27 km^2 still present in 1989–91. The total fresh weight decreased from about 46 000 t to approx. 9000 t (Zens et al., 1997; Herlyn et al., 1999).

A new survey of the mussel stock of the Lower Saxonian tidal flats was carried out in spring 1996 to document the further development.

The exact causes for the decrease are still unknown, but some factors have been discussed to be potentially stock reducing: predation by seabirds, parasites, macroalgal cover, blooms of *Phaeocystis* spp., pollutants, blue mussel fishery, dredging and dumping of sediments, as well as storms and ice rafting (Michaelis et al., 1995; Herlyn, 1996, 1999; Hilgerloh et al., 1997; Hilgerloh & Herlyn, 1996; Petri & Vareschi, 1997; Zens et al., 1997; Herlyn et al., 1999). To estimate the impact of mussel fishery on the development of young mussel beds, 16 newly established beds were investigated in summer and autumn 1995. These settlements originated from an intensive spatfall in summer 1994. Half of the beds were fished for seed mussels. They were located at sites of former stable mussel beds with a high potential to develop to the biotope mature mussel bed (Millat & Herlyn, 1999). Mature mussel beds are able to survive for decades (Linke, 1939, 1954; Bahr, 1950; Dankers & Koelemaij, 1989; Obert & Michaelis, 1991). A more detailed description of this investigation on fishery impact has been given by Herlyn et al. (1999). Up to now, only Schirm (1991) has studied direct effects of mussel fishery on blue mussel beds of the tidal flats along the Wadden Sea coast of the Netherlands, Germany and Denmark. He estimated that fishery has a smaller effect on the further development of a mussel bed than other causes for mortality.

Methods

The investigations were carried out in the Wadden Sea of Lower Saxony, southern North Sea. The locations of the studied mussel beds are given in Figure 1.

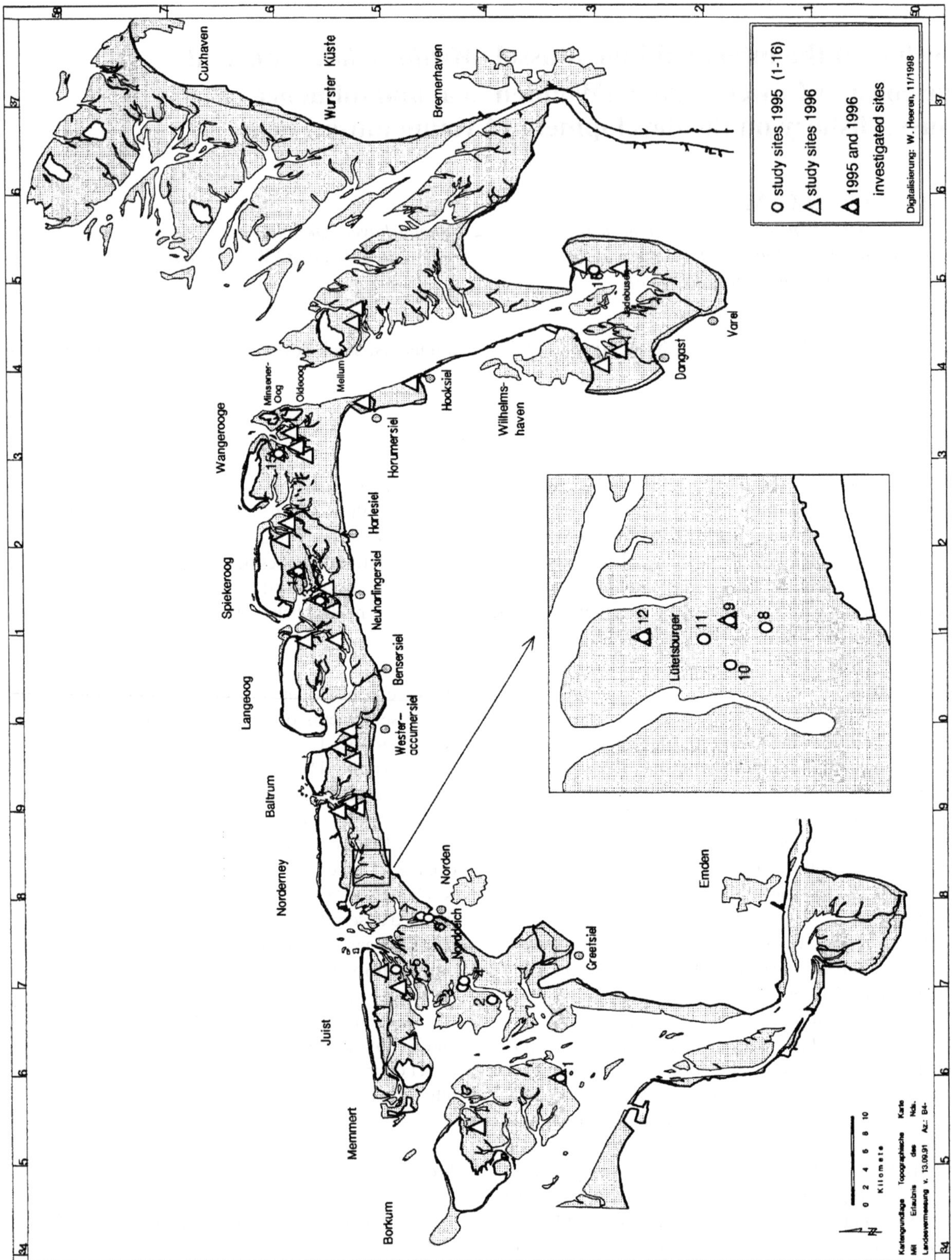

Figure 1. Investigation area with the study sites.

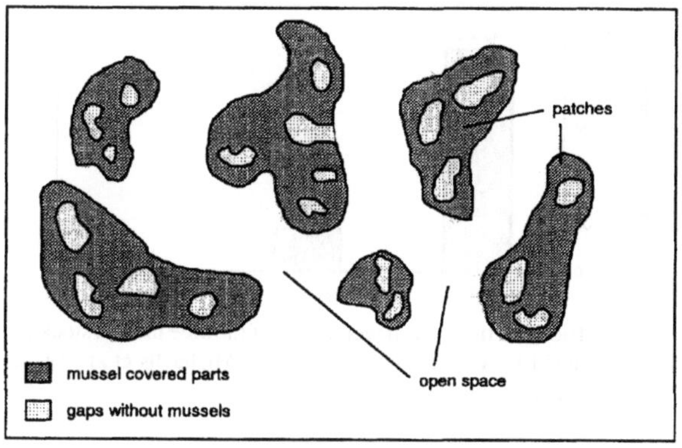

Figure 2. Schematic diagram of a mature intertidal mussel bed, which is divided in mussel covered patches and open spaces without mussels.

Survey 1996

Remote sensing

Position and extension of mussel beds and those locations of mature beds without actual mussel cover were determined by aerial photographs (true colour aerial photographs, scale 1:15 000, 5th and 6th June 1996). The data were analysed using a geographic information system (program: IDRISI 4.0 (Eastman, 1992)).

Ground truth

An intensive ground truth (incl. sampling) of 34 positions of mussel beds (Figure 1) was carried out from the end of April to the beginning of July 1996. The ground truth was used to assess the percentage of the actually populated mussel beds in the surveyed region. No samples were taken along the coast between Bremerhaven and Cuxhaven (Wurster Küste) as the mussel stock of that region has been extinguished since 1994 (Zens et al., 1997). A mature mussel bed consists of mussel covered small beds and open spaces without mussels (Linke, 1954) (Figure 2). In this publication, the mussel-covered small beds are called 'patches'. Within the patches there are areas without mussels, too. For this reason, the area covered by mussels occupies only a part, often only a fraction of the whole mussel bed. In mature mussel beds, three variables were determined for obtaining reliable quantitative approaches of biomass (Michaelis et al., 1995):

1. the total area covered by a bed (a (km^2)),
2. the cover (c (%)), i.e. the area of mussel covered patches in relation to the total mussel bed area,
3. the proportion (c' (%)) of the area covered by mussels within the patches.

The mussel-covered part x of a mussel bed is determined as follows:

$$x = a * c * c' * 10\,000^{-1}.$$

For the assessment of cover, proportion and biomass, 34 sites of *Mytilus edulis* were investigated (Figure 1). The mean values were used for the calculation of the whole mussel stock of the Lower Saxonian tidal flats.

Cover and proportion of the mussel beds were measured in the field using a transect method.

Cover

The share of patches and open spaces was measured by counting steps along parallel transects over the total extent of the mussel bed. These values were converted proportionally to the area of the whole bed.

Proportion

The mussel-covered share of the patches was determined along five parallel transects each on 12 different patches with a tape measure. These 12 patches were located along the transects for the measurement of the cover and were distributed evenly over the whole mussel bed area.

Sampling

Twelve random samples of 177 cm^2 each were taken on mussel covered pats on the 12 patches which were investigated for the proportion to determine the biomass (total fresh weight). Only the mussel layer without the sediment below was taken from the patches. The samples were sieved in the field using a sieve with 1 mm mesh size. The fresh weight of the cleaned mussels was determined immediately in the laboratory.

Influence of blue mussel fishery

For the estimation of the impact of blue mussel fishery on the development of young mussel beds, fished and non-fished young mussel beds were compared in summer and autumn 1995. Information of the Staatliches Fischereiamt Bremerhaven is used to classify fished and non-fished beds. The investigated beds were of the same age (1994 spatfall) and were located at sites of former mature mussel beds.

For this comparison, the cover and the proportion of the beds were measured in the field. In the case of

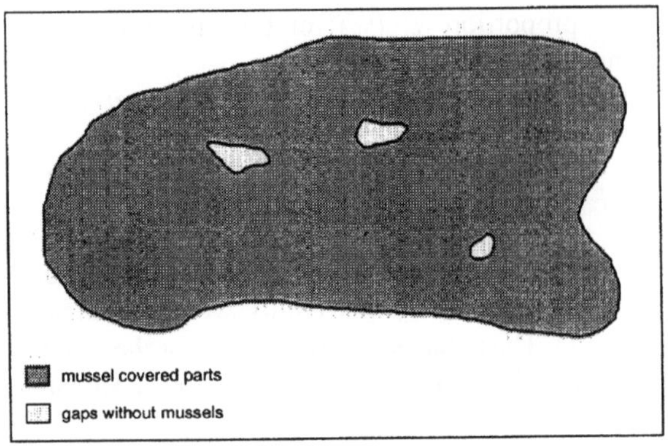

Figure 3. Schematic diagram of a young intertidal mussel bed.

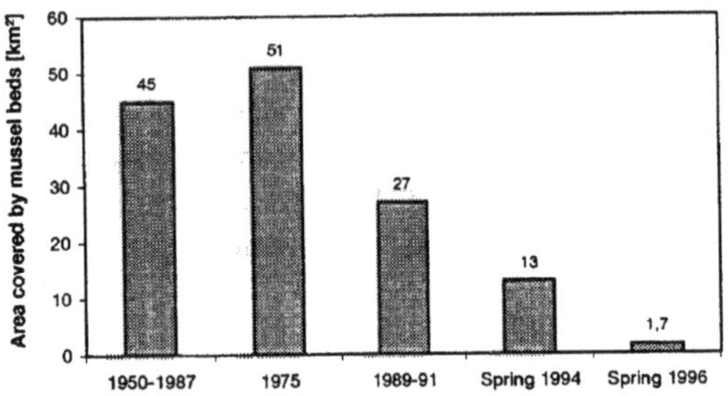

Figure 4. Total area (km^2) covered by blue mussel beds in the tidal flats of Lower Saxony (1950–1987 and 1989–1991: Michaelis et al., 1995, 1975: Dijkema, 1989; 1994: Zens et al., 1997).

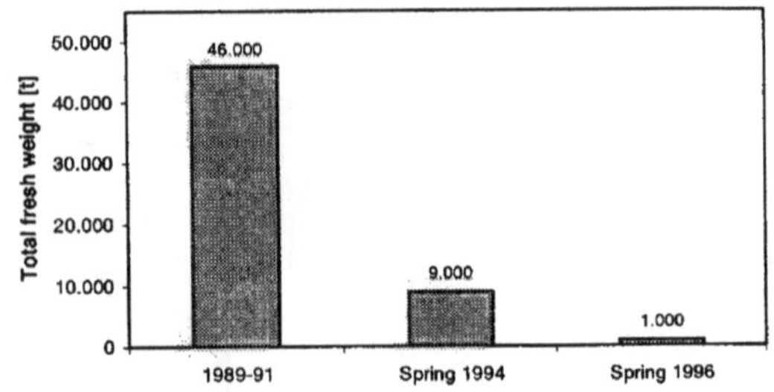

Figure 5. Biomass (total fresh weight (t)) of the blue mussel stock in the tidal flats of Lower Saxony (1989–1991: Michaelis et al., 1995, 1994: Zens et al., 1997).

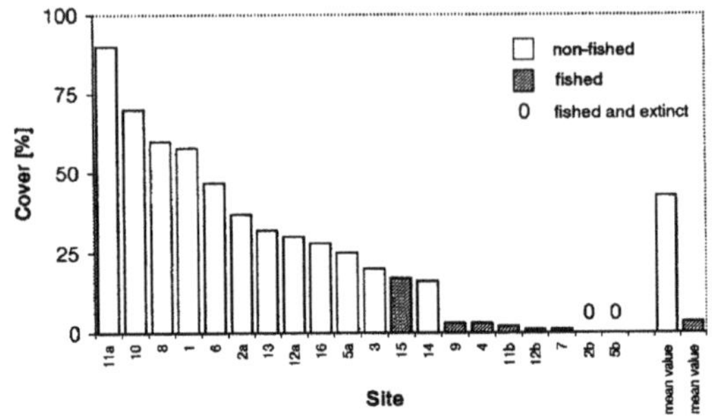

Figure 6. Cover of fished and non-fished young mussel beds and bed areas (spatfall 1994) in autumn 1995. For locations see Figure 1; (a) non-fished areas of partly fished beds, (b) fished areas of partly fished beds. The mean cover of the fished and non-fished beds and bed areas differs significantly (U-test: a = 0.05; $p = 0$; $U = 1.0$).

only partly fished mussel beds, cover and proportion were determined for both the fished and the non-fished areas.

A non-fished young mussel bed differs in structure from the described mature bed (Figure 2; Linke, 1954): the mussels cover nearly the total area of the bed without showing the typical structure of mature beds with heaps and open spaces. But as shown in Figure 3, there are small parts without mussels. The mussel covered area of the young mussel beds was measured in the field using the transect method, too.

Statistics

To analyse the results of the investigation of the fished and non-fished mussels beds, the Wilcoxon, Mann–Whitney U-test was used.

Results

Survey 1996

In spring 1996, most of the investigated mussel bed positions were devoid of living mussels. Only at nine of the 34 positions studied mussels were present and only three of them showed the typical structure of mature mussel beds. The total area covered by mussel beds was about 1.7 km^2 (Figure 4), this means a loss of nearly 95% of the mussel bed area since 1989–91.

The decrease of the biomass was even more pronounced because most of the remaining mussel beds were in a poor condition. The total fresh weight of the intertidal mussel stock amounted to about 1000 t (Figure 5), this is equivalent to 2% of the biomass in 1989–91.

Influence of blue mussel fishery

Cover (percentage of mussel containing patches in relation to total bed area)

The cover of the fished mussel beds and mussel bed areas 3.4 ± 5.6% (mean value ± sd) was significantly lower than of the non-fished 42.8 ± 22.6% (mean value ± sd) (Figure 6). Only the cover of one fished mussel bed (no. 15) reached a value comparable to the non-fished ones (Figure 6). In this case, the period between fishery activities and ground truth investiga-

Table 1. Time between first fishing and ground truth of the fished mussel beds and bed areas as well as their cover

Site	First fishing	Ground truth	Month between first fishing and ground truth	Cover [%]
2	Nov. 1994	Aug. 1995	9	0
4	Jan./Feb. 1995	Aug. 1995	7.5	3
5	Nov. 1994	Sept. 1995	10	0
7	Oct./Nov. 1994	Dec. 1995	13.5	1
9	May 1995	Nov. 1995	6	3
11	May 1995	Nov. 1995	6	2
12	Dec. 1994	Nov. 1995	11	1
15	May/June 1995	Oct. 1995	4.5	17

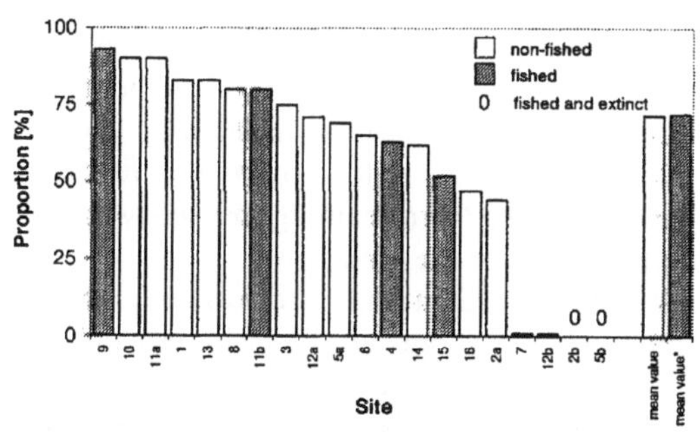

Figure 7. Proportion of fished and non-fished young mussel beds and bed areas (spatfall 1994) in autumn 1995. For locations see Figure 1; (a) non-fished areas of partly fished beds, (b) fished areas of partly fished beds. The mean proportion of the remained fished and non-fished beds and bed areas does not differ significantly (*U*-test: a = 0.05; $p = 0.069$; $U = 24.5$). *For the calculation of the mean values of the fished beds and bed areas only the data >5% were considered.

tion was only 4.5 months, whereas in the other cases six to up to 14 months went by (Table 1).

Proportion (the area covered by mussels within the patches)

The proportion of the non-fished beds and bed areas amounted to 71.6 ± 15.2%. The fished beds and bed areas showed two groups: at four sites, mussels were extinguished or near to extinction (with a proportion of 1%) (Figure 7). The second group of the fished mussel beds and mussel bed areas had a mean proportion of 72.0 ± 18.1%, which did not differ significantly from the mean value of the non-fished (71.6 ± 15.2%).

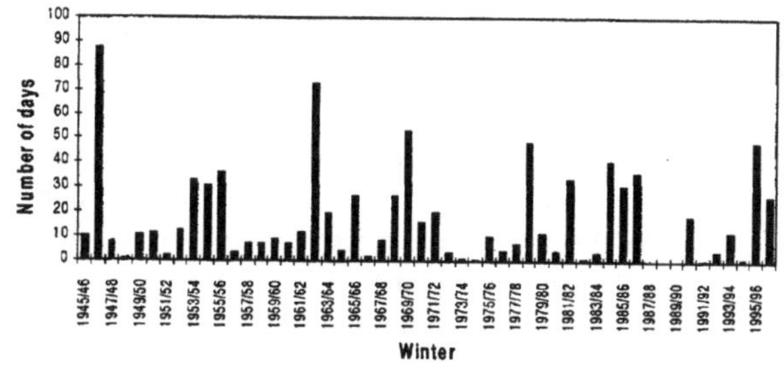

Figure 8. Number of days with ice coverage at the German North Sea coast in winters from 1945/1946 to 1996/1997. (Mean values of 13 stations: Koslowski (1987) and data of the Bundesanstalt für Seeschiffahrt und Hydrographie, Hamburg).

Discussion

According to recent results, blooms of *Phaeocystis* (Petri & Vareschi, 1997), macroalgal cover (Herlyn, 1996; Anonymus, 1999) and parasites (Zens, 1999) can be excluded as essential factors responsible for the decline of *Mytilus edulis* (Herlyn et al., 1999).

The mussels of the Wadden Sea coast are highly contaminated by pollutants (Buchwald, 1990; Jonkers & Everts, 1992; Bakker, 1994), but the knowledge about direct effects on stock development is limited. It is remarkable that the area showing the most pronounced decrease (Wurster Küste, Figure 1) (Michaelis et al., 1995; Zens et al., 1997) is exposed to the heavily polluted river Weser (Buchwald , 1990). More information about the contamination of mussels and effects of pollution can be found in Widdows & Donkin (1992). The knowledge about the effects of dredging and dumping of sediments on the mussel stock is scanty too. After Hagena (pers. comm.,1994), a singular mussel bed near the island Langeoog is disturbed by direct effects of dumping. According to ongoing investigations, such direct effects are not the rule. To elucidate the role of these factors which also include the remobilisation of pollutants (e.g. TBT) from harbour sediments to the Wadden Sea, further research is necessary.

During the last decades, increased populations of mussel eating birds (eider, herring gull, oyster catcher; Exo, 1994) have intensified predation pressure on the mussel stock. Recent investigations show the importance of this factor for the mussel population (Nehls, 1989; Hilgerloh, 1996, 1997; Hilgerloh et al., 1997), but further research is necessary to quantify its role for the whole area of the Wadden Sea. Seed mussel fishery competes with the increased stocks of birds for mussels of the same size classes (Herlyn et al., 1999).

For the development between 1987 and 1995, ice rafting was not a decisive factor, because the winters from 1987/88 to 1994/95 were remarkably mild. The last phase of the noticed decrease (Figure 4) was influenced mainly by effects of the winter 1995/96, which belonged to the five severest ice winters since 1945 (Figure 8). During that winter, ice rafting led to the erosion of most of the remaining mussel beds (Figures 4 and 5).

Severe storms are also known to have a potential influence on the development of mussel beds (Hagena, 1991; Michaelis, 1991; Reise, 1991; Nehls & Thiel, 1993), but there has been no change either of storm frequency (Schmidt & Von Storch, 1993) or storm flood frequency (Niemeyer et al., 1995) in the German Bight in the past 50 years. Recently, the number of wind floods increased, but the combination of frequency and intensity ranks behind the storm flood situation in the 1970s (Niemeyer et al., 1995). If this would have had lasting effects on the mussel beds, the decrease of the stock should have begun in the 1970s and not in the 1980s. On the other hand, it maybe that the proneness of the mussel beds to storms has increased (Herlyn, 1996).

The main result of the comparison of fished and non-fished mussel beds is that mussel fishery led to complete loss or heavy decrease of most of the fished young beds and bed areas (Figure 6). One year after recruitment, seven of eight fished beds and bed areas were lost or next to extinction, whereas none of the unfished beds were destroyed. The majority of mussels which were not fished and were left on the fished beds and bed areas disappeared in the period after fishery took place. Evidently, the losses caused by indirect effects are higher than the direct taking (Herlyn et al., 1999). Only some solitary patches within fished beds and bed areas remained (Figures 6 and 7). We assume that the majority of the patches disappeared completely because mussel fishery destroyed the structure of the fished parts (Figure 6), thereby disturbing or preventing development into mature mussel beds. These results confirm the hypothesis that fished beds are damaged in their structure and are presumably less resistant to storms and ice rafting (Obert & Michaelis, 1991).

Our comparison of the fished and non-fished mussel beds contradicts the hypothesis of the Lower Saxonian fishermen and the Staatliches Fischereiamt Bremerhaven (W. Hagena pers. comm., 1994) that mussel fishery has a positive effect on fished beds. They argue that the survival rate of young mussels increases because the mortality depending on density is reduced by fishery impact. If this would be the case, the proportion of the not extinguished fished beds and bed areas would be lower than of the non-fished beds (Figure 7) because the remaining mussels would also settle in the trails of the dredges.

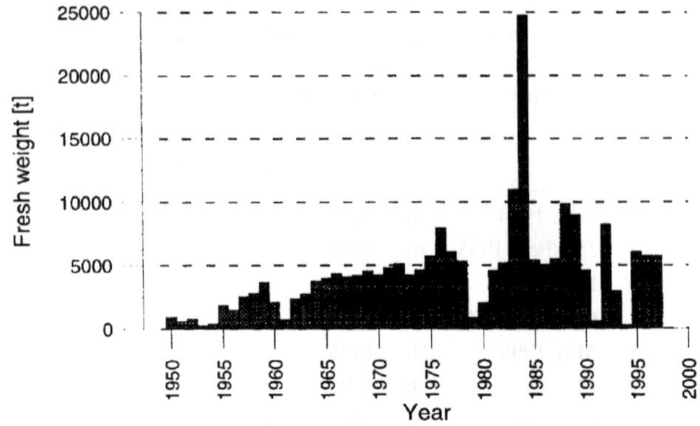

Figure 9. Annual landings (fresh weight) of mussels (*Mytilus edulis*) at the coast of Lower Saxony from 1950 to 1997 (after data of Staatliches Fischereiamt, Bremerhaven).

Until this investigation, there was only the study of Schirm (1991) about the direct impact of mussel fishery on mussel beds. He estimated that the influence of fishery is not a decisive factor for the development of mussel beds. The results of Schirm (1991) are not in contradiction to the present results because he investigated only one mussel bed in a way which differed from the way mussel fishery acts normally upon natural young beds. Furthermore, Schirm (1991) investigated the bed from April to October, and this short and mild period was presumably not suitable to record indirect effects – especially those resulting from severe storms and ice rafting. For this reason, his results cannot be transferred to young mussel beds which were fished in the usual way to obtain seed mussels like in this investigation (Herlyn et al., 1999).

The intense fishery in 1983 (11 000 t) and 1984 (24 700 t) has put an extraordinary stress on the population, and the gaining of seed mussels from tidal flats may endanger the wild beds as observed in the Netherlands (Obert & Michaelis, 1991).

The question of the exact relation of the beginning mentioned stock reducing factors is still unanswered, because of their mutual influence, other factors affecting the stock development and the natural high dynamic of the intertidal mussel stock. Nevertheless, the mussel fishery activities of the last decades can be assessed (Herlyn et al., 1999).

Especially during times of decreasing mussel stock, it is important to use the mussel stock in a restricted way. It violates the principle of sustainable management to intensify mussel fishery (Figure 9) while the stock development is declining. This is also valid if mussel fishery were not one of the main stock reducing factors.

Summary

Since the middle of the 1980s, the blue mussel stock at the tidal flats of Lower Saxony has declined. In spring 1996, the mussel beds of the region covered an area of 1.7 km^2 with a biomass (total fresh weight) of about 1000 t, this means a loss of about 95% of the mussel bed area and 98% of the biomass since 1989–91.

While the last phase of the decrease was affected by the ice winter 1995/96, ice rafting was not the decisive factor in the years after 1987/88 with mild winters.

Mussel fishery resulted in a heavy decrease or even complete loss of most of the investigated fished young beds and bed areas at sites of former mature beds. This decrease was not only restricted to the fished amounts because high losses were also caused by indirect effects. Until the next spatfall, the development of mature beds at the concerned sites is disturbed or prevented.

Acknowledgements

We are indebted to Dr N. Dankers (IBN-DLO, Texel), Dr S. Dittmann (Terramare, Wilhelmshaven), U. Fooken (Terramare, Wilhelmshaven), M. Grotjahn (Aqua-Marin, Norden), I. Harms (Terramare, Wilhelmshaven), J. Kahle (Terramare, Wilhelmshaven), H.-H. Kramer (NLÖ – Forschungsstelle Küste, Norderney), Dr H. Michaelis (NLÖ – Forschungsstelle Küste, Norderney), R. Zühlke (Aqua-Marin, Norden) and the crew of the research vessel 'MS Buise', A. Beekmann and E. Eilers (Niedersächsisches Landesamt für Ökologie – Forschungsstelle Küste, Norderney). This study is supported by the Ökosystemforschung Niedersächsisches Wattenmeer (publication no. 358) and the Niedersächsische Wattenmeer-Stiftung.

References

Anonymus, 1999. Überwachung der niedersächsischen Küstengewässer – Routineuntersuchungen 1997. Dienstber. Forschungsstelle Küste Norderney, March 1999, Norderney: 83 pp.

Bahr, K., 1950. Geschichte der ostfriesischen Miesmuschelfischerei. Ostfriesland-Kalender 1950: 53–56.

Bakker, J. F., 1994. Schadstoffe in Miesmuscheln. In Lozán, J. L. et al. (eds), Warnsignale aus dem Wattenmeer. Blackwell Wiss.-Verl., Berlin: 132–143.

Buchwald, K., 1990: Nordsee – Ein Lebensraum ohne Zukunft? Verlag Die Werkstatt, Göttingen: 552 pp.

Dankers, N. & K. Koelemaij, 1989. Variations in the mussel population of the Dutch Wadden Sea in relation to monitoring of other parameters. Helgoländer wiss. Meeresunters. 43: 529–535.

Dijkema, K. S., 1989. Habitats of the Netherlands, German and Danish Wadden Sea. Res. Inst. Nature Managem., Texel: 31 pp.

Eastman, J. R., 1992. User's Guide, IDRISI Version 4.0 rev. 1, Worcester.

Exo, K.-M., 1994. Bedeutung des Wattenmeeres für Vögel. In Lozán, J. L. et al. (eds), Warnsignale aus dem Wattenmeer. Blackwell Wiss.-Verl., Berlin: 87–90.

Hagena, W., 1991. Herz- und Miesmuschelfischerei aus der Sicht der niedersächsischen Fischereiverwaltung. Schriftenreihe Schutzgemeinschaft Deutsche Nordseeküste e.V.: Probleme der Muschelfischerei im Wattenmeer: 53–59.

Hagena, W., 1994. Miesmuschelfischerei in Niedersachsen. Paper presented at meeting of the National Parc Advisary committee, 19 April 1994, Wilhelmshaven: 1–3. Nationalpark-Beiratssitzung, 14 July 1994, Wilhelmshaven: 3 pp.

Herlyn, M., 1996. Zur Bestandssituation der Miesmuschelbänke des niedersächsischen Wattenmeeres. Mitteilungen aus der NNA January 1996, Schneverdingen: 56–61.

Herlyn, M., 1999. Zur Entwicklung des Miesmuschelbestandes der niedersächsischen Watten. In Umweltbundesamt & Nationalparkverwaltung Niedersächsisches Wattenmeer (eds), Umweltatlas Wattenmeer Bd. 2 – Wattenmeer zwischen Elb- und Emsmündung. Verlag Eugen Ulmer, Stuttgart: 70–71.

Herlyn, M., G. Millat & H. Michaelis, 1999. Einfluß der Besatzmuschelentnahme auf die Entwicklung eulitoraler Neuansiedlungen von *Mytilus edulis* im niedersächsischen Bereich des Wattenmeeres. Dienstber. Forschungsstelle Küste Norderney, September 1999, Norderney: 27 pp.

Hilgerloh, G., 1996. Miesmuscheln (*Mytilus edulis*) in der Nahrung von Eiderenten (*Somateria mollissima*) auf den Nordseeinseln Langeoog und Spiekeroog). Acta ornithoecol. 3 (3): 239–243.

Hilgerloh, G., 1997. Predation by birds on blue mussel *Mytilus edulis* beds of the tidal flats of Spiekeroog (southern North Sea). Mar. Ecol. Prog. Ser. 146: 61–72.

Hilgerloh, G. & M. Herlyn, 1996. Auswirkungen der Prädation durch Vögel auf eine junge Miesmuschelbank. Verh. dt. zool. Ges. 84: 306.

Hilgerloh, G., M. Herlyn & H. Michaelis, 1997. The influence of predation by herring gulls *Larus argentatus* and oystercatchers *Haematopus ostralegus* on a newly established mussel *Mytilus edulis* beds in autumn and winter. Helgoländer wiss. Meeresunters. 51: 173–189.

Jonkers, D. A. & J. W. Everts, 1992. Derivation of micropollutant risk levels for the North Sea and Wadden Sea. Ministries of Housing, Physical Planning and Environment and Transport, Public Works and Water Management, Distribution Center VROM, Zoetermeer, Rep. 1992/3: 74 pp.

Koslowski, G., 1987. Der Eiswinter 1987/88 im deutschen Küstengebiet zwischen Ems und Trave. Dt. hydrogr. Z. 40: 215–219.

Linke, O., 1939. Die Biota des Jadebusenwattes. Helgoländer wiss. Meeresunters. 1: 201–348.

Linke, O., 1954. Die Bedeutung der Miesmuscheln für die Landgewinnung im Wattenmeer. Natur u. Volk 84: 253–261.

Michaelis, H., 1991. Veränderungen im niedersächsischen Wattenmeer. Schriftenreihe der Schutzgemeinschaft Deutsche Nordseeküste e.V.: Probleme der Muschelfischerei im Wattenmeer: 16–25.

Millat, G. & M. Herlyn, 1999. Documentation of intertidal mussel bed (*Mytilus edulis*) sites at the coast of Lower Saxony. Senckenbergiana marit., 29 (Suppl.): 83–93, 4 figs., 2 tabs.; Frankfurt a. M.

Michaelis, H., B. Obert, I. Schultenkötter & L. Böcker, 1995. Die Miesmuschelbestände der niedersächsischen Watten. Ber. Forschungsstelle Küste Norderney 40: 55–70.

Nehls, G., 1989. Occurrence and food consumption of the common eider, *Somateria mollissima*, in the Wadden Sea of Schleswig-Holstein. Helgoländer wiss. Meeresunters. 43: 385–393.

Nehls, G. & M. Thiel, 1993. Large-scale distribution patterns of the mussel *Mytilus edulis* in the Wadden Sea of Schleswig-Holstein: Do storms structure the ecosystem? Neth.J. Sea Res. 31: 181–187.

Niemeyer, H. D., R. Kaiser & D. Glaser, 1995. Sturmfluthäufigkeiten zwischen Ems und Weser von 1946–1994 – Pegel: Emden, Borkum, Norderney, Bremerhaven. Dienstber. Forschungsstelle Küste Norderney (unpublished): 7 pp.

Obert, B. & H. Michaelis, 1991. History and ecology of the mussel beds (*Mytilus edulis* L.) in the catchment area of a Wadden Sea tidal inlet. In Elliot, M. & J. P. Ducrotoy (eds), Estuaries and Coasts: Spatial and Temporal Intercomparisons. Olsen & Olsen: 185–194.

Petri, G. & E. Vareschi, 1997. Utilisation of *Phaeocystis globosa* colonies by young *Mytilus edulis*. Arch. Fish. Mar. Res. 45: 77–91.

Reise, K., 1991. Wattenökologie – Zustand und Trends. Schriftenreihe der Schutzgemeinschaft Deutsche Nordseeküste e.V.: Probleme der Muschelfischerei im Wattenmeer: 9–15.

Schirm, B., 1991. Der unmittelbare Einfluß gezielter Fischerei auf die Lebensgemeinschaft von Miesmuschelbänken im Wattenmeer. Diplomarbeit, Universität Kiel (unpublished): 62 pp.

Schmidt, H. & H. Von Storch, 1993. German Bight storms analysed. Nature 365: 791.

Widdows, J. & P. Donkin 1992. Mussels and environmental contaminants: bioaccumulation and physiological aspects. In Gosling, E. (ed.), The Mussel *Mytilus*: Ecology, Physiology, Genetics and Culture. Elsevier, Amsterdam: 383–424.

Zens, M., 1999. Der Einfluß von Parasiten auf Vitalität und Bestandsentwicklung der Miesmuschel (*Mytilus edulis*). Dienstber. Forschungsstelle Küste Norderney, May 1999, Norderney: 40 pp.

Zens, M., H. Michaelis, M. Herlyn & M. Reetz, 1997. Die Miesmuschelbestände der niedersächsischen Watten im Frühjahr 1994. Ber. Forschungsstelle Küste Norderney 41: 141–155.

MIX
Papier aus verantwortungsvollen Quellen
Paper from responsible sources
FSC® C105338

If you have any concerns about our products,
you can contact us on
ProductSafety@springernature.com

In case Publisher is established outside the EU,
the EU authorized representative is:
Springer Nature Customer Service Center GmbH
Europaplatz 3, 69115 Heidelberg, Germany

Printed by Libri Plureos GmbH
in Hamburg, Germany